Food safety assurance in the pre-harvest phase

Food safety assurance and veterinary public health

– *volume 1* –

Food safety assurance in the pre-harvest phase

edited by:
Frans J.M. Smulders
John D. Collins

Wageningen Academic
Publishers

CIP-data Koninklijke Bibliotheek
Den Haag

paperback

Subject headings:
Veterinary public health
Food safety
Preventive medicin

First published, 2002

Wageningen Academic Publishers

The Netherlands, 2002

Printed in The Netherlands

Contents

Food Safety Assurance and Veterinary Public Health: The Pre-Harvest Phase

The objectives of this book series is to systematically address those public health hazards, the prevention and control of which is a function of veterinary public health. A longitudinal integrated approach has been chosen, in line with modern approaches to food quality and safety assurance programmes and relying upon the input of recognised experts engaged at the cutting edge of research on the safety of foods of animal origin. The programme is targeted towards an audience of both established senior scientists in this field and more junior colleagues now embarking on a career in food safety and veterinary public health. A feature of the programme is the creation of a platform for the formal and informal exchange of views between the experienced scientist and those newly entering this field of endeavour.

The resolution of many of the problems encountered in food safety requires a multidisciplinary approach and requires the expertise of researchers and practitioners involved in different elements of food safety and food quality assurance, as well as monitoring and surveillance and other aspects of risk analysis at each stage "from farm to fork". This approach requires an in-depth knowledge of all the food safety factors that apply to the pre-harvest, processing and post-harvest phases of the of the food chain, so as to provide the basis for recognising and implementing effective intervention measures and risk management strategies that protect the consumer against foodborne hazards in the light of current challenges.

The involvement of leading experts in the field in this publication series, including those involved in both academic and industrial research, as well as in decision-making at the world food trade level, or as members of multidisciplinary medical teams at local or regional level, will ensure that current issues will be addressed in depth and in a manner that will encourage the active participation of the younger participants in the proceedings.

The recent White Paper on Food Safety published by the Commission of the European Union has focussed attention on the longitudinal nature of the food chain and upon the need to approach food safety issues on an integrated basis. The document addresses the need for such an approach if food borne hazards, such as zoonoses and other foodborne infections and intoxications associated with foods of animal origin, together with the adoption of a formal risk analysis approach to each hazard, are to be prevented or controlled. Surveillance, monitoring and management of the safety aspects of various phases of food production, processing and distribution have attracted the Commission's special attention and again call for a multidisciplinary approach by all sectors concerned. The Commission, in its White Paper has alerted the food industry of its own responsibility for ensuring the safety of its products at all stage of the food chain and of the need for competency in all aspects of food safety assurance. The current perception of the problems associated with the transmissable spongiform encephalopathies (TSE's) such bovine spongiform encephalopathy (BSE), and more recently, dioxin contamination of animal feedstuffs and their potential consequences for the vulnerable consumer, provide ample evidence of the need for a rational, unbiased approach to the adoption of evidence-based intervention measures that are effective and feasible and that offer the necessary degree of protection for the public. Such a need can only be addressed with the continuous support of the scientific community.

The issues to be addressed are not restricted to the food product alone and relate very much to the condition of primary production, and extend to such matters as animal welfare and the protection of the environment. Scientists share these concerns and recognise that a wide range of issues require to be addressed. These include real, or in some cases, perceived, hazards as represented by the occurrence in food of chemical contaminants, residues of medicinal products including antimicrobial agents, natural and synthetic hormones, environmental contaminants such as the heavy metals and radio isotopes, well recognised pathogenic microorganisms and their toxic products, foodborne parasites and, possibly, genetically modified microorganisms. The relative importance of each of these categories varies from country to country from time to time, though the foodborne pathogenic microorganisms have attracted most attention in the past, as evidenced by the design and nature of the classical forms of food and in particular meat inspection systems in the past, have been adopted in most countries.

In many countries, with the elimination of many of the major animal diseases that formed a threat to public health through the consumption of contaminated meat and milk products, the efficacy of the tradition methods of control, based on final product inspection, have recently been widely questioned both by the veterinary profession and the authorities responsible for consumer protection. In the case of a number of the pathogens of current concern, such as *Campylobacter* spp. and *Cryptosporidium* spp., and some foodborne viruses, for instance, our knowledge of the exact routes of infection is incomplete due to a lack of sufficient quantitative epidemiological data in particular. The adoption of the longitudinal integrated approach to food safety addresses this deficit and provides evidence of the nature of such transmission by its effectiveness or otherwise in preventing human disease caused by these foodborne hazards. Such a response calls for the multidisciplinary approach and it is this objective that the series of Euroconferences has set out to achieve.

In this, the first in a series of three publications, the topic of interest is "Food Safety in the Pre-Harvest Phase". The elements that jeopardise the safety of foods of animal origin include factors relating to the food animals' environment, including their water supply, related for instance to the effects of industrial pollution and exposure to pathogenic microorganisms, of changes in animal husbandry and harvesting practices and in particular the risks associated with contaminated feedstuffs, as well as the untoward effects of antimicrobial usage resulting in the emergence of resistant microorganisms which in themselves constitute a further hazard for the consumer. All these issues, and others, are addressed by Public Health strategists representing the European Union and the United States of America and are discussed against the background of world trade as well as local market demands. These proceeding serve as a record of those discussions and provide a basis for further discussion by all interested parties.

The remaining publications in this series will address Safety Assurance during Food Processing and Risk Management Strategies: Monitoring and Surveillance, "Farm to Fork".

Vienna / Dublin, September, 2002

Frans J.M. Smulders
John D. Collins

Keynote contributions

The EU food safety action plan

Willem Daelman
European Commission, Health and Consumer Protection Directorate General, Rue de la Loi 2000, B-1049 Brussels, Belgium

1. Introduction

At the meeting of the European Parliament on 5 October 1999, Mr. Prodi, President of the European Commission, has announced that food safety is a clear and urgent priority and that the Commission would present a White Paper on Food Safety reflecting that priority.

2. The White Paper on Food Safety

On the 12th of January 2000, the Commission presented its White Paper on Food Safety. This White Paper basically covers two aspects. On the one hand, it presents an exhaustive action plan to make the European legislation more coherent and more comprehensive. On the other hand, it envisages the establishment of an independent European Food Authority with particular responsibilities for both risk assessment and risk communication on food safety issues.

2.1 The Basic Principles Underlying the Legislative Action Plan

In the past, EU food safety measures have been mostly developed on a sectoral basis. The increasing integration of economies, developments in farming and food processing, and new handling and distribution patterns clearly require a new more integrated approach. The White Paper on Food Safety outlines a comprehensive range of actions to make EU food legislation more coherent, understandable and flexible, to promote better enforcement of that legislation, and to provide greater transparency to consumers. It contains a detailed action plan with an indicative timetable, but most importantly, it reaffirms clearly the basic principles upon which the Commission intends to base its future proposals for food safety legislation.

The EU legislation needs to be clear on the *responsibilities* of the various players in the food production chain. Legislation will be proposed to clarify that responsibility for placing safe food on the market rests with the economic operators; that the competent authorities of the Member States are in charge of monitoring and enforcing this responsibility through national surveillance and control systems; and finally that the Commission shall concentrate on evaluating the ability of competent authorities to carry out these tasks through audits and inspections.

The Commission will continue to reinforce its *farm to table policy*, covering all sectors of the food chain, including feed production, production on the farm, food processing, storage, transport and retail sale. The Commission has even extended the concept of food safety, by placing nutrition within the scope of the document. Consumers should be offered not only safe food, but also healthy food, and be given the possibility to make better informed choices about the food they consume.

The Commission also places great importance on the possibility to withdraw a product from the market, in case a safety problem arises. In order better to protect consumers, but also to avoid more disruption of the market than is really necessary, adequate procedures to facilitate identification of the source of the problem are needed. Improved *traceability* of food components throughout the food production chain is what is required. It involves many practical difficulties, but it is the price to pay to respond rapidly and without hesitation to a real or potential food safety crisis.

In line with international agreements, the European Union must base its food policy on the three components of *risk analysis: risk assessment* (consisting of scientific advice and information analysis), *risk management* (consisting of regulation and control) and *risk communication*.

Also, in the EU decision making process, *other legitimate factors* relevant for the health protection of consumers and for the promotion of fair practices in food trade can be taken into account. The definition of the scope of such legitimate factors is presently being studied at international level. Examples of such other legitimate factors are environmental considerations, animal welfare, sustainable agriculture, consumers' expectation regarding product quality, fair information and definition of the essential characteristics of products and their process and production methods. Where appropriate, the *precautionary principle* will be applied in risk management decisions. It covers cases where scientific evidence is insufficient, inconclusive or uncertain and preliminary scientific evaluation indicates that there are reasonable grounds for concern that health could be at risk. The guidelines for the application of this principle are described in the Communication of the Commission on the Precautionary Principle, which was adopted on 2 February 2000.

Finally, the importance of *transparency* towards all stakeholders is underlined in the White Paper. All steps in policy making will be taken in full openness. Such an approach will allow consumers to follow and understand the development of new measures and the improvements which they bring. But transparency should go beyond passive transmission of information. It should be a two-way process, allowing consumers and all other stakeholders to feed into the process. It should also have a supplementary, pro-active dimension towards certain vulnerable groups of the population, to warn them about the inherent and unavoidable risks of some types of food, in an area where zero risk does not exist.

2.2 The European Food Authority

The most innovative element of the White Paper on Food Safety relates to the creation of the European Food Authority. The responsibilities of the European Food Authority would consist of the preparation and provision of scientific advice, the collection and analysis of information required to underpin both that advice and the Community's decision making process, the monitoring and surveillance of developments touching upon food safety issues (including rapid alert systems) and the communication of its findings to all interested parties. This corresponds to the functions of risk assessment and risk communication, two of the three components of risk analysis.

Once scientific advice is available and if a risk has been identified, a decision has to be taken with respect to the way in which this risk will be managed. Various strategies to manage risks are available. They include the establishment of lists of products which are either authorised or prohibited, the setting of maximum limits for contaminants, the application of specific production methods or the application of good practices. Risk management decisions can be taken on the basis of scientific advice alone, but in some cases other factors are also relevant. The final decision - touching upon the daily life of Europe's consumers and citizens - must be taken by a politically accountable body. This is the reason why, in the White Paper on Food Safety, the inclusion of risk management in the mandate of the Authority is not considered appropriate. The Commission has taken the view that any extension of the responsibilities of the Authority in the risk management area should only be considered in the light of the experience with its functioning and the confidence gained in its operation, including the possible need to change the Treaty.

The White Paper lists the criteria which will govern the functioning of the European Food Authority: *independence, excellence and transparency*. The existing situation, where scientists involved in scientific advice must respect strict rules concerning their independence, must continue. To allow the Authority to act as a point of scientific excellence and reference, and to resolve disputes on scientific issues, it will need rapidly to establish its international pre-eminence. This involves not only the rapid, open, presentation of the findings and recommendations of the Authority, but also implies that the processes followed in reaching them are as open and transparent as possible, in order to respond to the fundamental right of access of citizens as laid down in the Treaty. The new Authority will also have a key role in communicating with the public at large about food safety issues, a much neglected area at present.

The creation of a network of scientific contacts with national scientific agencies and institutions, with the Authority at its centre, will ensure that all concerned become associated with the analytical process, and have a clearer understanding and greater acceptance of the basis for the scientific opinions that are generated.

To set up such an Authority requires legislation. The White Paper foresees that the Commission will make its proposal for setting up an independent European Food Authority as a matter of priority.

3. Issues with an Impact for the Veterinary Public Health Sector

3.1 Proposals for new rules on food hygiene

On July 14, the Commission adopted a set of proposals which have as main objective to modernise and upgrade the existing food hygiene rules. These proposals will replace the extensive requirements that have been developed over the last thirty years as a scattered response to the needs of the internal market. The proposals are announced in the Commission's White Paper on Food Safety and will introduce consistency and clarity throughout the food production chain.

3.2 The Basic Principles Underlying Food Hygiene

The recast, consisting of a merging of all existing hygiene rules (17 Directives) together with the introduction of new principles, offers an excellent opportunity to fix uniform and simplified hygiene requirements applicable to *all food*. The result is a proposal that integrates a number of important requirements that must be respected by food businesses in order to achieve the level of hygiene that is required to produce safe food.

Firstly, food hygiene must be ensured along the food chain, from *farm to table*. At present, this continuity is not systematically ensured. This concept is new for several commodities and sectors since for example primary production has not been covered in the past in a horizontal way. Rules for ensuring hygiene at farm level have therefore been introduced

In addition, the proposals are based on the principle that food business operators have the *primary responsibility* for food safety. The HACCP (Hazard Analysis Critical Control Point) system is proposed as a management tool to be implemented by food business operators to ensure food safety. It prescribes a number of logical steps to be followed by operators throughout the production cycle in order to allow - through hazard analysis - the identification of points where control is critical with regard to food safety. It allows focusing the hazards that that are specific to a particular food business.

Another instrument playing an important role in food hygiene will be the use of *codes of good hygienic practice* to be established by the food sectors. They must give guidance to food business operators on food safety. The advantage of such codes is that they are sector specific, and that they can enter into details that cannot be adequately covered by legislative provisions. They will be complementary to the basic hygiene rules prescribed in Community legislation.

Food business operators must also ensure *traceability* of food and food ingredients, and must put in place product withdrawal procedures in case of food emergencies. For that purpose, they shall keep adequate records which must enable them to identify the supplier of ingredients and raw materials used in their operation.

The basic hygiene to be respected by food business operators (cleanliness of premises, personal hygiene, equipment requirements, food transport etc.) remain as before.

In addition to the basic principles, it is proposed that specific rules apply in order to control the very specific hazards that may influence the safety of certain commodities such as meat, milk and other products of animal origin.

3.3 Flexibility

The proposals introduce a certain level of flexibility that is believed essential in order to take account of particular situations. This is in particular the case with regard to the implementation of the HACCP-system, traditional ways of preparing certain food, and for certain small enterprises. It is clearly stated that this flexibility must not compromise food safety.

3.4 The Review of Meat Inspection Procedures

Traditional meat inspection techniques were developed at the beginning of the century in order to control the hazards in meat present at that time. These hazards were mostly characterised by gross pathological changes which are detectable by organoleptic inspection. This technique has been successful insofar that it has enabled to eliminate or reduce considerably certain hazards (glanders, tuberculosis).

Whilst certain hazards have been reduced or have disappeared, other hazards have emerged. These are in particular of a microbiological and chemical nature. They can be present in animals without provoking pathological changes and are therefore not always detectable by using traditional organoleptic techniques. Another approach is required. For that purpose, a policy document (Document SANCO/364/2000) has been developed by the Commission services.

This document is based on the principle that future meat inspection systems should be defined in terms of priorities with regard to human health protection. There is general international agreement that food-borne microbiological hazards resulting from contamination of meat with enteric pathogens need particular attention, in view of the increased prevalence reported for such hazards. On the other hand, the present organoleptic meat inspection system has been accused of being old fashioned and maintaining a number of inspection techniques that are not proportionate to the hazards for which they were designed originally. They are therefore not risk-based.

It seems therefore necessary that future meat inspection policies be re-oriented so as to:
a) develop a strategy to ensure increased food safety by reducing microbiological hazards.
b) undertake a profound re-assessment based on a risk analysis of the classical meat inspection methods.

In re-organising the inspection system, a number of factors should be taken into account:
- the homogeneity of groups of animals raised under identical conditions: this is already the case for poultry, but could be extended to homogeneous lots of pigs and veal calves. The scientific committee is preparing a report with regard to pigs; this could result in the detailed inspection of a limited number of animals from a homogeneous group;

- the availability of health information from the farm. This aspect is also covered under the poultry meat Directive. The need for health information to follow the animal from the farm to the abattoir is also embedded in the "recast exercise";
- avoiding incision and palpation techniques since they are likely to increase microbiological contamination;
- the availability of new inspection techniques e.g. serological tests for trichinosis or cysticercosis; new technical device for mechanical inspection of poultry carcases;
- a possible regionalisation where scientific evidence exists that it is unlikely that a hazard occurs in a particular region (e.g. trichinosis, hydatidosis);
- new inspection techniques must not only address food safety aspects, but also accommodate animal health and animal welfare needs.

A number of the above elements have already been included in a study of the Scientific Committee on a simplified inspection procedure for slaughter pigs.

HACCP is likely to play a crucial role in a revised meat inspection system. With the introduction of HACCP, food operators must assume full responsibility for food safety by assuring that the health objectives are attained, whilst the competent authority monitors whether the food operators have put all measures in place to obey the given objective and whether the practical implementation of these measures result in the attainment of the objective (audit).

4. Conclusions

With the White Paper on Food Safety and the action resulting therefrom a move is made to integrate the different policies on food safety so as to come to a global approach covering all food.

The recast and the simplification of the hygiene rules is one example where such an integration is being achieved. Another area where such integration is being developed is the one on food controls, as spelled out in Action 4 referred to in the Annex to the White Paper on Food Safety.

This global approach does not mean however that there is no room for more specific hygiene or official control requirements. Where more specific requirements are needed, they should be limited to those that are essential to ensure food safety.

Approaches to pre-harvest food safety assurance

John N. Sofos
Department of Animal Sciences, Colorado State University, Fort Collins, Colorado 80523-1171, USA

Summary

Animals may be contaminated or are asymptomatic carriers of chemical residues and pathogenic bacteria and, thus, serve as sources of subsequent meat contamination or contamination of other foods through contaminated manure and water. Microbial contaminants especially pathogenic bacteria of enteric origin such as *Escherichia coli* 0157:H7 and *Salmonella* are of major concern because they can cause adverse effects on human health and compromise food safety. Thus, there is a need to control pathogenic microorganisms in animals and their products in order to enhance the safety of our food supply. Progress has been made in developing interventions for pathogen control following harvest, but pre-harvest pathogen control has major constraints. Approaches to pre-harvest pathogen control should be based on research addressing pathogen ecology and risk analysis of animal management, handling, feeding and shipping for slaughter practices. It is important to realize, however, that control or management of food safety risks should be based on an integrated effort and approach that addresses all sectors, from the producer through the processor, distributor, packer, retailer, food service worker and consumer. Reduction of pre-harvest pathogen prevalence may lead to a reduced probability that errors occurring in subsequent parts fo the food chain will lead to foodborne illness.

Keywords: food safety assurance, pre-harvest, animal products, hazard control

1. Introduction

Animal products, including carcasses and fresh meat, are easily contaminated with microorganisms and support their growth if not properly handled, processed and preserved. A variety of sources, including air, water, soil, feces, feed, hides, intestines, lymph nodes, processing equipment, utensils and humans, contribute to the microbial contamination of the sterile muscles of healthy animals during slaughter, fabrication and further processing and handling (Bell, 1997; Gill, 1998; Sofos, 1994). Extensive contamination, or abusive conditions of handling and storage that allow microbial proliferation, increase the potential for presence of pathogenic bacteria and formation of toxins, and may lead to product spoilage and public health problems (Sofos, 1994; Sofos et al., 1999a). According to the United States Centers for Disease Control and Prevention (CDC), foodborne bacterial pathogens caused the following estimated numbers of cases per 100,000 people in 1999: *Campylobacter* 21, *Salmonella* 12.4, *Escherichia coli* 0157:H7 2.8, *Yersinia enterocolitica* 1, *Listeria monocytogenes* 0.5 and *Vibrio* 0.3 (www.cdc.gov). However, in terms of fatalities, the picture is different, with *L. monocytogenes* causing the highest percentage of deaths.

An outbreak of *Escherichia coli* 0157:H7 in the western United States in 1992-1993 was associated with consumption of undercooked ground beef patties at restaurants of a major chain and led to development of illness in several hundred people and four deaths (Bell et al., 1994; Sofos and Smith, 1993). This highly publicized outbreak may be considered as the beginning of intensified public scrutiny on food safety that has led to major developments, including the complete change of a meat inspection system which was in place since the early 1900s. The new United States Meat and Poultry Inspection Regulation was established in 1996 (FSIS-USDA, 1996a) and requires federally inspected slaughtering plants to: (1) establish sanitation standard operating procedures; (2) implement the hazard analysis critical control point (HACCP) system of process control (NACMCF, 1998); and, (3) apply microbial testing for *Escherichia coli* counts and *Salmonella* incidence as criteria of HACCP verification and pathogen reduction, respectively. In implementing the HACCP principles, slaughtering plants must address hazards defined as physical or foreign objects, natural toxins (e.g., mycotoxins), zoonotic diseases (e.g., bacteria, parasites), other microbial and chemical contamination, as well as food additives and product decomposition.

The incidence on animals before slaughter of pathogens such as *E. coli* 0157:H7, appears to be increasing in recent years (Elder et al., 2000), which necessitates employment of approaches for their control to enhance food safety. In efforts to increase pressure on the meat processing industry to control the incidence of *E. coli* 0157:H7 and other pathogens in meat, the Food Safety and Inspection Service (FSIS) of the United States Department of Agriculture (USDA) has been enforcing a zero tolerance policy for visible soil on carcasses during slaughter and has declared *E. coli* 0157:H7 an adulterant (i.e., a prohibited contaminant) in fresh ground beef and other non-intact fresh beef cuts (FSIS-USDA, 1993). Testing of fresh beef for this pathogen has resulted in several, highly publicized, product recalls from the marketplace. In addition, these food safety concerns led to the major overall revision of the United States Meat and Poultry Inspection Regulations, mentioned above. Furthermore, publicity over these and other food safety issues has led to initiatives such as: (1) the United States National Food Safety Initiative and associated programs or activities such as the FoodNet(r) and PulseNet(r) foodborne illness surveillance networks; (2) the FightBac(r) and Thermy(r) educational programs; (3) emphasis on risk assessment studies and evaluations; and, (4) an increase in federal funding for food safety research and education issues (http://www.foodsafety.gov).

The meat processing industry, specifically the animal slaughtering sector, has also undertaken initiatives and efforts to comply with consumer demands for food safety, customer specifications, and regulatory requirements included in the new meat and poultry inspection regulations (FSIS-USDA, 1996a). The objective of the meat processing industry is to improve operations through implementation of HACCP requirements, employment of various carcass decontamination interventions (Smulders and Greer, 1998; Sofos and Smith, 1998), and verification of compliance with regulatory or customer requirements through microbial testing.

The producers of animals have contributed to the overall effort of improving food safety by supporting development and applying quality assurance programs developed through their associations, and by financially supporting research and development studies on microbial reduction interventions applied during animal slaughter. All segments, including, regulators, consumers, health authorities, research scientists and the industry agree that efforts should be made to reduce incidence and eliminate or control pathogenic bacteria at all stages of the food chain. The objective of this chapter is to present an overall picture of problems and approaches to pre-harvest food safety assurance in the United States without presenting a detailed review of research on the subject.

2. Statement of the problem

Live animals are often highly contaminated, or are asymptomatic carriers of pathogenic bacteria (Fedorca-Cray et al., 1998; Hancock et al., 1997a,b; Letellier et al., 1999; Skjerve et al., 1998), and can serve as sources of subsequent meat contamination. Animal cleanliness is influenced by climate, geographic location, method of transportation and holding conditions. For example, animals raised on pastures may carry more bacteria of soil origin, while microorganisms of intestinal origin may be more common on carcasses from animals finished in feedlots (Sofos, 1994; Sofos et al., 1999a). Pathogenic bacteria of animal enteric origin such as *E. coli* O157:H7, other enterohemorrhagic serotypes, and *Salmonella* are of special interest because of their association with live animals and their major adverse effects on human health (Armstrong et al., 1996; Buchanan and Doyle, 1997; Mead et al., 1999). Thus, microbial contaminants are a major concern because they originate from the live animals or their environment and may contaminate carcasses or colonize the plant environment and serve as sources of cross-contamination of carcasses, meat and resulting products (Sofos et al., 1999a). Carcass and environmental cross-contamination with *E. coli* O157 and other pathogens may occur during slaughter. Chapman et al. (1993) isolated *E. coli* O157 from 30% of 23 carcasses of rectal swab positive cattle and 8% of 25 carcasses of rectal swab negative cattle.

Fecal shedding of *E. coli* O157:H7 by cattle is widespread and intermittent (Wells et al., 1991), and is affected by season and age of the animal. Shedding peaks in spring and summer months (Hancock et al., 1997a,b; Mechie et al., 1977; Wells et al., 1991), is more likely in younger animals (Harmon et al. 1999; Garber et al., 1995, 1999), and may be influenced by diet (Dargatz et al., 1997; Hancock et al., 1994; Herriot et al., 1998). Shedding of *E. coli* O157:H7 has been reported to be more likely after than before weaning (Corrier et al., 1990; Garber et al., 1995; Zhao et al., 1995). Elder et al. (2000) conducted a survey to estimate the frequency of *E. coli* O157 in feces, on hides and carcasses of fed cattle in the United States. The pathogen was present in 72% of fecal samples from 29 lots of animals tested and in 38% of hide samples. Of 30 lots of carcasses sampled, 87% were positive at pre-evisceration, 57% after evisceration and 17% after processing. Prevalence of *E. coli* O157 was 28% on feces, 11% on hides, and on carcasses it was 43% at pre-evisceration, 18% at post-evisceration and 2% after processing (Table 1).

The reduction of prevalence between pre-evisceration and post-processing indicates that sanitary procedures and decontamination interventions applied during slaughter reduce contamination (Dorsa, 1997, Smulders and Greer, 1998; Sofos and Smith, 1998). The results

Table 1. Prevalence of Escherichia coli *0157 on fecal, hide and carcass samples of fed beef cattle taken at slaughter (July-August, 1999).*

			Carcass		
Samples	Feces	Hide	Pre- evisceration	Post- evisceration	Post- processing
Lots Sampled	21/29 (72)[a]	11/29 (38)	26/30 (87)	17/30 (57)	5/30 (17)
Total Samples	91/327 (28)	38/355 (11)	148/341 (43)	53/332 (18)	6/330 (1.8)

Elder et al. (2000)
[a]Positive/tested (%)

also showed that fecal and hide prevalence were significantly correlated with carcass contamination, which suggests that control of *E. coli* 0157 on live cattle may be useful in food safety enhancement. Previous surveys found a lower prevalence of *E. coli* 0157 on fed cattle (Gansheroff and O'Brien, 2000), which may have been due to limitations of testing methodologies employed in earlier studies. A study by Wells et al. (1991) found that the prevalence of shedding *E. coli* 0157:H7 was 2.3, 3 and 0.15% in calves, heifers and adult cattle, respectively. Another enteric pathogen of concern is *Salmonella*. Wells et al. (2001) collected fecal samples from representative cows in 91 dairies and 97 cull dairy cow markets in 17 states and recovered *Salmonella* from 5.4% of milk cows, 18.1% of milk cows expected to be culled within 7 days, and 14.9% of culled dairy cow markets. Herds positive for *Salmonella* were more common in the period of May through July and in herds from southern states. Thus, there is a need to control pathogenic microorganisms in animals and their products in order to enhance the safety of our food supply. In order to develop management practices that will help in the control of the prevalence of pathogens in animals and their products, there is a need to determine risk factors. Factors to be considered for application at the pre-harvest stage include animal fasting, feeding and stressing practices such as those applied during animal confinement and transportation, amount of roughage and other dietary components, animal cleanliness, etc. (Cray et al., 1998; Dargatz et al., 1997; Diez-Gonzalez et al., 1998; Hadley et al., 1997; Herriot et al., 1998; Jordan and McEwen, 1998; Sofos et al., 1999a; van Donkersgoed et al., 1997).

3. Animal Slaughtering Industry Approaches To Food Safety

Extensive studies in the 1990s confirmed and expanded previous findings that carcass decontamination by various interventions (e.g., hot water, steam-vacuuming, organic acid rinses or steam) during animal slaughter reduces contamination (Dorsa, 1997; Smulders and Greer, 1998; Sofos, and Smith, 1998; Sofos et al., 1999a). Results indicated that commercial decontamination procedures in beef slaughtering plants reduced overall contamination and pathogen levels (FSIS-USDA, 1996b; Gorman et al., 1995; Hardin et al., 1995; Kochevar et al., 1997; Phebus et al., 1997; Reagan et al., 1996; Sofos et al., 1999b) and increased the probabilities of passing regulatory *E. coli* performance criteria (Sofos et al., 1999c) and *Salmonella* pathogen reduction performance standards (Sofos et al., 1999d) set in the new United States Meat and Poultry Inspection regulations (FSIS-USDA, 1996a) compared to levels of contamination before application of carcass decontamination interventions. Other studies have provided data demonstrating the value of multiple decontamination treatments applied in sequence in the laboratory (Graves Delmore et al., 1998) and in commercial plants (Bacon et al., 2000). Data released by USDA-FSIS (http://www.fsis.usda.gov/ophs/haccp//salmraw.htm) for prevalence of *Salmonella* in commercial plants indicated that the performance standards were met for carcasses of all species in the period 1998-99 in large plants, while in small plants only hog carcasses failed to meet their performance standard (Table 2).

Table 2. *Prevalence (%) of Salmonella in meat products (USA) in 1998-2000.*

		Large Plants		Small Plants	
Product	Regulatory Performance Standard	1998-1999	1999-2000	1998-1999	1999-2000
Broilers	20.0	10.9	8.0	16.3	13.6
Hogs	8.7	6.5	3.7	18.2	13.5
Cows/Bulls	2.7	0	0	2.3	1.7
Steers/Heifers	1.0	0	0.3	0.2	0.2
Ground Beef	7.5	4.8	9.1	4.3	4.7
Ground Chicken	44.6	4.4	8.8	18.9	17.7
Ground Turkey	49.9	36.4	29.8	27.0	30.2

(http://www.fsis.usda.gov/ophs/haccp/salmraw.htm)

Salmonella prevalence was lower for most species in the period 1999-00 compared to 1998-99. Overall, percentages of samples meeting the *Salmonella* standards in 1998-00 were in the range of 82 to 100 (Table 3).

Thus, it appears that interventions applied during animal slaughter and dressing are effective in reducing the level of contamination introduced on carcasses as their hide, fleece or feathers are removed at the initial stages of the slaughtering process. In general, it is believed that carcass decontamination interventions contribute to the production of carcasses with lower levels of contamination, and that reduced incidence of enteric pathogens helps in meeting regulatory requirements during slaughter (Sofos et al., 1999a) and should have an impact in the overall reduction of incidence of foodborne illness.

4. Responsibilities Of The Pre-harvest Sector

Animal producers are often pressured to implement pre-harvest pathogen control strategies. Reasons for employment of pre-harvest pathogen control strategies on animals

Table 3. Percentage of samples meeting Salmonella *performance standards (USA) in 1998-2000.*

	Large Plants		Small Plants		Combined
Product	1998-1999	1999-2000	1998-1999	1999-2000	
Broilers	91	95	79	87	92
Hogs	71	95	47	63	82
Cows/Bulls	100	100	77	83	84
Steers/Heifers	100	100	100	100	100
Ground Beef	90	82	87	87	87
Ground Chicken	NC	100	100	100	100
Ground Turkey	91	91	83	67	82
Total	88	94	84	84	88

(http://www.fsis.usda.gov/ophs/haccp/salmraw.htm)

include (Hancock and Dargatz, 1995; Hancock et al., 1997c): (1) the fact that animals and their environment constitute the main reservoir for most foodborne pathogens; (2) the increasing number of foodborne illness episodes associated with foods derived from animals or water and foods derived from plants which become cross-contaminated with animal waste; (3) DNA fingerprinting having shown that pathogen isolates may match from the farm through foodservice; and, (4) the fact that animals are usually asymptomatic carriers of human foodborne pathogens.

In recent years, the United States Department of Agriculture (USDA) Food Safety and Inspection Service (FSIS) together with other government agencies have undertaken several initiatives as part of a farm-to-table food safety assurance strategy. These initiatives include: (1) the implementation of the in-plant Pathogen Reduction and HACCP rule (FSIS-USDA, 1996a); (2) a risk assessment effort; (3) consumer education activities (FightBac(r) and Thermy(r)); (4) food product transportation and storage issues; (5) retail food safety issues; (6) foodborne illness surveillance efforts (FoodNet(r) and PulseNet(r)); (7) international trade issues; and, (8) endorsement of HACCP-compatible practices in animal production (http://www.fsis.usda.gov/oppde/ap/Presentations). In order to address pre-harvest food safety issues, the USDA-FSIS has established the Animal Production Food Safety (APFS) group under the Office of Policy, Program Development and Evaluation (OPPDE). This group was formed to provide a mechanism for the development of food safety programs at the pre-harvest level (http://www.fsis.usda.gov/oppde/ap/default.htm). The mission of the APFS staff is to: (1) provide assistance and leadership to foster the research needed to develop voluntary science-based food safety practices, and verification procedures for food animal production that will reduce the risk of chemical, physical and microbial hazards entering the food chain; (2) develop and maintain cooperative relationships and state-based partnerships to support FSIS food safety initiatives; (3) communicate information needed by the animal production sector to assist them in meeting reasonable, science-based requirements of animals at the receiving stage of processing; and, (4) work cooperatively with information multipliers to promote adoption of good production practices by producers and suppliers, which result in the safest and best quality animals being presented to meat and poultry slaughtering and egg processing plants implementing HACCP systems (http://www.fsis.usda.gov/oppde/ap/default.htm).

As indicated on its internet web page (http://www.fsis.usda.gov/oppde/ap/Presentations) USDA-FSIS follows a voluntary intervention, instead of a regulatory, approach in the implementation of HACCP compatible practices in the animal production sector. The reason for this strategy is that FSIS has no statutory authority for on-farm activities. However, FSIS provides assistance and leadership for: (1) adoption of animal production food safety assurance practices; (2) encourages, guides and supports research and educational activities; (3) works with information multipliers for voluntary implementation of food safety assurance programs; (4) forms partnerships with states, producer groups, consumer representatives, universities, etc., to enhance implementation of pre-harvest food safety assurance practices; and, (5) recommends the implementation of food safety assurance programs based on the latest scientific principles and HACCP concepts, especially for use of pesticides, antibiotics and veterinary drugs. The goal stated by FSIS is to assist the animal production industry in implementing voluntary programs based on science and HACCP principles (http://www.fsis.usda.gov/oppde/ap/Presentations).

Although it has no statutory authority a the pre-harvest level, the regulatory authority of FSIS covers meat and poultry slaughtering and processing plants, which must follow the requirements of the new United States Meat and Poultry Inspection Regulation of 1996 (FSIS-USDA, 1996a). This regulation states that slaughtering plants "must conduct a hazard analysis to determine the food safety hazards reasonably likely to occur before, during and after entry into the establishment." Furthermore, plants are required to meet microbiological performance standards and are "responsible for preventing illegal, or violative residues from adulterating their meat and poultry products. In order to control potential physical, chemical or microbiological hazards, including those occurring before entry into their establishment (pre-harvest), slaughtering plants should develop proper preventive approaches, such as: (1) reject at-risk animals; (2) sort animals in groups according to risk; (3) conduct testing; (4) inspect producers and review their records; (5) demand implementation of quality assurance programs by producers; (6) obtain certification letters from producers; etc. In order for the animal slaughtering industry to be successful in meeting its regulatory requirements (FSIS-USDA, 1996a), it may apply pressure on the pre-harvest sector to employ HACCP-compatible practices that may include: (1) herd health management practices; (2) residue avoidance or control programs; (3) pathogen reduction strategies; (4) animal and premise identification; (5) recordkeeping and management of records; (6) quality production practices that reduce diseases and drug use; (7) quality assurance program certification; and, (8) good sanitation and hygiene practices during production, marketing, transportation and preparation of animals for slaughter. The animal industry associations of the United States, such as the National Cattlemen's Beef Association and the National Pork Producers Council, have developed Production Quality Assurance Programs and encourage their members to adopt and implement these programs (http://www.fsis.usda.gov/oppde/ap/default.htm).

In efforts to contribute to the enhancement of food safety through activities implemented pre-harvest, a National Conference on Animal Production Food Safety was held on 6-7 of September 2000 in Saint Louis, Missouri, United States (Anonymous, 2000). The conference was sponsored by various agencies of USDA including FSIS, Agricultural Research Service (ARS) Animal and Plant Health Inspection Service (APHIS), Agricultural Marketing Service (AMS) and Cooperative States Research, Education and Extension Service (CSREES) as well as the Center for Veterinary Medicine (CVM) and the Center for Food Safety and Applied Nutrition (CFSAN) of the Food and Drug Administration (FDA) of the Department of Health and Human Services (DHHS). The aim of the conference was to: (1) review the current status of food safety at the production level; (2) provide an update on industry quality assurance activities; (3) review ongoing research in support of animal production food safety practices; (4) discuss gaps in research to address food safety at the animal production level; and, (5) identify additional educational efforts needed to improve food safety at the animal production level (http://www.fsis.usda.gov). It is interesting to quote excerpts of the remarks of Thomas J. Billy, Administrator of FSIS and Chairman of the Codex Alimentarious commission prepared for the National Conference on Food Safety (6-7 September 2000, St. Louis, MO, USA). These excerpts indicate the FSIS approach to pre-harvest food safety assurance at that time (http://www.fsis.usda.gov/OA/speeches/tb_apfs.htm). The FSIS farm-to-table approach to food safety "includes encouraging the animal production community to implement management practices that improve food safety, setting new requirements for slaughter and processing plants, ensuring that products remain safe as they move through distribution channels, and educating consumers and other food handlers to prepare and store food safely. It was also stated in the remarks of Mr. Billy that "when we use the term 'farm-to-table,' we by no means see FSIS taking regulatory action on the farm. We have no regulatory authority in this area. Rather, FSIS supports research and educational activities that promote the adoption of voluntary, industry-implemented food safety and quality assurance programs. In addition, it was stated that FSIS has provided information to producers about HACCP and how its implementation might affect their ability to market their animals for slaughter, working closely with other federal, state and local government agencies. With HACCP clarifying industry's responsibility for food safety, slaughter plants are move seriously considering the potential hazards in incoming animals when developing their preventive plans to the extent to which they can be addressed during animal production. FSIS believes HACCP-compatible practices at the animal production level include producer record keeping, good hygiene, herd health management, residue avoidance, and, where appropriate pathogen reduction strategies." (http://www.fsis.usda.gov/OA/speeches/tb_apfs.htm). At the conclusion of the conference, Dr. Lester Crawford of the Center for Food and Nutrition Policy at Georgetown University (Washington, DC) summarized the recurring themes of the conference as follows: (1) producers expressed questions and concerns relative to mechanisms for determination and establishment of criteria for pre-harvest food safety approaches and associated costs and profits; (2) verification procedures and third party auditing of quality and safety assurance programs should be established; (3) there should be international cooperation and countries should be informed and involved as pre-harvest food safety policies are developed; (4) international organizations, such as the World Health Organization (WHO), Food and Agriculture Organization (FAO) and the World Trade Organization (WTO) must be involved and serve as resources in the process of establishing pre-harvest food safety assurance requirements and criteria; (5) a specific major food safety concern identified was the issue of antibiotic resistance development; and, (6) the necessity for reliable analytical methodology development was identified.

5. Chemical Residues and Antimicrobial Resistance Control Approaches

The issue of chemical residues in animal products is considered as part of the hazard identification and control process of the HACCP regulation. The approach of controlling chemical residues in the United States before the HACCP regulation (FSIS-USDA, 1996a) became effective was based on FSIS enforcement activities and the National Residue Monitoring Program (http://www.fsis.usda.gov/oppde/ap/Presentations/leman.htm). Under that approach, FSIS inspectors collect samples of animal products at slaughtering facilities, which are analyzed for violative residue concentrations, as established by the United States Environmental Protection Agency (EPA) for pesticides and by the FDA for animal drugs and environmental contaminants. The new inspection system, established by the 1996 regulation (FSIS-USDA, 1996a) provides a new framework for hazard control, including chemical hazards. Part 417.2 of Title 9 of the United States Code of Federal Regulations requires inclusion of chemical residues in the hazard analysis. If the hazard analysis identifies chemical residues as a food safety hazard reasonably likely to occur during production, a critical control point must be established to control the hazard; failure to do so would result in adulterated product and discontinuance of plant inspection, and, thus operation. Since responsibility for chemical residue prevention starts with animal production, the farm sector will have to be involved in chemical residue control (http://www.fsis.usda.gov/oppde/ap/Presentations/leman.htm).

An important food safety issue is the increase in antimicrobial resistance development in bacterial isolates (Anonymous, 1999; Brisabois et al., 1997; Davis et al., 1999; Ferber, 2000; Threlfall et al., 2000; Tollefson, 1999; Tollefson et al., 1999). To address this issue, a United States interagency Task Force on Antimicrobial Resistance developed a Public Health Action Plan to Combat Antimicrobial Resistance. The Task Force consisted of individuals from the CDC, FDA, National Institutes of Health (NIH), USDA, United States Department of Defense (USDD), United States Department of Veteran Affairs (USDVA), EPA, Agency for Health Care Research and Quality, Health Care Financing Administration, and Health Resources and Services Administration. This multi-agency Task Force was led by the CDC, NIH and FDA. The plan designates priorities, identifies responsible agencies and creates timelines to deal with the issue of antimicrobial resistance (http://www.cdc.gov/drugresistance/). The Task Force dealt first with domestic, rather than international, issues of antimicrobial resistance and identified four major components in the proposed plan of action: (1) surveillance; (2) prevention and control; (3) research; and, (4) product development. Under surveillance, CDC will work with state health departments and others to design and implement a plan that will define responsibilities for antimicrobial resistance surveillance at the national, regional, state, and local level, and to coordinate methodologies and activities of the various entities. In addition, CDC, FDA and USDA plan to develop systems to monitor patterns of use of antimicrobials in human medicine, agriculture and consumer products. Under prevention and control, the partners identified above will launch a national public education campaign to reduce abuse, misuse or overuse of antimicrobials in the health care system. Research efforts will be led by NIH to provide the research community with new information technologies, including genetic fingerprints for microorganisms, to identify targets for new diagnostics, treatments and vaccines that are needed to prevent the emergence and spread of resistant pathogens. Furthermore, NIH plans to develop clinical studies to evaluate new antimicrobials and novel approaches against resistant pathogens. The United States Department of Health and Human Services plans to create an Interagency Antimicrobial Product Development Working Group in order to address the area of product development by identifying and publicizing priority health needs for new products that prevent or treat resistant infections (http://www.hhs.gov/news).

6. Pre-harvest Microbial Pathogen Control Approaches

Although research efforts are in progress, currently, there are no scientifically established critical control or management points available to deal with foodborne pathogenic microorganisms at the pre-harvest level (Smith et al., 1997). The issue of reducing pathogens in animals is challenging because: (1) scientific information available is limited; (2) there are numerous and complicating variables involved; (3) specific pathogen reservoirs are not well identified; (4) presence of certain pathogens is ubiquitous; (5) pathogen detection methodologies at the farm level are still developing; (6) interventions are not well identified; (7) verification and auditing approaches need to be developed; and, (8) economic issues need to be addressed and resolved. However, pathogen control at the pre-harvest sector may be approached in various ways, including: (1) testing of live animals for presence of pathogens; (2) traceback of contaminated animals and eradication of positive herds; (3) visits and inspections of livestock operations at the farm level; (4) development and administration of animal vaccines; (5) competitive exclusion or inhibition approaches against target pathogens; (6) on-farm niche engineering practices; and, (7) application of pre-slaughter measures to eliminate, control or reduce contamination (Hancock and Dargatz, 1995; Hancock et al., 1997c).

Testing of live animals, or even batches of finished product such as ground beef, for presence of pathogens is a simplistic approach that may result in elimination of certain animals or product batches found to be positive without ensuring that animals or batches found to be negative are in effect free of pathogens. This testing and rejection approach has major drawbacks because: (1) there are limitations associated with sampling protocols; (2) extent of testing may be impractical and limited by costs; (3) testing methods are still under development and have inherent deficiencies; and, (4) pathogen incidence is either low, nonrandom, or unpredictable, which makes testing very costly and inefficient (AMSA, 1999; Brown et al., 2000; Gill, 2000). However, it should be noted that various organizations and groups of scientists agree that microbiological testing in process control is appropriate and necessary when used for validation, maintenance and verification of effective HACCP programs, but such testing should involve enumeration of indicator microorganisms rather than testing for the detection of pathogens, due to the above drawbacks.

Control of pathogens through pathogen traceback and eradication is also unlikely to be successful because the conditions for its success cannot be met for animal-associated pathogens, (Hancock and Dargatz, 1995; Hancock et al., 1997c). These conditions include: (1) a pathogen may be eradicated only when it has a single host species; (2) pathogen eradication programs are economically feasible only when the pathogen is present in a small portion of farms; (3) procedures for identification of carriers should be available, effective and economical; and, (4) long-term eradication costs may be high and the resolve for application of eradication or depopulation measures may be lacking. Farm animal sources of *E. coli* O157:H7, in addition to cattle, may include sheep, pigs and deer (Chapman and Ackryod, 1997; Chapman et al., 1997; Kudva et al., 1997). Studies found that fecal samples from feedlots may be over 60% positive for *E. coli* O157:H7 (Hancock et al., 1997b). Studies by the Veterinary Services of Animal and Plant Health Inspection Service (APHIS) of USDA have concluded that pens of cattle that had been on feed for less than 20 days were 3.4 more likely to have had a sample positive for *E. coli* O157:H7 compared to those on feed for longer periods of time (Anonymous, 1997). In addition, pens receiving barley in the ration were 2.75 times more likely to have a positive sample than those not receiving barley; pens of cattle with entry weights of 350 kg or more were less likely to have had positive samples; pens of at least 85% heifers were less likely to have a positive sample; and there was no association between positive fecal samples and ionophore use. Furthermore, the incidence of *E. coli* O157:H7 has been found to be widespread in dairy farms, but at a low prevalence rate (Hancock et al., 1994; Zhao et al., 1995). These facts indicate that the traditional eradication approaches of testing, quarantine and slaughter are impractical. Although animal and meat fingerprinting, identification and traceback approaches, procedures, methods and mechanisms are becoming available, their employment may prove impractical in tracing back to the farm a pathogen that may have been isolated from ground beef which may have originated from a lot of beef trimmings representing many animals or sources of product (Troutt and Osburn, 1997).

Control of pathogens through farm inspections is limited by lack of specific objectives to be accomplished by the inspection and which would lead to pathogen control on the resulting meat. Furthermore, this would be a major and costly task (Hancock and Dargatz, 1995; Hancock et al., 1997c). Control of pathogens, through animal vaccinations may appear promising and worthwhile for some pathogens, but extensive and costly research is needed before practical use of vaccines can be recommended. Competitive exclusion (CE) or inhibition of pathogens with probiotics, which appears promising for further research, may also be successful to some degree and for certain species of food animals, provided that additional research is completed for the development and establishment of competitive microbial flora in the animal intestine (Hancock and Dargatz, 1995; Hancock et al., 1997c).

Competitive exclusion has been studied extensively against *Salmonella* colonization in poultry (Hinton et al., 1990; Nisbet et al., 1993; Nurmi and Rantala, 1973). When young chicks are inoculated with a CE culture derived from *Salmonella*-free adult chickens, the CE culture is established in the digestive track early and inhibits colonization with pathogens. Inhibition of colonization is believed to be due to lower pH, competition for attachment

sites or nutrients, and production of acids including lactic and volatile fatty acids in the digestive tract, or other mechanisms (Bailey et al., 1998; Corrier et al., 1993, 1995; Fuller, 1997; Ha et al., 1994; Hume et al., 2001; Nisbet et al., 1993, 1996; Salminen, 1999; Sanders, 1999; Stavric et al., 1987; Zhao et al., 1998). Many studies have found that exposing newly hatched chicks to intestinal microflora of older chickens decreases cecal colorization with *Salmonella* (Corrier et al., 1993; Hinton et al., 1991, 1992; Nisbet et al., 1993, 1996). Studies have also shown that avian *Lactobacillus* cultures exert competitive exclusion on *Salmonella* adhesion to the crop mucosa and reduce cecal colonization (Soerjadi et al., 1981, 1982). Tellez et al. (2001) evaluated the effect of an avian-specific probiotic combined with *Salmonella* antibodies on the cecal colonization and organ invasion of *Salmonella* in broilers. The probiotic-treated birds had a significantly lower concentration of *Salmonella* cecal colonization compared to the control group. Organ invasion by *Salmonella* was also lower in the probiotic-treated group, while there was no effect of the probiotic on body weight. The results indicated that a combination of *Lactobacillus acidophilus*, *Streptococcus faecium*, and *Salmonella enteritidis*, *Salmonella Typhimurium* and *Salmonella Heidelberg*-specific antibiodies had a beneficial effect in reducing the colonization of *S.* Enteritidis in market-aged broilers (Tellez et al., 2001). Hume et al. (2001) used continuous flow (CF) chemostat cultures as models to determine the potential usefulness of undefined porcine cecal bacteria as CE cultures against colonization by *S. typhimurium*. The results indicated that in vitro CF cultures of porcine cecal bacteria inhibited growth of *S.* Typhimurium and that inhibition was not restricted by prior exposure of the cecal bacteria to chlortetracycline. It was concluded that CF cultures have potential for use as models to identify CE cultures against *Salmonella* in pigs. Probiotics have also been found effective in reducing *E. coli* 0157:H7 in cattle (Zhao et al., 1998). Additional research and development is needed for wider application of the CE approach in pathogen control at the pre-harvest level.

Reducing prevalence of *E. coli* 0157:H7 in animals such as cattle should contribute to lowering contamination of meat, water and other foods (Gansheroff and O'Brien, 2000). Strategies proposed to be considered for reduction of pathogens in cattle include modification of farm practices and the bovine diet, vaccinations and animal treatment with probiotics or lytic phages (Gansheroff and O'Brien, 2000; Kudva et al., 1999; Zhao et al., 1998). On farm niche engineering offers promise and should be investigated, but success may be limited. It is useful to determine farm management practices that may allow increased pathogen prevalence in order to develop intervention strategies for their control (Hancock and Dargatz, 1995; Hancock et al, 1997c). The objective of the on-farm niche engineering approach for pathogen control at the pre-harvest level is to modify nutritional and environmental variables with the potential of reducing pathogen colonization in animals (Hancock and Dargatz, 1995; Hancock et al., 1997c). Specific areas of interest include animal management practices, feed additives, feed management practices, water trough management, pen management, and manure handling practices. For example, Garber et al. (1999) reported that fecal samples collected during the summer months from dairy operations were more likely to be contaminated with *E. coli* 0157. In addition, this study (Garber et al., 1999) indicated that flushing farm alleyways with water to remove manure increased the likelihood for samples to test positive for *E. coli* 0157. Studies have demonstrated that fecal shedding of *E. coli* 0157:H7 by feedlot cattle is highest in the

period of time immediately after weaning (Dargatz et al., 1997; Hancock et al., 1997a,b). This may be due to a potentially rapid transmission of bacteria among animals caused by the stress of weaning, transportation, diet changes, or overall animal handling. Laegreid et al. (1999) reported results indicating that the higher prevalence of *E. coli* 0157:H7 in fall weaned calves which had been on feed for the shortest period of time may be due to more calves arriving at the feedlot already infected with the pathogen. Based on this, the authors suggested that control measures may need to be implemented in the cow-calf herd and early in the life of a calf. It was also stated that further research is needed to determine on-farm ecological factors that may be influencing within and between herd transmission of *E. coli* 0157:H7 as well as interventions to control pathogen incidence (Laegreid et al., 1998).

Diez-Gonzalez et al. (1998) suggested that short duration hay feeding could be a viable pre-harvest approach for enhancement of food safety. The results indicated that switching the animals from grain to hay feeding for four days resulted in reduction of the prevalence of acid-resistant *E. coli* in the feces. Feeding high grain diets may allow accumulation of undigested starch in the colon which can then be fermented to yield volatile fatty acids and a reduced pH which could enhance growth of acid-resistant *E. coli*. This could lead to the hypothesis that reducing the starch load could reduce the numbers of *E. coli* 0157:H7, an acid-resistant serotype (Diez-Gonzalez et al., 1998; Scott et al., 2000). The conclusions of the study by Diez-Gonzalez *et al.* (1998) became the subject of dispute by correspondence (Hancock et al., 1999; Russell, 2000; Russell et al., 1999). Hovde et al. (1999) found that cattle perorally administered *E. coli* 0157:H7 and fed roughage or hay shed the pathogen longer than animals fed concentrate diets, while shed bacteria were not different in acid tolerance. It should be noted that, hay feeding is not considered a practical approach in cattle feeding for replacing starch. However, Russell et al. (2000a,b) suggested that cattle feeding management practices may be manipulated to decrease the risk of contamination with *E. coli* 0157:H7, but more work is needed to verify these effects. The amount of starch could be reduced by including wet corn gluten and high moisture corn in the diet. Based on this, Scott et al. (2000) developed the hypothesis that manipulation of the finishing diet to reduce the amount of starch could cause an increase in colonic pH and a reduction in numbers of acid-resistant *E. coli*. The results indicated that manipulation of finishing diets either by substituting ingredients or limit-feeding increased colonic pH, but it was not associated with reduced population of acid-resistant *E. coli*. Thus, Scott et al. (2000) reported that manipulation of finishing diets did not reduce shedding of acid-resistant *E. coli* in feces. However, short duration hay feeding reduced shedding of acid-resistant *E. coli* in the feces. Manipulating the diet by limit-feeding of finishing diets did not affect total or acid-resistant *E. coli* populations. Altering dietary ingredients did not affect total *E. coli* populations; however, steers fed diets containing dry-rolled or high moisture corn had lower acid-resistant *E. coli* populations. Switching steers to alfalfa hay lowered both total and acid resistant *E. coli* populations.

Loy et al. (2001) reported that when corn bran and wet corn gluten feed were substituted for dry rolled corn, colonic pH of feedlot steers increased until volatile fatty acid concentration decreased. The corn bran and the wet corn gluten feed diet reduced acid resistant *E. coli* shedding. Restricting intake increased colonic pH and decreased volatile

fatty acid concentration, but did not affect acid-resistant *E. coli* shedding. Prevalence of *E. coli* 0157:H7 was not affected by diet or feed intake. Other studies found that calves fed a high roughage diet had lower mean rumen volatile fatty acid concentrations and higher rumen pH values than calves fed a high concentrate diet. However, the calves fed the high roughage diet did not have higher rumen populations of *E. coli* 0157:H7 and did not show higher or longer fecal shedding of the pathogen compared to calves fed the high concentrate diet (Tkalcic et al., 2000). There was a significantly positive correlation between fecal shedding and rumen volatile fatty acid concentration in calves fed the high concentrate diet. It was concluded that the impact of diet on fecal shedding of *E. coli* 0157:H7 is still unclear and may depend on dietary effects, on fermentation in the colon and on diet-induced changes in the resident microflora. The authors, however, presented in vitro data indicating that *E. coli* 0157:H7 may develop acid tolerance in the rumen of calves fed high concentrate diets, which could allow larger populations to survive passage through the acidic abomasum to proliferate in the colon and increase fecal shedding of the pathogen (Tkalcic et al., 2000). It should be noted that the gastrointestinal microenvironment is complex and variable between individual animals and does not allow predictions of the outcome of various manipulations (Tkalcic et al., 2000). However, identifying and understanding potential factors that may affect shedding and resistances of intestinal pathogens may lead to selection of possible measures that may lead to an overall control of pathogen shedding. For this to occur, there is a need for additional studies to evaluate the effect of factors such as diet on pathogen survival, acid tolerance and shedding.

Buchko et al. (2000) fed three groups of yearling steers 85% cracked corn, 15% whole cottonseed and 70% barley, or 85% barley. The steers were inoculated with 10^{10} CFU of *E. coli* 0157:H7 and the presence of the inoculated strain was followed in the rumen fluid and feces for a 10-week period. The organism was rapidly eliminated from the rumen of the animals on all three diets but persisted in the feces of some animals for up to 67 days after inoculation, suggesting that the bovine hindgut is the site of *E. coli* 0157:H7 persistence. The number of animals that were culture positive for *E. coli* 0157:H7 during the 10 week period was significantly higher for the barley fed group compared to the other two groups. It appears that control of fecal shedding of *E. coli* 0157:H7 by cattle requires a multifaceted approach (Buchko et al., 2000), especially because, in addition to shedding, there is recurrent exposure to environmental sources of the pathogen that may originate from animal shedding. Thus, potential diet modifications should be coupled with control of environmental sources of the pathogens.

Anderson et al. (2000) hypothesized that, since *E. coli* 0157:H7 and *S.* Typhimurium possess respiratory nitrate reductase which reduces chlorate to the lethal chlorite ion, while most anaerobic bacteria do not possess the enzyme, chlorate may selectively kill these two pathogens without affecting the potentially beneficial anaerobes. Their research data indicated that populations of the two pathogens were reduced from approximately 6 log CFU to $\leq$10 CFU after in vitro incubation (24h) in buffered (pH 6.8) ruminal contents containing 5 mM added chlorate. The bactericidal effect of chlorate against *E. coli* 0157:H7 and *S.* Typhimurium was concentration dependent and higher at pH 6.8 compared to 5.6. Chlorate had little effect on anaerobe counts. Based on these findings, the authors

(Anderson et al., 2000) discussed the potential use of chlorate as an intervention to reduce pathogen prevalence immediately pre-harvest.

Anderson et al. (2001) proceeded to evaluate the chlorate treatment in pigs immediately before harvest. The study involved infecting weaned pigs orally with 8 x 10^7 CFU of a novobiocin and nalidixic acid-resistant strain of *S.* Typhimurium. Then the animals were treated after 8 and 16 h by oral gavage (10 ml) with 0 or 100 mm sodium chlorate. The pigs were euthanized at 8 h intervals after treatment and samples were analyzed microbiologically. The results indicated a significant chlorate treatment effect in reducing cecal concentrations of *Salmonella* (Anderson et al., 2001). Additional work by the same research group (Anderson and Nisbet, 2001) has indicated that giving pigs and sows sodium chlorate before slaughter may inactivate *Salmonella* and *E. coli* 0157:H7. Additional work is needed, however, to confirm these results in large scale field studies and to establish parameters for potential practical use.

There is evidence that feed withdrawal before slaughter may increase incidence of *E. coli* 0157:H7 in the feces of cattle (Harmon et al., 1999). Results of studies indicated that *Salmonella* and *E. coli* growth in ovine and bovine rumen may increase during fasting, while *E. coli* 0157:H7 grow poorly in vitro in rumen contents from well-fed compared to those from fasted cattle (Garber et al., 1995; Rasmussen et al., 1993; Zhao et al., 1995). Evidence also exists that increased holding time before slaughter may be associated with higher numbers of *E. coli* and *Salmonella* in the rumen at slaughter (Grau et al., 1968). Prevalence of verotoxigenic *E. coli* was reported as higher in cattle at slaughter than on the farm (Suthienkul et al., 1990). However, Harmon et al. (1999) presented results indicating that feed withdrawal and associated changes in rumen pH and volatile fatty acid concentrations had little effect on fecal shedding and rumen proliferation of *E. coli* 0157:H7 in cattle already shedding the pathogen. In addition, Van Donkersgoed et al. (1999) concluded that fasting may not be an important risk factor for increased fecal shedding of *E. coli* 0157:H7 in slaughter cattle under existing commercial transport and processing conditions in Western Canada.

Smith et al. (2001) reported that *E. coli* 0157:H7 prevalence did not differ between cattle feedyards, but did vary widely within feedyards; muddy pens were more likely to have higher prevalence than normal pens. Faith et al. (1996) concluded that there may be multiple sources of *E. coli* 0157:H7 on dairy farms, with animal drinking water being one of them. *Escherichia coli* 0157:H7 may survive for relatively long periods of time in the environment and farm practices such as the spreading of feces based fertilization may facilitate spread of the pathogen (Bolton et al., 1998). Swanenburg et al. (2001) reported that *Salmonella* was isolated in 70 to 90% of lairage samples from two pig slaughterhouses when pigs were present. The level was reduced to 25% and 10% following usual cleaning and improved cleaning and disinfecting, respectively. Based on the results, the authors (Swanenburg et al., 2001) concluded that a waiting period in the lairage of more than 2 h constitutes a substantial risk for slaughter pigs to become contaminated with *Salmonella*. Smith et al. (2000) found that coliform bacteria in water and biofilm were reduced by 99% and 99.9%, respectively, after draining, scrubbing, chlorine disinfecting

for 15 min, and refilling of feedlot water tanks. Levels of coliforms, however, returned to pre-treatment levels in 24-48 h following treatment and drinking by cattle.

The preslaughter external fecal load on the live animal has been reported as an important determinant of carcass contamination levels (Ridell and Korkeala, 1993). Pathogens from the hide and gastrointestinal tract are then transferred to the carcass during slaughter (Gill et al., 1976). Byrne et al. (2000) reported that thorough washing may be a suitable method for decontamination of animal hides before slaughter. Thus, one, seemingly obvious, approach that may contribute to the reduction of external animal contamination, and subsequently, carcass contamination is to clean or wash the hide of the animals before slaughter and dressing. Bolton et al. (1998) suggested that animals must be cleaned preslaughter to reduce fecal matter that can serve as a vehicle of carcass contamination with pathogens such as *E. coli* O157:H7. Pre-slaughter washing of sheep has been practiced in New Zealand (Biss and Hathaway, 1996), while, partial or complete, washing of cattle before slaughter has been used in Australia and by some plants in the United States. Individual operations have evaluated, or applied interventions, such as removal (by cutting or shearing) of hair and fecal tags from the exterior of the animals or washing of animals before slaughter, but in many instances the results are generally less than promising (Gill, 1998). In general, animal washing before slaughter has variable influence on carcass contamination. Furthermore, application of the procedure may be limited by climate, type of animal, and availability of facilities (Sofos and Smith, 1998; Sofos et al., 1999a). United States regulatory guidelines require cattle to be dry, or at least not dripping, when they are slaughtered (Reed, 1996), which can be a constraint when animal washing is considered before slaughter. However, when animals are wet or excessively soiled, slaughter speeds should be reduced to minimize accidental transfer of contamination from the exterior of the animals onto the carcass or the plant environment. In addition, modifications in the steps involved in hide removal, or in equipment used for hide removal, may help in minimizing transfer of contamination onto the carcass surface (Hadley et al., 1997). One approach that may help in the reduction of carcass contamination with pathogens may be to process highly contaminated or infected animals separately from cleaner or pathogen-free herds (Gill, 1998). This approach, however, may be impractical in some systems of animal production, marketing, distribution, and slaughtering, or for control of more than one type of pathogenic microorganisms on the same animals. Nevertheless, highly soiled animals are an important potential source of plant contamination, and presentation of clean animals for slaughter is desirable because it reduces the likelihood of pathogen presence and transfer onto carcasses (Bolton et al., 1998). However, poor sanitation, hygiene and manufacturing practices during slaughtering, fabrication and processing can lead to excessively contaminated meat, even when less heavily soiled animals are processed (Sofos et al., 1999a).

Other hide decontamination techniques include shaving and chemical dehairing which appear to be labor intensive and costly (Schnell et al., 1995; Van Donkersgoed et al., 1997). A patented process (Bowling and Clayton, 1992), for chemical dehairing of cattle early during slaughter, has been proposed for use, with the objective of removing hair, mud, manure and other external contaminants before hide removal, and, thus, to minimize carcass and plant contamination from these sources (Sofos and Smith, 1998; Sofos et al., 1999a).

The process was applied experimentally at the post-exsanguination stage in a commercial beef slaughtering operation and the resulting carcasses were compared with those from conventionally (not dehaired) processed animals (Schnell et al., 1995). It was found that dehairing reduced visible contaminants on the carcasses and the amount of knife-trimming needed to meet regulatory inspection requirements. Application of the dehairing process to hide samples in laboratory experiments (Castillo et al., 1998; Graves Delmore et al., 1997) caused significant reductions in inoculated *E. coli* 0157:H7, *Salmonella* spp. and *L. monocytogenes*. In addition to inactivation, however, the dehairing process also resulted in injured bacterial cells (Graves Delmore et al., 1997), which may be of concern during subsequent product storage, if they repair their injury. Overall, it can be postulated that the bacterial status of dehaired carcasses could be improved in facilities designed for the exclusive processing of dehaired animals (Sofos and Smith, 1998). It is anticipated that removing the dirt, feces, and hair in a separate room and prior to removing the hide should decrease the occurrence of pathogens on beef carcasses. It should be noted, however, that contamination of the resulting meat will also depend on plant design, good manufacturing practices, sanitation and hygienic practices, and overall avoidance of environmental cross-contamination. The issue of waste (hydrolyzed hair and dehairing chemical residues of sodium sulfide and hydrogen peroxide) disposal needs to be resolved before this technology can be adopted (Sofos and Smith, 1998; Sofos et al., 1999a).

Research data on contribution of pre-harvest contamination levels to resulting meat contamination, as well as on the usefulness of pre-harvest interventions in reducing meat contamination are scarce or conflicting. Studies by Duffy et al. (2000) found that pen wetness, use of bedding in pens, and wool length had only minor effects on resulting microbial contamination levels on lamb pelts and carcasses (Tables 4 and 5). Data from Australia (Table 6) indicated that wool length and crutching (none, keyhole, traditional, or traditional and belly) did not influence lamb carcass microbial counts (Bobbit et al., 2000).

Table 4. Pen wetness and lamb bedding (straw/corn shucks, 15 cm) effect on external pelt and lamb carcass surface contamination (log CFU/cm²).

	APC		ECC	
	Pelt	Carcass	Pelt	Carcass
Pen condition				
Dry	6.2[b]	4.5[b]	2.9[a]	1.8[a]
Wet*	6.4[a]	5.1[a]	2.4[b]	1.9[a]
Bedding				
No	6.5[a]	4.7[b]	2.7[a]	1.9[a]
Yes	6.2[b]	5.0[a]	2.6[a]	1.8[a]

Duffy et al. (2000); Number of samples per group: 74-100

*Muddy (water on for 12/24 hours)

APC: aerobic plate count; ECC: *Escherichia coli* count

Table 5. Wool length effect on external pelt and lamb carcass surface contamination (log CFU/cm²).

	APC		ECC	
	Pelt	Carcass	Pelt	Carcass
Unshorn	6.2[b]	5.1[a]	2.5[b]	2.0[a]
Shorn (5 d)*	6.4[a]	4.6[b]	2.7[a]	1.6[b]
Shorn (30 d)*	6.4[a]	4.7[b]	2.7[a]	1.9[ab]

Duffy et al. (2000); Number of samples per group: 46-64
*Shorn: 5 or 30 days before slaughter
APC: aerobic plate count; ECC: *Escherichia coli* count

Table 6. Effect of wool length and crutching practice on lamb carcass microbial counts (log CFU/cm²).

Wool lengt (mm)	Crutch	TCP	ECC
0-5	None	3.8	-1.8
	Keyhole	4.1	-1.7
	Traditional	4.1	-1.5
	Traditional & Belly	3.6	-1.9
50-75	None	3.8	-1.6
	Keyhole	3.9	-3.0
	Traditional	3.9	-2.0
	Traditional & Belly	4.2	-2.1

Bobbitt et al. (2000)
TPC: total plate count; ECC: *Escherichia coli* count

There is evidence that sanitation and hygienic practices as well as carcass decontamination interventions reduce contamination levels on carcasses irrespective of levels of contamination on live animals before slaughter. Rowland *et al.* (2000) found no correlation between visual contamination on the live animal and the microbial levels on beef carcasses, while various animal cleaning treatments resulted in no significant differences in bacterial counts. However, changes in carcass dressing procedures guided by appropriate microbiological data may result in consistent reductions in the microbiological contamination of carcasses (Gill and McGinnis, 1999). Bacon et al. (2000) studied contamination levels in eight beef slaughtering operations. The results indicated that the high cattle hide (8.2 to 12.5 log CFU/100 cm^2 total plate counts; 5.9 to 7.5 log CFU/100 cm^2 *Escherichia coli* counts) and dehided carcass counts (6.1 to 9.1 and 2.6 to 5.3, respectively) were reduced to 3.8 to 7.1 and 1.0 to 3.0 in washed carcasses and to 2.3 to 5.3 and <0.9 to 0.9 in chilled carcasses, respectively (Table 7).

Table 7. Bacterial counts (log CFU/100cm^2; ranges for 8 plants) on cattle hides and resulting carcasses.

Sample	TPC	ECC
Cattle hide	8.2 – 12.5 (0)	5.9 – 75 (0)
Dehided carcass	6.1 – 9.1 (0)	2.6 – 5.3 (1)
Washed carcass	3.8 – 7.1 (2)	1.0– 3.0 (167)
Chilled carcass	2.3 – 5.3 (62)	<0.9 – 0.9 (315)

Numbers in parenthesis: Number of samples (out of 320 tested) below the detection limit (2.2 for TPC and 0.9 for ECC)
Bacon et al. (2000)
TPC: total plate count; ECC: *Escherichia coli* count

The variability present in the above results is due to plant variation, indicating the importance of plant practices in reducing carcass contamination. Nevertheless, presentation of cleaner animals for slaughter should reduce extent of total microbial contamination and the likelihood of pathogenic bacteria being transmitted to carcasses or being established in the plant environment and cross-contaminating meat products (Sofos et al., 1999a).

As indicated, knowledge on pre-harvest measures to control pathogen presence on food animals is limited (Hancock and Dargatz, 1995; Hancock et al., 1997c). Furthermore, the limited information available does not indicate any specific promising approaches that would enhance food safety of resulting animal products. Additional research is needed to examine potential pre-harvest interventions for pathogen control. Areas in need of research include: (1) evaluation of environmental variables with the objective of minimizing animal hide contamination; (2) examination of dietary factors, such as type of feed (e.g., hay, concentrates) or related factors that may minimize intestinal colonization and pathogen shedding; (3) determination of animal handling practices before slaughter that minimize pathogen shedding; and, (4) reducing external animal contamination (e.g., evaluation of methods with the objective of cleaning or dehairing) at the stage of exsanguination.

7. Conclusions

It is obvious that approaches to pre-harvest pathogen control have major constraints. A pathogen of major recent concern, *E. coli* 0157:H7 may serve as an example to point out such constraints: (1) a documented source of the pathogen are cattle farms and feedlots; (2) the pathogen may persist for long periods in the farm environment but not on the same animals; (3) its persistence on animals varies with season; (4) cattle carriers may not show signs of illness; and, (5) it is isolated from environmental sources such as water and water troughs (Hancock and Dargatz, 1995; Hancock et al., 1997a). These facts complicate the issue of applying, pre-harvest measures to enhance the safety of resulting food products. However, as indicated by Elder et al. (2000), "the association between fecal prevalence and

carcass contamination indicates a role for control in cattle on the farm... Unfortunately, no effective control methods are currently available for producers to use...development of such control methods remains an area of active research. Areas of needed research include: (1) pathogen ecology; (2) risk analysis for dairy, beef, steers, heifers, poultry, seafood, at the farm or feedlot level; (3) pre-slaughter risk factors associated with animal management and handling; (4) animal and feed management practices; (5) bacterial attachment and detachment in the rumen; and, (6) animal stressing associated with handling and shipping. Differences in one or more of these parameters may be responsible for variation in pathogen prevalence rates associated with animals from different lots or farms, as indicated by results reported by Elder et al. (2000), Bacon et al. (2000) and Sofos et al. (1999c).

Overall, however, pre-harvest control of foodborne pathogens is useful because presence of pathogens on animals and their environment leads to animal product contamination, as well as contamination of other foods through contaminated manure and water. Reduction of pathogen incidence in the farm and on animals should reduce the likelihood of contamination of animal and plant food products and water. However, scientifically defined and verified critical control points or management practices at the pre-harvest level are presently unavailable. Ongoing research should continue and be expanded to define risks and develop effective and practical controls for such risks. It should be noted, however, that elimination of pathogenic microorganisms at the pre-harvest level is unlikely; nevertheless, their reduction and management should be sought. As pathogen control strategies are developed and implemented at the pre-harvest level, their verification, auditing and economic aspects should also be considered. Finally, it is important to realize that control or management of food safety risks from foodborne pathogenic microorganisms should be part of an integrated effort and approach that includes all sectors, from the producer through the processor, distributor, packer, retailer, food service worker, and consumer. Reduction of pre-harvest pathogen prevalence may lead to a reduced probability that errors occurring in subsequent parts of the food chain, through cooking and preparation for consumption, will lead to foodborne illness.

References

AMSA (American Meat Science Association), 1999. The Role of Microbiological Testing in Beef Food Safety Programs, The Scientific Perspective. Consensus of the 1999 symposium. American Meat Science Association, Kansas City, Missouri, USA. 16 p.

Anderson, R.C. and Nisbet, D.J., 2001. Feeding sodium chlorate to livestock to kill *Salmonella* and *E. coli*. Agricultural Research. March. Page 19.

Anderson, R.C., Buckley, S.A., Calloway, T.R., Genorese, K.J., Kubena, L.F., Harvey, R.B. and Nisbet, D.J., 2001. Effect of sodium chlorate on *Salmonella* Typhimurium concentrations in the weaned pig gut. J. Food Prot. 64, 255-258.

Anderson, R.C., Buckley, S.A., Kubena, L.F., Stanker, L.H., Harvey, R.B. and Nisbet, D.J., 2000. Bactericidal effect of sodium chlorate on *Escherichia coli* O157:H7 and *Salmonella* Typhimurium DT104 in rumen contents in vitro. J. Food Prot. 63, 1038-1042.

Anonymous. 2000. National Conference on Animal Food Safety. Federal Register 65(160), 50176-50177.

Anonymous., 1999. Antimicrobial resistance, an ecological perspective. Report from the American Academy of Microbiology, Washington, DC. 14p.

Anonymous., 1997. Factors associated with *Escherichia coli* 0157:H7 in feces of feedlot cattle. Info Sheet, October, Veterinary Services, Animal and Plant Health Inspection Service, United States Department of Agriculture, Fort Collins, Colorado, USA.

Armstrong, G.L., Hollingsworth, J. and Morris, Jr. J.G., 1996. Emerging foodborne pathogens: *Escherichia coli* 0157:H7 as a model of entry of a new pathogen into the food supply of the developed world. *Epidemiol. Rev.* 18, 29-51.

Bacon, R.T., Belk, K.E., Sofos, J.N., Clayton, R.P., Reagan, J.O. and Smith, G.C., 2000. Microbial populations on animal hides and beef carcasses at different stages of slaughter in plants employing multiple-sequential interventions for decontamination. J. Food. Prot. 63, 1080-1086.

Bailey, J.S., Blankenship, L.C., Stem, N.J., Cox, N.A. and McHan, F., 1988. Effect of anticoccidial and antimicrobial feed additives on prevention of *Salmonella* colonization of chicks treated with anaerobic cultures of chicken feces. Avian Dis. 32, 324-329.

Bell, B.P., Goldoft, M., Griffin, P.M., Dans, M.A., Gordon, D.C., Tarr, P.J., Bartleson, C.A., Lewis, J.H., Barrett, T.J., Wells, J.W., Baron, R. and Kobayashi, J., 1994. A multistate outbreak of *Escherichia coli* 0157:H7-associated bloody diarrhea and hemolytic uremic syndrome from hamburgers, the Washington experience. JAMA 272, 1349-1353.

Bell, R.G., 1997. Distribution and sources of microbial contamination on beef carcasses. J. Appl. Microbiol. 82:292-300.

Biss, M.E. and Hathaway, S.C., 1996. Effect of pre-slaughter washing of lambs on the microbiological and visible contamination of the carcasses. Vet. Record 138:82-86.

Bobbit, J., Phillips, M., Haines, H., Rowland, D., Whitehouse, J., Barlow, S., Isaac, J., Kondekas, N., Isgro, D. and Coates, K., 2000. Preparation of clean livestock for slaughter: effect of crutching practices on the microbiological quality of lamb carcasses. Proc. Int. Congress Meat Sci. Technol. 46, 404-405.

Bolton, D.J., Byrne, C. and Sheridan, J.J., 1998. The survival characteristics of *Escherichia coli* 0157:H7 and transfer from the contaminated hide to the carcass during slaughter. Proceed. Int. Congress Meat Sci. Technol. 44, 351-352.

Bowling, R.A. and Clayton, R.P., 1992. Method for dehairing animals. United States Patent 5,149,295.

Brisabois, A., Cazin, I., Breuil, J. and Collatz, E., 1997. Surveillance of antibiotic resistance in *Salmonella*. Surveillance of antibiotic resistance in *Salmonella*. Surveillance Report. Eurosurveillance. 2(3), 19-20.

Brown, M.H., Gill, C.O., Hollinsworth, J., Nickelson II, R., Seward, S., Sheridan, J.J., Usborne, W.R. and Zink, D., 2000. The role of microbiological testing in systems for assuring the safety of beef. Int. J. Food Microbiol. 62, 7-16.

Buchanan, R.L. and Doyle, M.P., 1997. Foodborne disease significant of *Escherichia coli* 0157:H7 and other enterohemorrhagic *Escherichia coli*. Food Technol. 51(10), 69-76.

Buchko, S.J., Holley, R.A., Olson, W.O., Gannon, V.P.J. and Veira, D.M., 2000. The effect of different grain diets on fecal shedding of *Escherichia coli* 0157:H7 by steers. J. Food Prot. 63, 1467-1474.

Byrne, C.M., Bolton, D.J., Sheridan, J.J., McCowell, D.A. and Blair, I.S., 2000. The effects of preslaughter washing on the reduction of *Escherichia coli* 0157:H7 transfer from cattle hides to carcasses during slaughter. Lett. Appl. Microbiol. 30, 142-145.

Castillo, A., Dickson, J.S., Clayton, R.P., Lucia, L.M. and Acuff, G.R., 1998. Chemical dehairing of bovine skin to reduce pathogenic bacteria and bacteria of fecal origin. J. Food Prot. 61, 623-625.

Chapman, P.A. and Ackryod, H.J., 1997. Farmed deer as a potential source of verocytotoxigenic-producing *Escherichia coli* 0157. Vet. Record 20, 314-315.

Chapman, P.A., Siddons, C.A., Cerdan-Malo, A.T. and Hawkins, M.A., 1997. A 1-year study of *Escherichia coli* 0157 in cattle, sheep, pigs and poultry. Epidemiol. Infect. 119, 245-250.

Chapman, P.A., Siddons, C.A., Wright, D.J., Norman, P., Fox, J. and Crick, E., 1993. Cattle as a possible source of verocytotoxin-producing *Escherichia coli* 0157 infections in man. Epidemiol. Infect. 111, 439-447.

Corrier, D.E., Nisbet, D.J., Hollister, A.G., and DeLoach, J.R., 1995. Control of *Salmonella typhimurium* colonization in broiler chicks with a continuous-flow characterized mixed culture of cecal bacteria. Poult. Sci. 74, 916-924

Corrier, D.E., Nisbet, D.J., Hollister, A.G., Scanlan, C.M., Hargis, B.M. and DeLoach, J.R., 1993. Development of defined cultures of indigenous cecal bacteria to control salmonellosis in broiler chicks. Poult. Sci. 72, 1164-1168.

Corrier, D.E., Purdy, C.E., and DeLoach, J.R., 1990. Effects of marketing stress on fecal excretion of *Salmonella* spp. in feeder calves. Am. J. Vet. Res. 51, 866-869.

Cray, Jr., W.C., Casey, T.A., Boswath, B.T. and Rasmussen, M.A., 1998. Effect of dietary stress on fecal shedding of *Eschereichia coli* 0157:H7 in calves. Appl. Environ. Microbiol. 64, 1975-1979.

Dargatz, D.A., Wells, S.J., Thomas, L.A., Hancock, D.D. and Garber, L.P., 1997. Factors associated with the presence of *Escherichia coli* 0157:H7 in feces of feedlot cattle. J. Food Prot. 61, 466-470.

Davis, M.A., Hancock, D.D., Besser, T.E., Rice, D.H., Gay, J.M., Gay, C., Gearhart, L. and DiGiacomo, R., 1999. Changes in antimicrobial resistance among *Salmonella enterica* serovar Typhimurium isolates from humans and cattle in the Northwestern United States, 1982-1997. Emerging Infectious Diseases 5, 802-806.

Diez-Gonzalez, T., Calloway, T.R., Kizoulis, M.G., and Russell J.B., 1998. Grain feeding and the dissemination of acid-resistant *Escherichia coli* from cattle. Science. 281, 1666-1668.

Dorsa, W.J., 1997. New and established carcass decontamination procedures commonly used in the beef-processing industry. J. Food Prot. 60, 1146-1151.

Duffy, E.A., LeValley, S.B., Belk, K.E., Sofos, J.N. and Smith G.C., 2000. Pre-harvest management practices, good manufacturing practices during harvest, and microbiological quality of lamb carcasses. Dairy, Food and Environ. Sanit. 20:753-762.

Elder, R.O. Keen, J.E., Siragusa, G.R., Barkacy-Gallagher, G.A., Koohmaraie, M. and Laegreid, W.W., 2000. Correlation of enterohemorrhagic *Escherichia coli* 0157 prevalence in feces, hides, and carcasses of beef cattle during processing. Proc. Nat. Acad. Sci. 97, 2999-3003.

Faith, N.G., Shere, J.A., Brosch, R., Arnold, K.W., Ansay, S.E., Lee, M.-S, Luchansky, J.B. and Kaspar, C.W., 1996. Prevalence and clonal nature of *Escherichia coli* 0157:H7 on dairy farms in Wisconsin. Appl. Environ. Microbial. 62, 1519-1525.

Fedorca-Cray, P.J., Dargatz, D.A., Thomas, L.A. and Gray, J.T, 1998. Survey of *Salmonella* serotypes in feedlot cattle. J. Food Prot. 61, 525-530.

Ferber, D., 2000. Superbugs on the hoof. Science 288, 792-794.

FSIS-USDA (Food Safety and Inspection Service-United States Department of Agriculture), 1993. Immediate actions: cattle clean meat program. FSIS Correlation Packet, Interim Guidelines for Inspectors. Food Safety and Inspection Service, United States Department of Agriculture, Washington, DC.

FSIS-USDA (Food Safety and Inspection Service-United States Department of Agriculture), 1996a. Pathogen reduction: hazard analysis and critical control point (HACCP) systems; final rule. Fed. Regist. 61, 38806-38989.

FSIS-USDA (Food Safety and Inspection Service-United States Department of Agriculture), 1996b. Achieving the zero tolerance performance standard for beef carcasses by knife trimming and vacuuming with hot water and steam; use of acceptable carcass interventions for reducing carcass contamination without prior agency approval. Notice of Policy Change. Fed. Regist. 61, 15024-15027.

Fuller, T., 1999. Probiotics for farm animals. In Probiotics, G.W. Tannock (ed.). Pages 15-23. Biddles Ltd., Guildford and King's Lynn, UK.

Gansheroff, L.J. and O'Brien, A.D., 2000. *Escherichia coli* 0157:H7 in beef presented for slaughter in the U.S.: higher prevalence rates than previously estimated. Proc. Nat. Acad. Sci. 97, 2959-2961.

Garber, L.P., Wells, S.J., Hancock, D.D., Doyle, M.P., Tuttle, J., Shere, J.A. and Zhao, T., 1995. Risk factors for fecal shedding of *Escherichia coli* 0157:H7 in dairy calves. J. Am. Vet. Med. Assoc. 207, 46-49.

Garber, L., Wells, S., Schroeder-Tucker, L. and Ferris, K., 1999. Factors associated with fecal shedding of verotoxin-producing *Escherichia coli* 0157 on dairy farms. J. Food Prot. 62, 307-312.

Gill, C., 2000. Microbial testing: Why it's so badly misunderstood. Meat Marketing Technology, October, pp. 62-63

Gill, C.O. and McGinnis, J.C., 1999. Improvement of the hygienic performance of the hindquarters skinning operations at a beef packing plant. Int. J. Food Microbiol. 51, 123-132.

Gill, C.O., 1998. Microbiological contamination of meat during slaughter and butchering of cattle, sheep and pigs. In *The Microbiology of Meat and Poultry*. A Davies and R. Board, Eds. pp. 118-157, Blackie Academic & Professional, London.

Gill, C.O., McGinnis, J.C., Badoni, M., 1976. Use of total *Escherichia coli* counts to assess the hygienic characteristics of a beef carcass during processing. Int. J. Food Microbiol. 31, 181-196.

Gorman, G.M., Sofos, J.N., Morgan, J.B., Schmidt, G.R. and Smith, G.C., 1995. Evaluation of hand-trimming, various sanitizing agents and hot water spray-washing as decontamination interventions for beef brisket adipose tissue. J. Food Prot. 58, 899-907.

Grau, F.H., Brownlie, L.E. and Roberts, E.A., 1968. Effect of some preslaughter treatments on the *Salmonella* population in the bovine rumen and ? . J. Appl. Bacteriol. 31, 157-163.

Graves Delmore, L.R., Sofos, J.N., Schmidt, G.R. and Smith, G.C., 1998. Decontamination of inoculated beef with sequential spraying treatments. J. Food Sci. 63, 8990-893.

Graves Delmore, L.R., Sofos, J.N., Schmidt, G.R. and Smith, G.C., 1997. Inactivation of pathogenic bacteria by the chemical dehairing process proposed for use on beef carcasses before slaughter. Presented at the 50th Reciprocal Meat Conference, June 29-July 2, Iowa State University, Ames, Iowa. American Meat Science Association, Chicago, Illinois, USA.

Ha, S.D., Ricke, S.C., Nisbet, D.J., Corrier, D.E. and DeLoach, J.R., 1994. Serine utilization as a potential competition mechanism between *Salmonella* and a chicken cecal bacterium. J. Food Prot. 57, 1074-1079.

Hadley, P.J., Holder, J.S. and Hinton, M.H., 1997. Effects of fleece soiling and skinning method on the microbiology of sheep carcasses. Vet. Rec. 140, 570-574.

Hancock, D.D., Besser, T.E., Gill, C. and Hovde, C., 1999. Cattle, hay and *E. coli*. Science. 284, 51-52.

Hancock, D. and Dargatz, D., 1995. Implementation of HACCP on the farm. Hazard Analysis and Critical Control Point (HACCP) Symposium. Presented in association with the 75th Annual Meeting of the Conference of Research Workers in Animal Diseases, November 12, Ramada Congress Hotel, Chicago, IL 6p.

Hancock, D.D., Besser, T.E., Kinsel, M.L., Tarr, P.I., Rice, D.H. and Paros, M.G., 1994. The prevalence of *Escherichia coli* 0157:H7 in dairy and beef cattle in Washington state. Epidemiol. Infect. 113, 199-207.

Hancock, D.D., Besser, T.E., Rice, D.H., Herriot, D.E. and Tarr, P.I., 1997a. A longitudinal study of *Escherichia coli* 0157 in fourteen cattle herds. Epidemiol. Infect. 118, 193-195.

Hancock, D.D., Rice, D.H., Thomas, L.A., Dargatz, D.A., and Besser, T.E., 1997b. Epidemiology of *Escherichia coli* 0150:H7 in feedlot cattle. J. Food Prot. 60, 462-465.

Hancock, D.D., Lyon, T.V. and Besser, T.E., 1997c. Cow/calf Management Update: Feasibility of preharvest food safety control. The Compendium, September, pp. 5200-5207, 5240.

Hardin, M.D., Acuff, G.R., Lucia, L.M., Oman, J.S. and Savell, J.W., 1995. Comparison of methods for decontamination from beef carcass surfaces. J. Food Prot. 58, 368-374.

Harmon, B.G., Brown, C.A., Tkalcic, S., Mueller, P.O.E., Parks, A., Jain, A.V., Zhao, T. and Doyle, M.P., 1999. Fecal shedding and rumen growth of *Escherichia coli* 0157:H7 in fasted calves. J. Food Prot. 62, 574-579.

Herriott, D.E., Hancock, D.D., Ebel, E.D., Carpenter, L.V., Rice, D.H. and Besser, T.E., 1998. Association of herd management factors with colonization of dairy cattle by shiga toxin-positive *Escherichia coli* 0157. J. Food Prot. 61, 802-807.

Hinton, A., Jr., Corrier, D.E., Spates, G.E., Norman, J.O. Ziprin, R.L., Beier, R.C. and DeLoach, J.R., 1990. Biological control of *Salmonella typhimurium* in young chicks. Avian Dis. 34, 626-633.

Hinton, Jr., A., Corrier, D.E. and DeLoach, J.P., 1992. In vitro inhibition of *Salmonella typhimurium* and *Escherichia coli* 0157:H7 by an anaerobic gram-positive coccus isolated from the cecal contents of adult chickens. J. Food Prot. 55, 162-166.

Hinton, A., Jrs., Corrier, D.E., Ziprin, R.L., Spates, G.E. and DeLoach, J.R., 1991. Comparison of the efficacy of cultures of cecal anaerobes as inocula to reduce *Salmonella typhimurium* colonization in chicks with or without dietary lactose. Poult. Sci. 70, 67-73.

Hovde, C.J., Austin, P.R., Cloud, K.A., Williams, C.J. and Hunt, C.W., 1999. Effect of cattle diet on *Escherichia coli* 0157:H7 acid resistance. Appl. Environ. Microbiol. 63, 3878-3886.

Hume, M.E., Nisbet, D.J., Buckely, S.A., Ziprin, R.L., Anderson, R.C. and Stanker, L.H., 2001. Inhibition of in vitro *Salmonella* Typhimurium colonization in porcine cecal bacteria continuous-flow competitive exclusion cultures. J. Food Prot. 64, 17-22.

Jordan, D and McEwen, S.A., 1998. Effect of duration of fasting and short-term high-roughage ration of the concentration of *Escherichic coli* biotype 1 in cattle feces. J. Food Prot. 61, 531-534.

Kochevar, S.L., Sofos, J.N., Bolin, R.R., Reagan, J.O. and Smith, G.C., 1997. Steam-vacuuming as a pre-evisceration intervention to decontaminate beef carcasses. J. Food Prot. 60, 107-113.

Kudva, I.T., Hatfield, P.G. and Hovde, C.J., 1997. Characteristics of *Escherichia coli* 0157:H7 and other shiga toxin-producing *E. coli* serotypes isolated from sheep. J. Clinic. Microbiol. 35, 892-899.

Kudva, I.T., Jelacic, S., Tarr, P.I., Youderian, P. and Hovde, C.L., 1999. Biocontrol of *Escherichia coli* 0157 with 0157-specific bacteriophages. Appl. Environ. Microbiol. 65, 3767-3773.

Laegreid, W.W., Elder, R.O. and Keen, J.E., 1999. Prevalence of *Escherichia coli* 0157:H7 in range beef calves. Epidemiol. Infect. 123, 291-298.

Letellier, A., Messier, S. and Quessy, S., 1999. Prevalence of *Salmonella* spp. and *Yersinia enterocolitica* in finishing swine at Canadian abattoirs. J. Food Prot. 62, 22-25.

Loy, T., Wilson, C., Bailey, D., Blackford, M., Kloffenstein, T., Moxley, R., Smith, D. and Ysunts, S., 2001. Influence of restricted intake and reduced dietary starch on colonic pH and *E. coli* prevalence. Nebraska Beef Report. pp. 86-88.

Mead, P.S., Slutsker, L., Dietz, V., McCary, L.F., Bresee, J.S., Shapiro, C., Griffin, P.M and Tauxe, R.V., 1999. Food-related illness and death in the United States. Emerging Infectious Diseases 5, 607-624.

Mechie, S.C., Chapman, P.A. and Siddons, C.A., 1997. A fifteen month study of *Escherichia coli* 0157:H7 in a dairy herd. Epidemiol. Infect. 118, 17-25.

NACMCF (National Advisory Committee on Microbiological Criteria for Foods), 1998. Hazard Analysis and Critical Control Point Principles and Application Guidelines. J. Food Prot. 61, 1246-1259.

Nisbet, D.J., Corrier, D.E. and DeLoach, J.R., 1993. Effect of mixed cecal microflora maintained in continuous culture and of dietary lactose on *Salmonella typhimurium* colonization in broiler chicks. Avian Dis. 37, 528-535.

Nisbet, D.J., Corrier, D.E., Ricke, S.C., Hume, M.E. and DeLoach, J.R., 1996. Maintenance of a competitive exclusion culture by continuous-flow fermentation. J. Food Prot. 59, 1279-1283.

Nurmi, E. and Rantala, M., 1973. New aspects of *Salmonella* infection in broiler production. Nature 241, 210-211.

Phebus, R.K., Nutsch, A.L., Schafer, D.E., Wilson, R.C., Riemann, M.J., Leising, J.D., Kastner, C.L., Wolf, J.R. and Prasai, R.K., 1997. Comparison of steam pasteurization and other methods for reduction of pathogens on surfaces of freshly slaughtered beef. J. Food Prot. 60, 476-484.

Rasmussen, M.A., Cray, W.C., Jr., Casey, T.A. and Whipp, S.C., 1993. Rumen contents as a reservoir of enterohemorrhagic *Escherichia coli*. FEMS Microbiol. Lett. 114, 79-84.

Reagan, J.O., Acuff, G.R., Buege, D.R., Buyck, M.J., Dickson, D.S., Kastner, C.L., Marsden, J.L., Morgan, J.B., Nickelson II, R., Smith, G.C. and Sofos, J.N., 1996. Trimming and washing of beef carcasses as a method of improving the microbiological quality of meat. J. Food Prot. 59, 751-756.

Reed, C.A., 1996. Foodborne illness prevention before slaughter? Yes! JAVMA 208, 1366.

Ridell, J. and Korkeala, H., 1993. Special treatment during slaughtering in Finland of cattle carrying an excessive load of dung: meat hygiene aspects. Meat Sci. 35, 223-228.

Rowland, D., Bobbit, J., Phillips, M., Haines, H., Whitehouse, J., Barlow, S., Isaac, J., Kondekas, N., Isgro, D. and Coates, K., 2000. Preparation of clean livestock for slaughter: effect of cleaning practices on the microbiological quality of beef carcasses. Proc. Int. Congress Meat Sci. Technol. 46, 678-679.

Russell, J.B., 1999. Letter to the Editor. Science 284, 52-53

Russell, J.B., 2000. Hay, grain and *E. coli* revisited. Amer. Soc. for Microbiol. News 66(1), 1.

Russell, J.B., Diez-Gonzalea, F. and Jarvis, G.N., 2000a. Effects of diets shifts on *Escherichia coli* in cattle. J. Dairy Sci. 83, 863-873.

Russell, J.B., Diez-Gonzalez and Jarvis, G.N., 2000b. Potential effect of cattle diets on the transmission of pathogenic *Escherichia coli* to humans. Microbes and Infection 2, 45-53.

Salminen, S., 1999. Probiotics: scientific support for use. Food Technol. 53, 66.

Sanders, M.E., 1999. Probiotics. Food Technol. 53:67-77.

Schnell, T.D., Sofos, J.N., Littlefield, V.G., Morgan, J.B., Gorman, B.M., Clayton, R.P. and Smith, G.C., 1995 Effects of postexsanguination dehairing on the microbial load and visual cleanliness of beef carcasses. J. Food Prot. 58, 1297-1302.

Scott, T., C. Wilson, D. Bailey, T. Klopfenstein, T. Milton, R. Moxley, D. Smith, J. Gray and Hungerford, L., 2000. Influence of diet on total and acid resistant *E. coli* and colonic pH. Nebraska Beef Report pp. 39-41.

Skjerve, E., Lium, B., Nielsen, B. and Nesbakken, T., 1998. Control of *Yersinia enterocolitica* in pigs at the herd level. Int. J. Food Microbiol. 45, 195-203.

Smith, D., Blackford, M., Younts, S., Moxley, R., Gray, J., Hungerford, L., Milton, T. and Klopfenstein, T., 2001. The relationship of the characteristics of feedlot pens to the percentage of cattle shedding *Escherichia coli* 0157:H7 within the pen. Nebraska Beef Report. pp. 81-84.

Smith, D., Milton, T., Moxley, R., Gray, J., Hungerford, L., Bailey, D., Scott, T and Klopfenstein, T., 2000. Cleaning coliform bacteria from feedlot water tanks. Nebraska Beef Report pp. 77-79.

Smith, R.A., Griffin, D.D. and Dargatz, D.A., 1997. The risks and prevention of contamination of beef feedlot cattle: the perspective of the United States of America. Rev. Sci. Technol. Off. Int. Epiz. 16, 359-368.

Smulders, F.J.M. and Greer, G.G., 1998. Integrating microbial decontamination with organic acids in HACCP programs for muscle foods: prospects and controversies. Int. J. Food Microbiol. 44, 149-169.

Soerjadi, A.S., Stehman, S.M., Snoeyenbos, G.H., Weinack, O.M. and Smyser, C.F., 1981. The influence of *Lactobacilli* on the competitive exclusion of paratyphoid salmonellae in chickens. Avian Dis. 25, 1027-1033.

Soerjadi, A.S., Rufner, R., Snoeyenbos, G.H. and Weinack, O.M., 1982. Adherence of salmonellae and native gut microflora to the gastrointestinal mucosa of chicks. Avian Dis. 26, 576-584.

Sofos, J.N., 1994. Microbial growth and its control in meat, poultry and fish. pp. 359-403, In: *Quality Attributes and Their Measurement in Meat, Poultry, and Fish Products*. A.M. Pearson and T.R. Dutson (Eds.). Blackie Academic and Professional, Glasgow, United Kingdom.

Sofos, J.N. and Smith, G.C., 1993. The headache of the United States meat industry: *E. coli* 0157:H7. Meat Focus Int. 2, 317-325.

Sofos, J.N. and Smith, G.C., 1998. Nonacid meat decontamination technologies: model studies and commercial applications. Int. J. Food Microbiol. 44, 171-188.

Sofos, J.N., Belk, K.E. and Smith, G.C., 1999a. Processes to reduce contamination with pathogenic microorganisms in meat. Proceedings of the International Congress of Meat Science and Technology (Yokohama, Japan). 45(2), 596-605

Sofos, J.N., Kochevar, S.L., Bellinger, G.R., Buege, D.R., Hancock, D.D., Ingham, S.C., Morgan, J.B., Reagan, J.O. and Smith, G.C., 1999b. Sources and extent of microbiological contamination of beef carcasses in seven United States slaughtering plants. J. Food Prot. 62, 140-145.

Sofos, J.N., Kochever, S.L., Reagan, J.O. and Smith G.C., 1999c. Extent of beef carcass contamination with *Escherichia coli* and probabilities of passing United States regulatory criteria. J. Food Prot. 62, 234-238.

Sofos, J.N., Kochever, S.L., Reagan, J.O. and Smith G.C., 1999d. Incidence of *Salmonella* on beef carcasses as related to the United States meat and poultry inspection regulations. J. Food Prot. 62, 467-473.

Stavric, S., Gleeson, T.M., Blanchfield, B. and Pivnick, H., 1987. Role of adhering microflora in competitive exclusion of *Salmonella* from young chicks. J. Food Prot. 50, 928-932.

Suthienkul, O., Brown, J.E., Seriwatana, J., Ttenthongdee, S., Sastravaha, S. and Echeverita, P., 1990. Shiga-like toxin producing *Escherichia coli* in retail meats and cattle in Thailand. Appl. Environ. Microbiol. 56, 1135-1139.

Swaneburg, M., Urlings, H.A.P., Keuzenkamp, D.A. and Snijders, J.M.A., 2001. *Salmonella* in the lairage of pig slaughterhouses. J. Food Prot. 64, 12-16.

Tellez, G., Petrone, V.M., Escorcia, M., Morishita, T.Y., Cobb, C.W. and Villasenor, L., 2001. Evaluation of avian-specific probiotic and *Salmonella Enteritidis*, *Salmonella Typhimurium*, and *Salmonella Heidelberg*-Specific antibodies on cecal colonization and organ invasion of *Salmonella Enteritidis* in Broilers. J. Food Prot. 64, 287-291.

Threlfall, E.J., Ward, L.R., Frost, J.A. and Willshaw, G.A., 2000. The emergence and spread ofo antibiotic resistance in food-borne bacteria. Int. J. Food Microbiol. 62, 1-5.

Tkalcic, S., Brown, C.A., Harmon, B.G., Jain, A.V., Mueller, E.P.O. Mueller, Parkes, A., Jakobsen, K.L., Martin, S.A., Zhao, T. and Doyle, M.P., 2000. Effects of diet on rumen proliferation and fecal shedding of *Escherichia coli* 0157:H7 in calves. J. Food Prot. 63, 1630-1636.

Tollefson, L., 1999. Antibiotic use in animals: controlling the human health impact. J. AOAC Int. 83, 245-254.

Tollefson, L., Fedorka-Cray, P.J. and Angulo, F.J., l999. Public health aspects of antibiotic resistance monitoring in the USA. Acta Vet. Scand. Suppl. 92, 67-75.

Troutt, H.F. and Osburn, B.I., 1997. Meat dairy cows: possible microbiological hazards and risks. Rev. Sci. Tech. Off. Int. Epiz. 16, 405-414.

VanDonkersgoed, J., Graham, T. and Gannon, V., 1999. The prevalence of verotoxins, *Escherichia coli* 0157:H7, and *Salmonella* in the feces and rumen of cattle at processing. Can. Vet. J. 40, 332-338.

VanDonkersgoed, J., Jericho, K.W.F., Grogan, H. and Thorlakson, B., 1997. Preslaughter hide status of cattle and the microbiology of carcasses. J. Food Prot. 60, 1502-1508.

Wells, J.G., Shipman, L.D., Greene, K.D., Sowers, E.G., Green, J.H., Cameron, D.N., Downers, F.D., Martin, J.L., Griffin, P.M., Ostroft, S.M., Potter, M.E. Tauxe, R.V. and Wachmuth, I.K., 1991. Isolation of *Escherichia coli* serotype 0157:H7 and other shiga-like-toxin-producing *E. coli* from cattle. J. Clin. Microbiol. 29, 985-989.

Wells, J.J., Fedorka-Gray, P.J., Dargatz, D.A., Ferris, K. and Green, A., 2001. Fecal shedding of *Salmonella* spp. by dairy cows on farm and at cull cow markets. J. Food Prot. 64, 3-11.

Zhao, T., Doyle, M.P., Shere, J. and Garber, L., 1995. Prevalence of enterohemorrhagic *Escherichia coli* 0157:H7 in a survey of dairy herds. Appl. Environ. Microbiol. 61, 1290-1293.

Zhao, T., Doyle, M.P., Harmon, B.G., Brown, C.A., Mueller, P.O.E. and Parks, A.H., 1998. Reduction of carriage of enterohemorrhagic *Escherichia coli* 0157:H7 in cattle by inoculation with probiotic bacteria. *J. Clin. Microbiol.* 36, 641-647.

Microbiological concerns associated with animal feed production

Servé Notermans and Henk Beumer
TNO Nutrition and Food Research Institute, Post Office Box 360, 3700 AJ Zeist, The Netherlands, E-mail notermans@voeding.tno.nl

Summary

Analysis of foodborne diseases demonstrates that the majority of food-related gastro-intestinal infections involve bacteria originating from foods of animal origin. In a minority of cases this concerns organisms which can be introduced to animals through feedstuffs. In this respect, *Salmonella* spp. are most important. It has clearly been proven that these may substantially contaminate animals through consumption of contaminated feed. However, up till now no effective legislation exists that mandates the absence of *Salmonella* in animal feed. Occasionally, animal feeds have caused severe health hazards in animals in The Netherlands. Examples include outbreaks of mycotoxicosis due to feeding of grains containing high amounts of certain types of unknown mycotoxins. Crisis management is the obvious strategy for those types of calamities. Feeding to animals of swill does not constitute an animal health hazard and neither does it represent even an indirect public health hazard, provided some essential hygiene practices and adequate heat processing are adhered to. However, the latter should be kept under strict control

Keywords: public health, animal health, microbiology, animal feed, risk analysis

1. Introduction

The slogan 'healthy animals, healthy humans' is often used to demonstrate that a clear relationship exists between the healthy state of animals and that of human beings. In the course of time it has been proven that transmission of diseases from animal to man and *vice versa* can only be effectively prevented by improving the health conditions of both populations. Prevention of transmission of zoonotic agents even becomes a bigger challenge since - due to an increase of the human population - there is also a concomitant increase in production animals. Feedstuffs play an important role in the health condition of production animals and consequently of humans.

To assess the indirect effect of feedingstuffs on human health an analysis was made of foodborne human diseases occurring in the Netherlands. It was estimated which proportion of foodborne disease was associated with ingestion of products of animal origin. Subsequently, an estimate was made of agents that are predominantly transmitted by animal feed. In addition, incidents of adverse animal health effects have been analysed as well as the possible consequences of feeding swill to production animals.

2. Human Foodborne Diseases

Recently the Dutch Health Council (2000) issued a report on foodborne diseases occurring in the Netherlands. Amongst others, the incidence of foodborne diseases, factors contributing to the occurrence of these diseases, groups at risk and preventive actions were discussed.

As far as incidence was concerned, the Health Council produced an overview of foodborne diseases based on both sentinel and populations studies, and on cases of foodborne diseases reported to the Dutch authorities. A summary of the results (average estimated values) is presented in Table 1. It is estimated that the results have an inherent variation of about a factor 2. They may be considered as the most reliable estimates possible so far.

Table 1. Occurrence of gastro-enteritis (and some other diseases) in men caused by pathogenic organisms transmitted by food. Adopted from the report of the Health Council in the Netherlands, 2000.

Causative agent	Average incidence rate in the general Population /year/10^6 population	Estimated food attributive factor
Bacteria		
Bacillus cereus	3,000	1
Campylobacter spp.	20,000*)	>0.9
Clostridium perfringens.	3,000*)	1
Escherichia coli VTEC 0157	< 70	0.5 - 0.9
Other pathogenic *E.coli.*	3,000*)	0,1 - 0,5
Salmonella spp.	8,000*)	>0.9
Shigella	370	0,1 - 0,5
Staphylococcus aureus	3,000	1
Yersinia enterocolitica.	350*)	>0.9
Listeria monocytogenes	< 7	>0.9
Clostridium botulinum	< 1	>0.9
Viruses		
Adenovirus 40/41	2,000*)	Presumptive**)
Astrovirus	2,000*)	Presumptive**)
Rotavirus	6,500*)	Presumptive**)
Small Round Structured Viruses	20,000*)	0,1 - 0,5
Parasites		
Cryptosporidum parvum	2,000*)	Unknown***)
Cyclospora spp	?	
Entamoebe histolytica	1,200*)	Presumptive**)
Giardia lamblia	2,000*)	Unknown***)

*) Estimates based on sentinel and population studies

**) Transmission by food is only assumed

***) Transmission is unknown

The sentinel study limited itself to causes of acute gastro-enteritis and consequently no data are available on foodborne diseases with other clinical manifestations such as those caused by toxoplasmosis and taeniasis. Actually, the incidence of the latter diseases it is unknown in the Dutch population.

In Table 1 the estimated food attributive factor is also presented. In the case of viral and parasitic diseases the food attributive factor is mostly low since the contact between men or between and men and animals plays a role as well. Table 1 reveals that on a yearly base 70.000 - 160.000 cases of foodborne diseases occur in the Netherlands per 10^6 population, 75% of which are caused by bacteria, with *Campylobacter* and *Salmonella* predominating.

3. Zoonotic Foodborne Diseases

Epidemiological analysis of foodborne diseases shows that the majority of cases of food-related gastro-enteritis involves bacterial infections contracted from foods of animal origin. For the Netherlands, the expert working group of the Dutch Health Council estimated 75% of foodborne diseases as being transmitted through products of animal origin. Based on these results, the most relevant organisms involved in foodborne diseases transmitted through products of animal origin are summarised in Table 2. *Campylobacter* spp., *Salmonella* spp, *Clostridium perfringens, Bacillus cereus* and *Staphylococcus aureus* are the organisms of major concern. It is estimated that each of these organisms yearly cause > 1000 zoonotic infections per 10^6 individuals.

3.1 Campylobacter spp.

Campylobacter jejuni is a Gram-negative, slender, curved and motile rod. It is a micro-aerophilic organism, which means that it has a requirement for reduced levels of oxygen. It is relatively fragile, and sensitive to environmental stress (e.g. to 21% oxygen, drying, heating, disinfectants, acidic conditions). The organism is especially sensitive to drying. Its dominant habitat is the intestinal tract of warm-blooded animals. Especially in the caecum of chickens > 10^6 *C. jejuni* may be present. The organism does not multiply at temperatures < 30 °C. Before 1972, when methods were developed for its isolation from faeces, it was believed to be primarily an animal pathogen causing abortion and enteritis in sheep and cattle. Surveys have shown that *C. jejuni* is the major cause of human bacterial diarrhoeal illness, not only in the Netherlands but in many other countries including the United States. Although *C. jejuni* is not carried by healthy individuals, it is often isolated from healthy cattle, chickens, birds and even flies. Sometimes it is present in non-chlorinated water sources such as streams and ponds.

C. jejuni frequently contaminates raw chicken meat. Surveys show that 20 to 100% of retail chicken meat is contaminated. This is not really surprising if one considers that many healthy chickens carry these bacteria in their intestinal tract. Raw milk is also a source of infection. The bacteria are often carried by healthy cattle and by flies on farms. Non-chlorinated water may also be a source of infection. However, properly cooking chicken, pasteurising milk, and chlorinating drinking water will eliminate the bacteria.

Table 2. Estimated incidence rate of foodborne diseases caused by products of animal origin (general incidence rate (table 1) x food attributive factor x 75% attributed to products of animal origin). Adopted from the report of the Health Council in the Netherlands, 2000.

Causative agent	Average incidence rate in the general population /year/10^6 population
Bacteria	
Bacillus cereus	2,250
Campylobacter spp.	15,000
Clostridium perfringens.	2,250
Escherichia coli VTEC 0157	< 40
Other pathogenic *E.coli*	675
Salmonella spp.	6,000
Shigella	80
Staphylococcus aureus	2,250
Yersinia enterocolitica.	260
Listeria monocytogenes	< 5
Clostridium botulinum	< 1
Viruses	
Adenovirus 40/41	Not transmitted by animals
Astrovirus	Not transmitted by animals
Rotavirus	Not transmitted by animals
Small Round Structured Viruses	Not transmitted by animals
Parasites	
Cryptosporidum parvum	Unknown but small
Cyclospora spp	Unknown but small
Entamoebe histolytica	Unknown but small
Giardia lamblia	Unknown but small

There are several routes for animals to become contaminated by *C. jejuni*. Watering of animals with untreated water is a dominant source. Contamination may also occur by feeding green crops since these may have become contaminated by birds excreting *C. jejuni*. In animal feeds like silage, cereals, or dry compound feeds, *Campylobacter* has rarely been isolated. This is in accordance with the above indicated characteristics of the organism.

3.2 *Salmonella* spp.

Salmonella is a rod-shaped, motile, non-spore forming and Gram-negative bacterium. *Salmonella* is predominantly present in the intestinal tract of animals. Several serotypes prevail. An overview of the dominant serotypes in production animals is presented in Table 3. The organisms are especially found in poultry and swine. Obviously, it is a ubiquitous organism, surviving long periods in natural environments and a.o. being resistant to dry

Table 3. The most important Salmonella *serotypes (in order of decreasing importance) found in production animals in the United Kingdom.*

Cattle	Calves	Pigs	Sheep	Poultry
S. typhimurium	*S. typhimurium*	*S. typhimurium*	S. typhimurium	*S. enteritidis*
S. dublin	*S. dublin*	*S. derby*	*S. arizona*	*S. typhimurium*
S. gold coast	*S. gold coast*	*S. kedegou*	*S. dublin*	*S. mbandaka*
S. agona	*S. agona*	*S. gold coast*	*S. montevideo*	*S. senftenberg*
S. enteritidis	*S. enteritidis*	*S. give*	*S. derby*	*S. thompson*
				S. virchow

Source: Report of CVL, New Haw, Addlestone, Surrey, KT15 3NB. Tel. + 44 1932 341111 Fax + 44 1932 349983

conditions. As a consequence, there is a wide environmental spread of the organism and hence it occurs in water, soil, insects, factory surfaces, kitchen surfaces, etc.. *Salmonella* can be present in raw poultry, raw seafoods, raw meats, eggs, milk and dairy products, fish, shrimps, frog legs, yeast, coconut, sauces and salad dressing, cake mixes, cream-filled desserts and toppings, dried gelatine, peanut butter, cocoa, and chocolate. It is also present in animal feed and raw materials for animal feed. An overview is presented in Tables 4a-d).

Table 4. Presence of Salmonella *in animal feed and ingredients for animal feed.*

4a. Presence of Salmonella *in processed animal feed in the United Kingdom.*

Year	Feed imported into UK	Feed produced in the UK
1995	7.2 %	2.0 %
1993	14.4 %	2.3 %
1991	22.0 %	3.5 %
1989	39.0 %	5.2 %

Source: Report of CVL, New Haw, Addlestone, Surrey, KT15 3NB. Tel. + 44 1932 341111 Fax + 44 1932 349983

4b. Presence of Salmonella *in processed animal feed in Belgium.*

Year	Number of samples tested	Feed produced in Belgium and UK
1995	217	9.7 %

Source: Report of Fierens, H. en Huyghebaert, A. (1996). Int. J. Food Microbiol. 31, 301-309

4c. Summary of contamination data of feed ingredients in UK feed mills.

Ingredients sampled		Number of Samples	Number *Salmonella*-positive	
Origin	Number		n (%)	Ingredient
Plant products	32	816	59 (7.2)	17*
Animal products	6	147	8 (5.5)	3
Total	38	963	67 (7.0)	20

* Cotton seed meal: 18 out of 41 Salmonella-positive
Source: UK, MAFF Survey 1992.)

4d. Results of monitoring programme on Salmonella *in complete feeds and animal feed ingredients in the Netherlands from 1986-1994.*

Year	Complete feeds and concentrates		Feed ingredients			
			Animal		Plant	
	number	% pos.	number	% pos.	number	% pos.
1986			303	8	85	7
1987	119	0	685	6	9	33
1988	56	0	546	6	17	18
1989	129	3	1084	3	88	0
1990/91	457	7	1337	5	144	5
1993	1059	2	1212	8	367	4
1994	2462	1	2026	7	1298	3

* Source: Beumer, H. (1996); Beumer, H. and A.F.B. van der Poel (1997)

Many transmission routes have been recorded, the route from contaminated environments to animals, from animal to animal products, and finally from these products to men ranking among the most important. To reduce the contamination of animal products with *Salmonella,* considerable efforts have been directed towards interrupting these routes of transmission. In the past, feed has undoubtedly contributed to animal contamination and has caused the environment to become contaminated. This in turn may have contributed to infections in animals. From the 1960's onwards fish and animal meal containing high numbers and a large variety of *Salmonella* serotypes, have been abundantly imported into European countries. After the impact of the presence of *Salmonella* in these feeds on environmental contamination was recognised many actions were taken to reduce feed contamination.

In an epidemiological study, carried out in Denmark, factors contributing to the risk of contamination of poultry flocks with *Salmonella* were assessed (Anonymous, 1996). This report listed the following factors (in order of relevance) as contributing significantly to the contamination of poultry with *Salmonella*: a) The origin of the day-old chicks, b) the company delivering the feed, c) the number of flocks present on the farm, d) the season (more positive flocks in autumn), and finally, e) presence of *Salmonella* in the previous flock(s).

Although several routes of transmission can be identified, animal feed is thought to be the dominant vector in the transmission of *Salmonella*. Therefore, continued efforts to further reduce *Salmonella* in poultry- and other animal feeds are fully justified. Literature contains a host of processing methods and control measures contributing to reduce or even eliminate *Salmonella* from animal feed (Beumer, 1996; Beumer and Van der Poel, 1997). In the United Kingdom and The Netherlands some success in reducing the *Salmonella* content in feed can be observed (Table 4).

3.3 *Clostridium perfringens*

C. perfringens is an anaerobic organism (unable to grow in the presence of free oxygen), Gram-positive, sporeforming rod. In contrast to *Salmonella* and *Campylobacter* the infective dose is rather high. Clinical symptoms are caused by ingestion of large numbers (greater than 10^{7-8}) vegetative cells. Toxin production in the digestive tract (or in test tubes) is associated with sporulation. This disease being a toxico-infection, high numbers of organisms have to be ingested before enough toxin is released in the intestinal tract.

C. perfringens is widely distributed in the environment and frequently prevails in the intestines of humans and many domestic animals. Investigations of raw meat demonstrated that meat is generally not contaminated with *C.perfringens* (Tschirdewahn et al., 1991). In all positive samples the number or *C.perfringens* was < 100 cfu/gram. Of the positives samples only a few contain the gene encoding the enterotoxin (Damme-Jongsten et al., 1989, 1990) which is the causative agent in foodborne *perfringens* clostridiosis.

Spores of the organism persist in soil, sediments, and areas prone to human or animal faecal pollution. The production of heat resistant spores contributes to the survival of the organism in animal feed. Yet, other sources are responsible for contaminating animals in a great many surroundings, whereby feed only plays a minimal role in the transmission.

3.4 *Bacillus cereus*

B. cereus is a large Gram-positive, facultative aerobic rod-shaped organisms, whose spores do not swell the sporangium. *B. cereus* may cause two types of food poisoning caused by two distinct metabolites. The diarrhoeal type of illness is caused by a large molecular weight protein, while the vomiting (emetic) type is believed to be caused by a low molecular weight, heat-stable peptide.

The symptoms of *B. cereus* diarrhoeal type food poisoning mimic those of *Clostridium perfringens* food poisoning. The onset of watery diarrhoea, abdominal cramps, and pain occurs 6-15 hours after consumption of contaminated food. Nausea may accompany diarrhoea, but vomiting (emesis) rarely occurs. The emetic type of food poisoning is characterized by nausea and vomiting within 0.5 to 6 h after consumption of contaminated foods. Occasionally, abdominal cramps and/or diarrhoea may occur as well. The symptoms of this type of food poisoning are similar to those caused by *Staphylococcus aureus* foodborne intoxication. The presence of large numbers of *B. cereus* (greater than 10^6 organisms/g) in a food product is indicative of active growth and proliferation of the organism and represents a potential hazard to health.

A wide variety of foods including meats, milk, vegetables, and fish have been associated with the diarrhoeal type food poisoning by *B.cereus*. The vomiting-type outbreaks have generally been associated with rice products; however, other starchy foods such as potato, pasta and cheese products have also been implicated. Food mixtures such as sauces, puddings, soups, casseroles, pastries, and salads have frequently been incriminated in *B.cereus* food poisoning outbreaks.

In dairy products *B. cereus* organisms are increasingly found, which have the capacity to grow at low temperatures (≤ 7°C). Important sources of infection are faecal materials from cows, the litter used and processing equipment, especially the equipment used on farms (Te Giffel, 1997). *B. cereus* is able to attach rapidly to surfaces of processing equipment; even in pasteurising equipment. Once attached, *B.cereus* will proliferate and continually contaminate food products. Fresh, rapidly cooked meat, eaten immediately, seems an unlikely vehicle of *B. cereus* food poisoning.

3.5 *Staphylococcus aureus*

S. aureus is a spherical bacterium (coccus) which on microscopic examination appears in pairs, short chains, or bunched, grape-like clusters. These organisms are Gram-positive. Some strains are capable of producing a highly heat-stable protein toxin (staphylococcal enterotoxin) that causes illness in humans. The onset of symptoms in staphylococcal food poisoning is usually rapid and in many cases acute, depending on individual susceptibility to the toxin, the amount of contaminated food eaten, the amount of toxin in the food ingested, and the general health of the victim. The most common symptoms are nausea, vomiting, retching, abdominal cramping, and prostration.

Foods that are frequently incriminated in staphylococcal food poisoning include meat and meat products, poultry and egg products, salads such as egg, tuna, chicken, potato, and macaroni, bakery products such as cream-filled pastries, cream pies, and chocolate eclairs, sandwich fillings and milk and dairy products. Foods that require considerable handling during preparation and that are kept at slightly elevated temperatures after preparation are frequently involved in staphylococcal food poisoning.

Staphylococci exist in air, dust, sewage, water, milk, and food or on food equipment, environmental surfaces, humans, and animals. Humans and animals are the primary

reservoirs. Additionally, *S.aureus* may become rapidly endemic in meat processing plants where they may exist for years (Notermans et al, 1982; 1991). It has been proven that, during slaughtering, chicken become predominantly contaminated by strains which are endemic in the processing plant.

3.6 *Yersinia enterocolitica*

Y. enterocolitica, a small rod-shaped, Gram-negative bacterium, is often isolated from clinical specimens such as wounds, faeces, sputum and mesenteric lymph nodes. However, it is not part of the normal human flora. *Y. pseudotuberculosis* has been isolated from the diseased appendix of humans.

Both organisms, *Y. enterocolitica* and *Y. pseudotuberculosis*, have often been isolated from such animals as pigs, birds, beavers, cats, and dogs. Only *Y. enterocolitica* has been detected in environmental and food sources, such as ponds, lakes, meats, ice cream, and milk. Most isolates have been found not to be pathogenic. There are 3 pathogenic species in the genus *Yersinia*, but only *Y. enterocolitica* and *Y. pseudotuberculosis* cause gastro-enteritis. Yersiniosis is frequently characterised by such symptoms as gastro-enteritis with diarrhoea and/or vomiting; however, fever and abdominal pain are the hallmark symptoms. *Yersinia* infections mimic appendicitis.

Strains of *Y. enterocolitica* can be found in meats (pork, beef, lamb, etc.), oysters, fish, and raw milk. The exact cause of the food contamination is unknown. However, the prevalence of this organism in the soil and water and in animals such as beavers, pigs, and squirrels, offers ample opportunities for it to enter our food supply. Poor sanitation and improper sterilisation techniques by food handlers, including improper storage, are most likely to contribute to contamination.

4. Other Relevant Microorganisms and Regulations

4.1 Relevant Microorganisms

Of the above zoonotic organisms *Campylobacter* and *Salmonella* have the largest public health consequences, *B. cereus, C. perfringens,* pathogenic *Escherichia coli* and *Yersinia* have lesser importance. Fact is that of all of these *Salmonella* is the most important organism contaminating animals through feed.

4.2 Regulations

The European Council Directive 92/117/EEC of 17 December 1992 (concerning measures for protection against specified zoonoses and specified zoonotic agents in animals and products of animals origin in order to prevent outbreaks of food-borne infections and intoxications) and the amendments made later, describes in its Annex III Section II the minimal requirements for animal feed stuffs. The wording here is as follows:

" When official samples are being taken on a holding or in cases of justified suspicion, sampling may be carried out on the compound feedingstuffs used to feed poultry.
Where a sample is positive for *Salmonella*, the competent authority shall carry out an investigation in order to:

a) identify the source of contamination, in particular by means of official samples taken at different stages of production;
b) examine the application of the rules and controls concerning the disposal and processing of animal waste and in particular those laid down in Directive 90/667/EEG;
c) establish procedures for good manufacturing practices and ensure compliance with recognised procedures..."

5. Special Cases

In the past, several feedborne incidents have been observed which have resulted in adverse health effects in animals. Some of these incidents and one case which could have led to such an incident, are described in the next paragraphs.

5.1 Botulism through Feeding of Spent Brewer's Grain

In 1976 and 1977 outbreaks of botulism, followed by the death of many animals, occurred in about thirty dairy farms in The Netherlands. The disease was caused by spent brewers' grain. Proteolytic *Clostridium botulinum* type B cells as well as botulinum toxin type B were found to be present in the brewers' grain (Breukink et al., 1978). Also, in 4,5 % of the cattle fed with the spent brewers' grain toxin type B was present.

Although, spent brewers' grain has been fed to animal for long periods, no cases of botulism have been reported upon feeding this product before and after this incident. In the brewery, after the heat extraction of malt, the remaining part of the grain is stored for a short period and is then transported to dairy farms. At these farms the spent brewers' grain is stored under anaerobic conditions (silage). During storage, normally a fermentation process occurs, decreasing the pH, and thus preventing spoilage. In the decribed case, the pH did not drop sufficiently, in all probability due to a lack of enough fermentable sugars present in the remaining grain.

How *C.botulinum* type B entered in the product remained unclear. It was observed that the storage tank of the spent brewers' grain at the brewery was not completely closed and infection may have been introduced by birds. Nevertheless, *C. botulinum* may be naturally present in malt in low numbers and the heat resistant spores are likely to survive the extraction process.

In the faeces of the cows large numbers of *C.botulinum* type B organisms (10^5 to 10^7/g) were detected. Also, after ceasing to feed the contaminated fodder, *C.botulinum* remained present in the faeces for more than 8 weeks. Consequently, the environment of the farms, especially the grass pastures, was probably contaminated by manure containing

C.botulinum spores. Investigations carried out on one of the farms demonstrated that cycles of *C.botulinum* had occurred. The cycles started by spreading of manure containing faeces of the affected cows. This spreading increased the contamination of the pastures. In grass silage's, prepared with wilted grass from these pastures the number of *C.botulinum* type B organisms increased, and toxin type B was produced. Feeding cows the contaminated silage fodder completed the cycle. It was observed that these cycles continued to exist for over 3 years (i.e. the period of investigation). Besides contamination of human foodstuffs (milk and meat), further contamination of the environment occurred. Incidentally, during the investigation it was also observed that fowl are important vectors in spreading *C. botulinum*. During this botulism episode and a period of 10 years thereafter, no human cases of botulism in The Netherlands have been reported.

5.2 Death by Mycotoxins

Aflatoxins
In 1960 more than 100,000 turkeys died in England some weeks before Christmas. Since initially no diagnosis could be made, the disease was indicated as the turkey 'X' disease. Later the cause was found to be feed containing Brazilian peanuts, which were infested heavily by *Aspergillus flavus* moulds. After extensive analysis of this feed, using among others thin layer chromatography, a series of fluorescent compounds were detected, later called aflatoxins, responsible for this outbreak. (Rustom, 1977).

Fumonisin
Fusarium moniliforme in mouldy feed, particularly corn feed, has been associated with horse and pig deaths since the 1970's. In horses, equine leukoencephalomalacia (ELEM) is the most frequently encountered disease associated with *F. moniliforme* (Kellerman *et al.*, 1990, Ross *et al.*, 1993, Wilson *et al.*, 1992). ELEM is characterised by liquefactive necrosis of the cerebral hemispheres. In pigs porcine pulmonary oedema was observed to develop after exposure to *F. moniliforme* in mouldy feed. Since those cases, legislation has been developed to reduce the fumonisin levels in feed to levels considered to be safe. Later, begin 1990, the agent causing the above indicated diseases was isolated and characterised (Gelderblom et al., 1991). It was indicated as fumonisin and found to be produced mainly by the moulds *F. moniliforme (F. verticillioides)*, *F. proliferatum*, and several other *Fusarium* species that grow on agricultural commodities in the field or during storage. Fumonisin has now been found worldwide, primarily in corn. More than ten types of fumonisins have been isolated and characterised. Of these, fumonisin B1 (FB_1), B2 (FB_2), and B3 (FB_3) are the major fumonisins produced. The most prevalent of these mycotoxins in contaminated corn is FB_1, which is believed to be most toxic (Thiel *et al.*, 1992, Musser and Plattner, 1997).

Based on the above indicated findings, the USA Food and Drug Administration has produced a document with recommended maximum levels for fumonisins in animal feed .

Although many production animals have been and continiue to be exposed to fumonisins, there are no reported disease cases in human beings after consumption of products of animal origin. Obviously, fumonisins being soluble in water and being excreted by the animals explains this.

5.3 Illegal Practices *(Source: FoodSafetyNet periode Juli-Augustus 2000).*

End of June 2000, Japanese consumers became diseased after drinking of pasteurised milk. On June 30, 1,152 patients asked for medical assistance with vomiting, nausea and diarrhoea as most important clinical symptoms. On July 6, the number of patients had increased to 10.780 and 159 of them were hospitalised. On July 11, the number of patients had increased to 14.000. Finally 14.555 patients became diseased. Epidemiological research made clear that the cause of the disease was present in the pasteurised milk originating from the 'Snow Brand Food Co Ltd'. This is the largest dairy plant in Japan. Laboratory investigations revealed the presence of *Stapylococcus aureus* enterotoxin type A (SEA) in the pasteurised milk. SEA is heat-stable and will easily survive the pasteurisation process.
More detailed investigations uncovered hygienic disorders in the dairy plant. The two most significant disorders were:

1. The use of movable connecting pipelines which were withdrawn from the automatic in-plant cleaning and disinfection system. Some of the pipelines had not been cleaned and disinfected for several weeks. In these pipelines high numbers of *S.aureus* were present. It turned out that these organisms produced SEA.
2. Use of return milk. Packages of return milk were opened by hand and stored in a separate tank (again not connected to the automatic in-plant cleaning and disinfection system). In addition, this tank was not adequately cooled. The stored return milk was mixed with fresh milk, pasteurised, packaged and retailed.

This case is an example of illegal activities and bad practices in the human food chain. The illegal activity was the use of return milk. The bad practices concerned the lack of proper cleaning and disinfection and the storage temperature. In this special case enterotoxigenic *S. aureus* multiplied and produced the heat stable SEA. Both men and many types of (young) animals are very sensitive for SEA. This especially applies to piglets (own observations).

Also, in animal feed production, return products are often used and an incident as described above, which diseased over 14,000 persons, may also occur in animal feed production. The three accidents (botulism through spent brewers' grain, production of unknown mycotoxins by moulds and the illegal practices in Japan) demonstrate that animal feed may occasionally cause severe adverse effects in animals. The probability of occurrence is rather low, however. Also, the damage caused is difficult to estimate.

Emergency management is the most obvious management strategy for those types of risks (Klinke and Renn, 2000). Emergency management is primarily directed to limitation of damage by increasing capacity building such as developing and promoting national programs of emergency protection.

6. Swill Feeding

Animals are fed with dry feedingstuffs such as concentrates, compound feeds and hay, wet feedingstuffs such as grass and silage and combinations of wet and dry feeding. Dry feedingstuffs can be fed in a dry state or as a slurry after addition of water. How animals are fed largely depends on the types of raw materials available and on economical parameters. As a consequence, there is also an interest in using many wet by-products of food production and returned foods, including kitchen remnants from restaurants etc. as animal feedingstuffs.

Processing of kitchen remnants from restaurants, indicated as 'swill', is at present under discussion since these types of products are thought to be involved in transmission of pathogenic agents such as classical- and African swine fever to animals. In addition, such kitchen remnants may easily become spoiled by growth of several types of bacteria and production of mycotoxins by growth of moulds. As a consequence, the feeding of swill is subject of political debate.

In a recent report of ID-Lelystad (Urlings et al., 2000) the microbiological risks of the feeding of swill was analysed. Starting point in the analysis was a heat treatment of swill of 133 °C for 20 minutes at a pressure of 3 bar. At such a temperature viruses and parasites will not survive. The same applies to bacteria and their heat resistant spores. Bacterial toxins such as botulinum toxins will be inactivated completely. However, toxins produced by *S. aureus* may not be inactivated. Also, it must be taken into account that after the heating process recontamination with bacteria should be avoided. Moulds will be killed in the process. Preformed mycotoxins, however, may not be inactivated by the heat treatment and require further attention.

As far as transmissible spongiform encephalopathies (TSE's) are concerned, there are no indications that the agent infects pigs via the oral route. In addition, the heat treatment process will inactivate TSE agents with a factor 10^3 (Opinion of SSC of the EU, March 26-27, 1999) up to 10^6 (Böhm amd Rieder, personal communication, 1999)

Based on these analytical results, ID-Lelystad has made the following recommendation for the safe use of swill for pigs. The most important ones are:

- The initial products to be used should be of a quality level in accordance with the requirements of the human food law.
- Immediately after obtaining the product it must be cooled down until processing occurs. The storage time should be limited.
- The product should be processed by heating at 130 °C at a pressure of 3 bar for 20 minutes. Heat treatment should preferably be carried out batchwise.
- Recontamination should be prevented by an adequate hygienic design of the equipment.
- Preservation of the product after heat treatment should be considered.
- Storage of the final product should be limited.

References

Anonymous, 1996. Preventive Veterinary Medicine 1996, 26, 223-2370.

Beumer, H., 1996. Quality and safety in the feed industry. Control of Salmonella and other pathogens in feed. Proceedings of the 4th international Feed Production Conference, Piacenza, February. 1996.

Beumer, H. and van der Poel, A.F.B., 1997. Effects on the hygienic quality of feeds. In: A.F.B. van der Poel (Ed.) Expander processing of animal feeds. Chemical, physical and nutritional effects. Wageningen Feed Processing Centre.

Breukink, H.K., Wagenaar, G., Wensink, T., Notermans, S. and Poulos, W.P., 1978. Food poisoning in cattle caused by ingestion of brewers' grain contaminated with Clostridium botulinum type B. Neth. J. Vet. Sci. 103, 303-311.

Damme-Jongsten, M. v., Wernars, K. and Notermans, S., 1989. Cloning and sequencing of *Clostridium perfringens* enterotoxin gene. Ant. Van leeuwenhoek 56, 181-190.

Damme-Jongsten, M. v., Haagsma, J. and Notermans, S., 1990. Testing strains of *Clostridium perfringens* type A isolated from dirrhoeic piglets for the presence of the enterotoxine gene. Vet. Rec. 126, 191-192.

Gelderblom, W. C. A., Kriek, N. P. J, Marasas W. F. O, and Thiel, P. G., 1991. Toxicity and carcinogenicity of the *Fusarium moniliforme* metabolite, fumonisin B1, in rats. Carcinogenesis 12:1247-1251.

Health Council, 2000. Foodborne infections in the Netherlands. Publication no 2000/09

Kellerman, T. S., Marasas, W. F. O., Thiel, P. G., Gelderblom, W. C. A., Cadwood, M. and Coetzer, J. A. W., 1990. Leukoencephalomalacia in two horses induced by oral dosing of fumonisin B1. Onderstepoort J. Vet. Res. 57:269-275.

Klinke, A. and Renn O., 2000. Precautionary Principle and Discursive strategies: Classifying and managing risks. Journal of Risk Research (In press).

Musser, S. M. and Plattner, R. D., 1997. Fumonisin composition in cultures of *Fusarium moniliforme, Fusarium proliferatum,* and *Fusarium nygami*. J. Agric. Food Chem. 45:1169-1173.

Notermans, S., Dormans, J.A.M.A. and Mead, G.C., 1991. Contribution of surface attachment to the establishment of microorganisms in food processing plants - a review. Biofouling, 5, 21-36.

Notermans, S., Dufrenne, J.B. and van Leeuwen, W.J., 1982. Contamination of broiler chickens by Staphylococcus aureus during processing: Incidence and origin. J. Appl. Bacteriol. 52, 257-280.

Opinion of SSC of the EU, March 26-27, 1999.

Ross, P. F., Ledet, A. E., Owens, D. L., Rice, L. G., Nelson, H. A., Osweiler, G. D., and Wilson, T. M. 1993. Experimental equine leukoencephlomalacia, toxic hepatosis and encephalopathy caused by corn naturally contaminated with fumonisins. J. Vet. Diagnostic Investigation 5:69-74.

Ruston, IYS. 1997. Aflatoxin in food and feed; Occurrence, legislation, and inactivation by physical methods. Food Chemistry. 59: 57-67.

Thiel, P. G., Marasas, W. F. O., Sydenham, E. W., Shephard, G. S. and Gelderblom, W. C. A. 1992. The implications of naturally occurring levels of fumonisins in corn for human and animal health. Mycopathologia 117:3-9.

Tschirdewahn, B., Notermans, S., Wernars, K.and Untermann, F., 1991. The presence of enterotoxigenic *Clostridium perfringens* strains in faeces of various animals. Int. J. Food Microbiol. 14, 175-178.

Urlings, H.A.P.. Oude Elferink and van Zijderveld, F.G. (2000). Veterinaire risico's van swill. Report of ID-DLO, Lelystad, The Netherlands.

Wilson, T. M., Ross, P. E., Owens, D. L., Rice, L. G., Green, S. A., Jenkins, S. J. and Nelson, H. A. 1992. Experimental reproduction of ELEM - a study to determine the minimum toxic dose in ponies. Mycopathologia 117:115-120.

Natural and industrial contaminants in animal feeds: emerging substances

Johanna Fink-Gremmels
Utrecht University, Faculty of Veterinary Medicine, Department of Pharmacology, Pharmacy and Toxicology, Yalelaan 16, 3508 TD Utrecht, The Netherlands; E-mail: J.Fink@vfft.vet.uu.nl

Summary

The contamination of animal-derived food products by chemicals is an issue of intensive debate. Animal feeds have been recognized as regular source of the contamination and despite of the established network of analytical control measures, residue-free products cannot be guaranteed. The latter is demonstrated by incidentally - but regularly - occurring cases of food contamination. This indicates the need for new strategies in quality control measures. A possible alternative to the analytical control for the presence-absence of individual substances in animal feeds, is the use of bio-indicators of exposure in the target animal species. This concept would allow the monitoring of livestock on individual farms, thus improving food safety.

Keywords: feed contaminants, indicators of exposure, industrial pollutants, mycotoxins, hormonally active substances.

1. Introduction

The implementation of a global market for agricultural products was one of the prerequisites for modern animal farming. Traditional animal husbandry depended on the local availability of feeds, hence limiting farm size, determining flock composition and finally limiting the productivity of animals on an individual farm. Traditional farming also reflected the skills of individual farmers to grow selected crops allowing the composition of the most appropriate diet for their livestock. These diets consisted of a few basic components only. In contrast, modern industrialised animal farming has access to a virtually unlimited number of feed components, allowing the composition of well balanced, tailor-made diets for individual (age, production) groups of animals. However, the obvious disadvantage is that the diverse feed components often originating from different sources and even different continents. They have been produced and harvested under unknown conditions, have been transported, shipped, preserved and processed by unknown procedures an may include recycling products from different technologies. As a consequence, farmers may be confronted with unexpected contaminants that exert detrimental effects on their livestock or give rise to undesirable residues in animal products, resulting in non-marketable food products. Animal feed production is driven not only by nutritional requirements of the target animal species but also by economic interest. The individual farmer purchases composed feed that generally fulfils the nutritional requirements of his animals. However, the exact composition remains unknown to him, and he has to rely on the quality concept provided by the feed industries. In turn, feed industries are guided - in Europe - by a number of Council Directives defining raw materials suitable for animal nutrition, hygienic standards and providing a list of prohibited ingredients in animal feeds (Council Directive 91/516/EEC and its amendments 92/508/EEC; 95/274/EC and 97/582/EC). Nevertheless, it soon became evident that even these detailed regulations that set maximal tolerable levels for undesirable contaminants could not cover the plethora of all potential feed contaminants under the diverse circumstances of feed mixing. Subsequently, one of the more recent EC Directives, addressing specifically the undesirable substances and products in animals feeds (1999/24/EC) requests that all feed components need to be of *"sound, genuine, and merchantable quality"*. The statement was made with the dual objective to stimulate the quality of substrates for animal nutrition whilst aiming, at the same time, to guarantee consumer safety. Only one year later, and in consideration of the recent negative experience with large scale dioxin contamination of batches of animal feeds and the emerging transmissible diseases (TSE/BSE), the EC indicated the need for additional measures in the White Paper of Food Safety (January, 2000). However, the indicated need for additional measures in animal production should consider a set of basic obstacles, jeopardising the immediate strategy of increased sampling and control procedures:

1. In a global market, the large number of undesirable contaminants prevents a substance-by-substance analytical control of feed ingredients.
2. In a global market the lack of harmonisation for risk assessment procedures results in unequal limits for permissible levels of chemical and bacterial contaminants.
3. The diversity in animal farming techniques in Europe, particularly the percentage of formulated (thus controllable) feed used on a farm *versus* the percentage of genuine feed materials present on the farm, results in different levels of quality control and thus end-product safety.

This implies that overall concepts following the principle of hazard analysis and hazard control must be implemented to improve feed and food quality. In consideration of the generally recognised major classes of chemical contaminants, concepts towards an integrated quality program should include the following considerations:

2. Heavy Metals

The adverse health effects of heavy metals, particularly cadmium, lead and mercury, have been known for centuries. Their ability to accumulate in certain edible tissues (particularly liver and kidneys) make them potential contaminants of foods from animal origin (Minura, 1995). Exposure to heavy metals varies considerably according the local geographic formation (Diels et al., 1999; Manz et al., 1999; Albering et al., 1999). Industrial processes contribute to the local environmental burden and the abundant use of heavy metal-containing products in crop protection and fertilisation may result in an undesirable increase in animal feeds as well (Giuffre-de-Lopez et al., 1997). Heavy metals are listed in the inventory of undesirable contaminants of both feed components, as well as animal-

Table 1. The Principles of Hazard Analysis and Control Measures for Chemical Contaminants.

Principle:*	Implementation:
Conduct a hazard analysis	Identification of geographical risks, risk materials
Determine the critical control points	Control of feed batches with a high risk
Establish critical limits	Acceptable limits for common feed contaminants
Establish monitoring procedures	Monitor farm animals at the farm level
Establish corrective actions	Evaluate the use of absorbents, detoxifying agents, decontamination procedures.
Establish verification procedures	Schedule analytical control of food products from farms with high risk factors
Establish record keeping and documentation procedures	Establish an integrated quality control program- Request records about all feed commodities (trace ability of contaminants)

* The National Advisory Committee of Microbiological Criteria in Foods, 1998/27, has proposed these principles for undesirable microbiological contaminants

derived food commodities (86/469/EEC). However, current legislation lacks an integrative approach and requires individual measurements rather than providing guidelines for strategic testing. A simple, and often cited example refers to formulated feed for veal calves, pigs and poultry. If the latter contain only trace (permissible) amounts of cadmium or lead, the short production period of these animals provides a natural guarantee that no harmful residues will occur in animal tissues. In turn, residue control should focus on herbivorous species (cattle, horses, wild game) with a long life span, using known geographic risk factors (Hecht, 1992). High-risk organs, such as kidneys from aged animals, should me omitted from human consumption as a preventative measure.

In addition to analytical control of organs from slaughter animals (lean muscle tissues contain very low concentrations of heavy metals), bio-indicators of exposure may contribute to the identification of highly polluted geographic areas. Heavy metals, particularly cadmium, are strong inducers of metallothionein (MT), a class of metal binding proteins involved in metal transport and metal sequestration in the animal body (Roesijadi, 1996). MT-induction may be a useful marker to monitor animal populations for exposure to heavy metals. In addition, biosensors based on other proteins, such as GST-SmtA and MerR have been identified and provide the possibility for rapid detection of various heavy metals at femtomolar levels (Bontidean et al., 1998; Corbisier, 1997). These tests should be adopted to assess exposure of production animals at the farm level.

3. Pesticides and Organic Pollutants

The initial triumph in the eradication of serious pests was turned into concern regarding the global burden of non-biodegradable components. Previously, prominent products such as DDT, dieldrin, chlordane, lindane and heptachlor are nowadays compounds of public apprehension. This concern is based on their detrimental effect towards marine and terrestrial organisms including mammals and man. Besides their tendency to bio-accumulate in the food chain, the nature of their toxic effects, comprising endocrine disregulation, immunotoxicity and genotoxicity, remains a source of serious concerns. In addition, the exposure *in utero* and the transmission via mother's milk to the neonatus, have gained increasing attention. As a consequence, the use of these compounds has been banned in most countries.

Organophosphorous pesticides (OP) and carbamates succeeded the first class of persistent chlorinated pesticides. The acute toxicity of these compounds, sharing a common mechanism of action (AChE-phosphorylation), was recognized immediately, but their biodegradability in the environment was considered a major advantage. Only recently, based on clinical findings from accidental intoxications, occupational exposure and experimental multi-generation studies in rodents, their potential to induce a delayed neurotoxicity syndrome (OPIDN) - at least at high doses – requested a re-evaluation of their potential adverse health effects (Clegg and Van Gemert, 1999; Zheng et al., 2000) and intensified the monitoring program (Mulchandani et al., 1999; Maroni et al., 2000, Nigg and Knaak, 2000).

The above-mentioned pesticides have had distinct beneficial effects in the control of pests. In contrast, structurally related compounds including polyhalogenated aromatic compounds have been generated by modern industry, neglecting initially not only their persistency in the environment but also their adverse effects to all forms of life. These polyhalogenated compounds, comprising polychlorinated biphenyls (PCBs), polybrominated biphenyls (PBBs), polychlorinated phenols and benzenes have been applied as plasticisers, flame-retardants, heat transfer fluids and dielectric fluids for transformers and capacitors. Other groups of polyhalogenated aromatics such as polychlorinated dibenzo-p-dioxins (PCDDs) and dibenzofurans (PCDFs) are formed as undesirable by-products during chemical synthesis or during combustion processes at municipal and chemical waste incinerators (Hutzinger et al., 1985). The toxic effects of these classes of compounds are diverse and range from chloracne and hepatotoxicity to systemic effects including immunotoxicity, neurotoxicity and neurobehavioral dysfunction, endocrine toxicity and carcinogenicity (for review see Safe, 1990 and 1994). All these compounds are highly lipophilic and extremely stable to hydrolysis, resulting in the above mentioned extremely slow biodegradation. The high lipophilicity explains the high oral bioavailability after ingestion with feed components and the disposition in liver and fatty tissue, as well as skin, causing bioaccumulation and biomagnification along the food chain (Van den Berg et al., 1994). Analytical monitoring of residues in edible tissues of production animals suffers not only from cost-intensive analytical procedures, but also from a lack of knowledge about biotransformation and bioaccumulation of specific congeners in target animal species. To circumvent these limitations, monitoring of wildlife and marine organisms is often conducted by an effect-related monitoring. Polyhalogenated aromatic compounds exert a broad variety of their biochemical and toxic effects via the aryl hydrocarbon (Ah) receptor (Probst et al., 1993). As part of this pathway, specific isoenzymes of the cytochrome P450 (CYP) enzyme family are induced. These enzymes are highly conserved throughout all animal species, and are also expressed in typical patterns in farm animals (Fink-Gremmels and Van Miert, 1996). Monitoring of production animals for elevated levels of CYP1A1 or CYP1A2 and CYP2B enzymes would provide an additional measure to recognise even accidental exposure at the farm level, as these measurements can be conducted by low-cost, high throughput (bio)assays.

4. Mycotoxins

Mycotoxins are naturally contaminants of plant-derived feed commodities and have been often regarded as neglected risk factors (Fink-Gremmels, 1999; Hinton, 2000). Research relating to their toxicity has been confined merely to Academia, and systematic toxicological investigations have been conducted only for a few prominent toxins. Mycotoxins are secondary metabolites of moulds and their actual synthesis is influenced by various environmental factors in the pre-harvest phase of plants, as well as during storage and transportation of crops (Nawaz et al., 1997; Scudamore et al., 1998). Mycotoxins comprise a group of chemically diverse compounds, with diverse toxicity based on different molecular mechanisms. Thus, no common bio-indicator of mycotoxin exposure exists and contamination of food and feed commodities require individual analytic tests (Coker et al., 1995; van Egmond, 1995). The most prominent mycotoxin, Aflatoxin B_1 has gained worldwide attention, and is regulated in most countries of the world (Verardi et al., 1995). Other toxins, such as the trichothecenes (including deoxynivalenon; DON), ochratoxins and related *Penicillium* toxins, as well as zearalenone and the fumonisins which have been clearly identified as adverse risk factors impairing animal health and productivity (Hollinger and Ekperigin, 1999; Le Bars and Le Bars, 1996; Shephard et al., 1996) and their impact on human health is a matter of continuous debate. Mycotoxins are rapidly biodegraded by microorganisms (ruminants), biotransformed and inactivated and thus have a relative short biological half life in the target animal without a tendency to accumulate in animal tissues. Thus, despite their detrimental effects to animal health and productivity, human exposure to mycotoxins from residues in animal tissues is low compared to the direct exposure from plant derived food commodities.

5. Hormonally Active Substances

The most recently acknowledged classes of compounds of concern are the hormonally active substances. Epidemiological data and mechanistic research identified diverse substances interfering with hormone synthesis or hormone-receptor interactions and signal transduction pathways. Commonly these compounds (examples are coumestrol, genistein, bisphenol A, DDT, fungistatic azoles, a.o.) are addressed as endocrine disrupting agents. Their effect on certain aquatic organisms and wildlife is well documented, however not completely understood in terms of the biological significance of the observed effects (for example on population kinetics). Moreover, their potential adverse effect on the human health and development is a matter of controversial debates (Vos et al., 2000) Hormonally active contaminants may impair animal productivity (see contamination of feeds with the mycotoxin zearalenone) and thus should be monitored by means of effects-based bioassays (receptor assays, reporter-gene assays; Diel et a., 1998)

6. Conclusions

Natural and industrial contaminants in the feed for farm animals represent one of the major risk factors for potentially hazardous (violating) residues in edible tissues and animal products, such as milk and eggs (the issue of veterinary drug residues has been discussed by Haagsma et al., 1999). The globalisation of the agricultural market makes it virtually impossible to monitor all feed commodities for the presence-absence of undesirable contaminants. This indicates the need for monitoring programs on the farm level, aiming to identify even accidental exposure of production animals. Milestones for an integrated quality program are:

1. the identification of geographical risks relating to feed quality (regional environmental burden with contaminants, climatic risk factors)
2. the identification (and declaration) of risk materials such as recycling products (fats, proteins)
3. the development of biosensors/bioindicators for the use in farm animals allowing a low-cost, high throughput screening of violating contaminants at the farm level.

Only an integrated quality control concept, which should be based on accepted principles in risk identification and exposure assessment, will meet the dual objective of guaranteeing animal health and welfare and improving food safety. The design of an advanced monitoring system requires a multidisciplinary co-operation, combining the knowledge of environmental sciences (Walker, 1998), toxicology and veterinary sciences with the skills of modern biotechnology, providing the probes for rapid effect-based biomonitoring (Maroni et al., 2000).

References

Albering, H.J., Van Leusen,S.M., Moonen, E.J., Hoogewerff, J.A., Kleinkans, J.C., 1999. Human health risk assessment: A case study involving heavy metals soil contamination after the flooding of the river Meuse during the winter of 1993-1994. Environ. Health Perspect. 107(1), 37-43.

Bontidean, I., Berggren, C., Johansson, G., Csoregi, E., Mattiasson, B., Lloyd, J.R., Jakeman, K.J., Brown, N.L., 1998. Detection of heavy metal ions at femtomolar levels using protein-based biosensors. Anal. Chem. 70(10), 4162-4169.

Bosveld, A.T.C. and Van den Berg, M. 1994. Biomarkers and bioassays as alternative screening methods for the presence and effects of PCDD, PCDF and PCB. Fresenius J. Anal. Chem. 348, 106-110.

Clegg, D.J., van Gemert, M., 1999. Expert panel report of human studies on chlorpyrifos and/or other organophophate exposures. J.Toxicol. Environ. Health. B. Crit. Rev. 2(3), 257-279.

Coker, R.D., Nagler, M.J., Blunden, G., Sharkey, A.J., Defize, P.R., Derksen, G.B., Whitker, T.B., 1995. Design for sampling plans for mycotoxins in foods and feeds. Nat. Toxins 3(4), 257-262, 280.

Corbisier, P. 1997. Bacterial metal-lux biosensors for a rapid determination of the heavy metal bioavailability and toxicity in solid samples. Res. Microbiol 148(6), 533.

Diel, P., Smolnikar, K., Michna, H. 1998. In vitro test systems for the evaluation of estrogenic activity of natural products. Planta Medica 65, 197-203.

Diels, L., De Smet, M., Hooyberghs, L., Corbisier, P. 1999. Heavy metals bioremediation of soil. Mol. Biotechnol. 12(2), 140-158.

Egmond, van H.P. 1995. Mycotoxins: regulations, quality assurance and reference materials. Food Addit. Contam. 12(3), 321-330.

Fink-Gremmels. 1999. Mycotoxins: their impact on human and animal health. Veterinary Quarterly 121: 115-120

Fink-Gremmels, J., Van Miert, A.S.J.P.A.M., 1996. Species Differences in Biotransformation: conventional models in unconventional animal species. Exp. Toxic Pathol. 48: S120-S127.

Haagsma, N., Ruiter, A., Van Miert, A.S.J.P.A.M., Fink-Gremmels, J., 1999. Veterinary Drug Residues. In F.J.M. Smulders (ed) Veterinary Aspects of Meat Production, Processing and Inspection. Publ. ECCEAMST, Utrecht, The Netherlands, pp 267-276

Giuffre-de-Lopez-Camelo, J., Ratto de Migues, S., Marban, L., 1997. Heavy metal input with phosphate fertilizers in Argentina. Sci. Total. Environ 204(3), 245-250.

Hecht, H., 1992. Schwermetalle in Rehen im Bayrischen Wald und Fichtelgebirge. Fleischwirtsch. 75, 790-799.

Hinton, M.H. 2000. Infections and intoxication associated with animal feed and forage which may present a hazard to human health. Vet. J. 159(2), 124-138.

Hollinger, K. and Ekperigin, H.E., 1999. Mycotoxicosis in food producing animals. Vet. Clin. North. Am. Food. Anim. Pract. 15(1), 133-165.

Hutzinger, O., Blumich M.J., Van den Berg, M., Olie, K., 1985. Source and fate of PCDDs and PCDFs: an overview. Chemosphere 14, 581-600.

Le Bars, J. and Le Bars, P.,1996. Recent acute and subacute mycotoxicoses recognized in France. Vet. Res. 27(4), 383-394.

Manz, M., Weissflog, L., Kuhne, R., Schurmann, G., 1999. Ecotoxicologicl hazard and risk assessment of heavy metals contents in agricultural soils of central Germany. Ectoxicol. Environ. Saf. 42(2), 191-201.

Maroni, M., Colosio, C., Ferioli, A., Fait, A., 2000. Biological monitoring of pesticide exposure: a review. Toxicology 143(1), 1-118.

Minura, K., 1995. *In:* Goyer, R.A and Cherian, M.G. (eds): Toxicology of Metals, Springer Verlag Berlin. pp 163-188

Mulchandani, P., Mulchandani, A., Kaneva, I., Chen, W., 1999. Biosensor for direct determination of organophosphate nerve agents. 1. Potentiometric enzyme electrodes. Biosens. Bioelectron. 14(1). 77-85.

Nawaz, S., Scudamore, K.A. and Rainbird, S.C., 1997. Mycotoxins in ingredients of animal feeding stuffs: I. Determination of Alternaria mycotoxins in oilseeds cake meal and sunflower seed meal. Food Addit. Contam. 14 (3), 249-262.

Nigg, H.N. and Knaak, J.B., 2000. Blood cholinesterase as human biomarker of organophosphorus pesticide exposure. Rev. Environ. Contam. Toxicol 163, 29-111.

Probst, M.R., Reisz-Porszasz, S., Agnunag, R.V., Ong, M.S., Hankinson, O. 1993. Role of the aryl hydrocarbon receptor nuclear translocator protein in aryl hydrocarbon (dioxin) receptor action. Mol. Pharmacol. 44, 511-518.

Roesijadi, G., 1996. Metallothionein and its role in toxic metal regulation. Comp. Biochem Physiol. 133C, 117-123.

Safe, S., 1990. Polychlorinated biphenyls (PCBs), dibenzo-p-dioxins (PCDDs), dibenzofurans (PCDFs) and related compounds: Environmental and mechanistic considerations which support the developments of toxic equivalent factors (TEFs). CRC Crit. Rev. Toxicol. 21, 51-88.

Safe, S. 1994. Polychlorinated biphenyls (PCBs): Environmental impact, biochemical and toxic responses and implications for risk assessment. CRC Crit.. Rev. Toxicol. 24, 87-149.

Scudamore, K., Nawaz, S., Hetmanski, M.T., 1998. Mycotoxins in ingredients of animal feeding stuffs: II. Determination of mycotoxins in maize and maize products. Food Addit. Contam. 15(1), 30-55.

Shephard, G.S., Thiel P.G., Stockenstrom, S., Sydenham E.M., 1996. Worldwide survey of fumonisins contamination of corn and corn-based products. J.A.O.A.C. Int. 70(3), 671-687.

Van den Berg, M., De Jongh, D., Poiger, H., and Olson, J.R., 1994. The toxicokinetics and metabolism of polychlorinated dibenzo-p-dioxins (PCDDs) aand dibenzofurans (PCDFs) and their relevance for toxicity. CRC Crit. Rev. Toxicol. 244, 1 – 74.

Verardi, G., De-Froidmont-Gortz, I., 1995. Overview on community legislation in the field on official control of mycotoxins in feedingstuffs. Nat. Toxins 3(4), 248-256, 280.

Vos, J.G., Dybing, E., Grein, H.A., Ladefogoed, O., Lambre, C., Tarazona, J.V., Brandt, I., Vethaal, A.D. 2000. Health effects of endocrine-disrupting chemicals on wildlife, with special reference to the European situation. Crit. Rev. Toxicol. 30(1), 71-133.

Walker, C.H. 1998. The use of biomarkers to measure the interactive effects of chemicals. Exotoxicol. Environ. Saf. 40(1-2), 65-70.

Zheng, Q., Olivier, K., Won, Y.K., Pope, C.N. 2000. Comparative cholinergic neurotoxicity of oral chlorpyrifos exposures in preweanling and adult rats. Toxicol. Sci. 55(1), 124-132.

The industrial production of safe animal feeds in Europe

Brian C. Cooke
Agricultural Consultant, UK, and FEFAC Past Chairman of Scientific and Legal Affairs Committee, rue de la Loi 223, Boite 3, B-1040 Brussels, Belgium; E-mail: brianccooke@lineone.net

Summary

The objective of animal feed production is to satisfy the nutritional requirements of the farm animal, whilst protecting animal health, human health and the environment. EU legislation lays down rules as to what materials can be used in the manufacturing process and various requirements especially labelling which relates to the manufacturing process and the finished product that is sold to the livestock producer. This legislation is discussed in detail. Of particular importance is the Establishments Directive 95/69 (EC, 1995(b)) which lays down basic rules relating to facilities and equipment, personnel, production, quality control, traceability and record keeping throughout the production and distribution process for compound animal feed. These legislative requirements are backed within the industry by Codes of Good Manufacturing Practice and operating instructions based upon the "FEFAC Guidelines for the implementation of a Code of Practice for the manufacture of animal feedstuffs" (FEFAC, 1999). This Code, together with the EU legislation refers to the use of the HACCP technique as a means of identifying hazards, critical control points and monitoring to avoid dangers to animal health, human health or the environment. Examples of the use of this technique are given in relation to the avoidance of cross-contamination of additives and medicines from one batch of feed to another, and to the control of Salmonella within the production processes. The adoption of Codes of Practice and the HACCP technique enables the European commercial feed compounder to ensure the production of safe animal feeds. The industry would welcome a similar approach to the production of feed on farm and in other parts of the world. To this end FEFAC hopes that Codex will introduce a detailed Code of Practice on an international basis.

Keywords: animal feeds, EU legislation, HACCP, cross contamination, *Salmonella* control

1. Introduction

The overall objective of animal feed production is to supply feeds with the correct nutritional balance to meet the requirements of the animals destined to receive the feed. However, this objective must be attained whilst giving equal importance to the protection of animal health, human health and the environment. These latter objectives are attained not only by the introduction of legislation by the EU, which is controlled by the Commission, Parliament and member states, but also by the application of Codes of

Practice and Good Manufacturing Practice on a voluntary and government-backed basis by the compound feed industry itself.

2. The EU Legislation

The legal basis for feed production applied directly to the 15 member states, but other countries applying for membership are in the process of adopting similar laws. Other non-EU countries of Europe, such as Norway, also have similar feed safety legislation. The EU legislation is in three forms: a) Regulations which have legal standing in all member states, b) Directives which must be incorporated into the national law of all member states, and c)
Commission decisions which modify annexes and details in existing regulations or directives.

Many Regulations and Directives have been introduced which control what can go into animal feeds, how they should be manufactured and what information must be given to the farmer about those feeds. The following list is not exhaustive but covers the main legislation of relevance to the safe manufacture of animal feeds.

3. Legislation Controlling the Materials which can be Used in Compounded Animal Feeds

3.1 The Additives Directive 70/524 (EEC, 1970)

This directive concerning additives in feeding stuffs has now been amended some five times and the 6th amendment is currently under discussion. This directive controls the use of all additives in feed; it specifies what can be used and, where necessary, it lays down minimum and maximum inclusion rates for the additive. It also lays down details as to how each additive should be labeled on the compound feed. It covers vitamins, trace elements, antibiotics, coccidiostats and antiblackhead drugs, growth promoters, antioxidants, colourants, emulsifiers, stabilisers, thickeners and gelling agents, binders, aromas and appetisers, preservatives, acidifiers, enzymes and micro-organisms.

All additives must be approved by entry into the annexes of this Directive before they can be used in animal feeds. This approval requires the presentation of a dossier covering the efficacy, purity and safety of the product to animals, man and the environment.

Examples of current maximum inclusion rates in complete compound feeds for some of the authorised additives are as follows:

(i)	Vitamin A	13,500 iu/kg except for young animals
(ii)	Vitamin D_3	2,000 iu/kg for pigs
		4,000 iu/kg for cattle, sheep and horses
		3,000 iu/kg for fish and poultry except

		5,000 iu/kg for turkeys and broilers
(iii)	Iron	1,250 mg/kg (600, 500 for horses)
(iv)	Cobalt	10 mg/kg (2)
(v)	Iodine	10 mg/kg except for horses @ 4 mg/kg
(vi)	Manganese	250 mg/kg (130, 150 for poultry)
(vii)	Zinc	250 mg/kg (120, 100 for pigs, 80 for cattle)
(viii)	Molybdenum	2.5 mg/kg
(ix)	Selenium	0.5 mg/kg
(x)	Copper	35 mg/kg except for sheep @ 15 mg/kg and higher levels for growing pigs (variable by species ranging from 15 for sheep to 30 for young pigs)

(xi) Some antioxidants e.g. BHA + BHT + Ethoxyquin @ 150 mg/kg
(xii) Some colourants e.g. in feed for egg production @ 80 mg/kg
xiii) Some binders e.g. bentonite @ 20,000 mg/kg

The maximum levels of some of the trace elements are currently under review in order to reduce their impact on the environment. Suggested new maxima are shown in brackets above. In the case of growth promoters and antibiotics since 1st September 1999, only 4 products are now authorised by the EU. These are: Salinamycin, Monensin Sodium, Flavomycin and Avilamycin. However, even these are banned in Norway, Sweden and Finland, whilst both Denmark and Holland are introducing a ban on their use.

3.2 The Undesirable Substances Directive 99/29 (EC, 1999)

This Directive on the fixing of maximum permitted levels for undesirable substances and products in feedingstuffs, controls the levels of many materials that might be used in animal feeds. Maximum levels are laid down for certain chemicals, including the heavy metals, e.g. arsenic, cadmium, fluorine, lead, mercury; certain mycotoxins e.g. Aflatoxin; natural anti-nutrients e.g. gossipol in cotton, hydracyanic acid in linseed and volatile mustard oil in rape. Certain toxic plant materials such as various mustard seeds and certain weed seeds such as those containing alkaloids are also controlled. It lays down maximum pesticide residues allowed in feed materials and in final compound feeds.

Examples of maximum inclusion rates of some of the listed Undesirable Substances in complete compound feed are as follows:

(i)	Arsenic	2 mg/kg	Except for fish @ 4 mg/kg	
(ii)	Cadmium	0.5 mg/kg	Except for cattle, sheep & goats @ 1 mg/kg	
(iii)	Lead	5 mg/kg		
(iv)	Mercury	0.1 mg/kg		
(v)	Nitrites	15 mg/kg		
(vi)	Fluorine	150 mg/kg	Except for:	100 mg/kg pigs 350 mg/kg poultry except chicks @ 250 mg/kg 50mg/kg for cattle, sheep & goats except when in milk @ 30 mg/kg

(vii)	Aflotoxin B_1	0.05 mg/kg	For cattle, sheep & goats except milking & young animals @ 0.01 mg/kg
		0.02 mg/kg	For pigs and poultry except for young animals @ 0.01 mg/kg
(viii)	Castor oil plant	10 mg/kg	Expressed as husk
(ix)	Free gossypol	20 mg/kg	For laying poultry, piglets, rabbits & horses
		60 mg/kg	For other pigs
		100 mg/kg	For other poultry
		500 mg/kg	For cattle, sheep & goats
(x)	Hydrocyanic acid	50 mg/kg	Except for chicks @ 10 mg/kg
(xi)	Ergot	1000 mg/kg	
(xii)	Volatile	150 mg/kg	(as allylisothiocyanate) except for adult cattle, sheep
	mustard oil		& goats @ 1000 mg/kg and adult pigs & poultry @ 500 mg/kg
(xiii)	Vinyl-thioxazolidine	1000 mg/kg	In poultry feed except for laying hens @ 500 mg/kg
(xiv)	Theobromine	300 mg/kg	Except for adult cattle @ 700 mg/kg N.B. For horses in training - **zero**
(xv)	Weed seeds containing alkaloids, glucoside and other toxic substances 3000 mg/kg except that Lolium and Datura have a maximum of 1000 mg/kg		
(xvi)	Apricot, Illipe, Bitter Almond and Mustard Seeds such as Indian, Chinese, Ethiopian and Black must not be present at a quantifiable level.		

Maximum levels for Dioxin in animal fat, fish and fish oil, vegetable oil, clays and compound feeds have recently been introduced or are currently under consideration.

3.3 The Certain Constituents Directive 82/471 (EEC, 1982)

This Directive concerns certain products used in animal nutrition. It controls the use of proteins produced from fermentation processes, such as bacterial proteins and yeasts, non-protein nitrogen sources such as urea, byproducts from amino acid production, amino acids and analogues of amino acids. It lays down not only labeling requirements for these constituents but also labeling requirements for the final feed containing these constituents. The presentation of a dossier proving the products' efficacy, purity and safety is required for any product before it can be authorised under this Directive.

3.4 The Feed Materials Directive 96/25 (EC, 1996)

This Directive on the circulation of feed materials lists the most commonly used feed materials in animal nutrition. It controls these materials whether they are fed directly to animals or as constituents of compound feeds. It lays down requirements for certain declarations on these individual ingredients, and defines their source and the processing through which the ingredients have passed.

3.5 Commission Decision 91/516 (EEC, 1991) - Establishing a List of Ingredients whose Use is Prohibited in Compound Animal Feeding Stuffs

This Decision currently bans the following:

(a) Faeces, urine and digestive tract contents
(b) Hide treated with tanning substances
(c) Treated seeds
(d) Wood & sawdust treated with wood preservative
(e) Sewage sludge
(f) Solid urban waste
(g) Untreated waste from eating places except vegetable material unsuitable for human consumption for reasons of freshness
(h) Packaging and parts of packaging

Following recent dioxin scares, the list of banned materials has or is being extended to include some or all of the following: i) Catering waste, ii) Domestic kitchen waste, iii) Fats & oils not produced in accordance with Directive 90/667, or iv) Vegetable oils not produced by companies operating a HACCP system to guarantee the nature, origin, traceability and safety of the material, and so certified by a national authority, and v) Filters used for refining processes.

3.6 The Pathogens Directive 90/667 (EEC, 1990 c)

lays down veterinary rules for the disposal of animal waste, for its placing in the market and for the prevention of pathogens in feedingstuffs, of animal or fish origin. It is under this directive that controls of salmonella, etc. are laid down. No comparable Directive, however, covers vegetable materials, although one is currently under discussion.

3.7 The Medicated Feedingstuffs Directive 90/167 (EEC, 1990 a)

laying down the conditions governing the preparation, placing on the market and the use of medicated feedingstuffs defines which medicines can be used and under what circumstances and at what levels. It also lays down the necessary labeling requirements for the compound feed containing medicines. These medicines are, of course, all under veterinary control and in member states they all have individual rules for the veterinary surgeon to write a prescription for the production of the animal feed containing the relevant medicine.

3.8 Directive 90/220 Control on the Use of Genetically Modified Organisms (EEC, 1990 b)

Directive 90/220 controls the authorisation for the growing of, import of, and use of GMO plants. Unfortunately its interpretation is not clear. In theory it only controls materials which contains genetically active material and thus any whole maize or soya from GMO varieties has to be authorised. The problem arises with raw materials such as extracted soya and maize gluten feed. Some national authorities, such as the UK and Holland, state that these materials do not contain genetically active material, because the processing fragments the DNA, and thus no authorisation is required. Other countries such as France take the contrary view and only allow material, even in by-product form, that has been specifically authorised. Currently only four GMO maize varieties, and one GMO soya variety are authorised in the EU, but many more varieties are authorised and grown in third countries, such as the US. Thus the European feed compounder is in a very awkward position with regard to knowing what he can legally use. However, a new Regulation on the circulation and use of genetically modified feed materials is currently under discussion and is likely to come into operation in 2001. This will require specific authorisation and labelling of all GMO feed materials and those derived from GMO varieties of crops even if they do not contain intact genetically modified DNA or protein. All these materials will be known as "Novel feed materials". The authorisation will require the presentation of a dossier, including a full assessment of the risk to human and animal health, and to the environment, appropriate safety and emergency response, a monitoring plan where necessary as well as precise instructions and conditions for use, labelling and packaging. It will lead to the phasing out by the end of 2004of any novel feed materials containing antibiotic resistant marker genes which have a current or potential use in medical or veterinary treatment.

3.9 Commission Decision 94/381 – BSE Controls (EC, 1994)

Under a number of legislative approaches various controls exist on the use of Meat and Bone Meal or mammalian protein in compound feeds.

i. In the EU mammalian Meat & Bone Meal is banned from use in ruminant feeds by Commission Decision 94/381, except where meat and bone is guaranteed to contain no ruminant protein.
ii. In the UK and Portugal the use of meat and bone is banned in all farm feeds.
iii. In Denmark and Ireland the use of Meat and Bone Meal is banned in all mills making ruminant feed.
iv. In France the use of fallen animals in Meat & Bone Meal is banned.

3.10 The Establishments Directive 95/69 (EC, 1995 b)

lays down the conditions and arrangements for approving and registering certain establishments and intermediaries in the animal feed sector. It has recently been extended to cover the approval/registration of manufacturers of additives and certain feed materials in non-EU countries. Initially this approval will be of the importer but ultimately will involve inspection and approval of establishments in third countries. It covers the production and use of additives, certain constituents and some feed materials high in undesirable substances. It, or some other Directive, is likely to be extended to require the authorisation of all manufacturing, including on-farm, of compound feed and some raw material suppliers such as those blending fats.

3.11 The Pesticides Residues Directive 86/362 (EEC, 1986)

The Pesticides Residues Directive 86/362 (EEC, 1986) fixes maximum levels for pesticide residues in and on cereals. It lays down maximum levels of various pesticides which may not be exceeded in cereals to be used in animal feeds.

4. Legislation Controlling the Manufacture of Compound Animal Feeds

4.1 The Additives Directive 70/524 (EEC, 1970)

This Directive is concerned with additives in feeding stuffs, not only controls the additives that can be used but also lays down specific labelling requirements for the final feed when certain additives are incorporated into that compound feed. For instance, the levels of certain vitamins and trace elements, and the presence of other additives such as antioxidants and colourants, has to be declared.

4.2 The Compounds Directive 79/373 (EEC, 1979)

on the marketing of compound feeding stuffs as amended, lays down general rules for the manufacture of all compound feeds for farm animals, pet animals and has now been extended to include wild animals, such as birds. It lays down a labelling requirement for compound feeds which not only includes nutritional parameters but also safety information, such as the species of animal to be fed, the age and type of animal within the species, together with feeding instructions. All the feed materials included in the compound feed have to be declared. Further it states that no compound feed shall be harmful to the health of animal or man.

4.3 The Undesirable Substances Directive 99/29 (EC, 1999)

on the fixing of maximum permitted levels for undesirable substances and products in feeding stuffs, controls the levels of many materials that might be used in animal feeds. Maximum levels in finished feed are laid down for certain chemicals, including the heavy metals e.g. arsenic, cadmium, fluorine, lead, mercury; certain mycotoxins e.g. Aflatoxin; natural anti-nutrients e.g. gossipol in cotton, hydracyanic acid in linseed and volatile mustard oil in rape. Certain toxic plant materials such as various mustard seeds and certain weed seeds such as those containing alkaloids are also controlled. It lays down maximum pesticide residues allowed in final compound feeds.

4.4 The Medicated Feedingstuffs Directive 90/167 (EEC, 1990a)

laying down the conditions governing the preparation, placing on the market and the use of medicated feedingstuffs defines which medicines can be used and under what circumstances and at what levels. It also lays down the necessary labelling requirements for the compound feed containing medicines. All these medicines are, of course, under veterinary control and in member states they all have individual rules for the veterinary surgeon to write a prescription for the production of the animal feed containing the relevant medicine.

4.5 The Dietetic Feeds Directive 93/74 (EEC, 1993)

on feedingstuffs intended for particular nutritional purposes lays down specific requirements for specific purposes in animal feeding, e.g. when animals have special nutritional status because of stress or some other cause. It specifies what materials can be used in that particular type of feed and at what level they can be included. It lays down specific labelling requirements for the finished feed.

4.6 The Establishments Directive 95/69 (EC, 1995 b)

lays down the conditions and arrangements for approving and registering certain establishments and intermediaries in the animal feed sector. It covers the production and use of additives, certain constituents and some feed materials high in undesirable substances and requires the authorisation or registration of all compound feed producers using certain additives, constituents and feed materials. It lays down some aspects of good manufacturing practice.

4.7 The Zoonosis Directive 92/117 (EEC, 1992)

concerns measures for protection against specified zoonoses and specified zoonotic agents, in animals and products of animal origin, in order to prevent outbreaks of food-borne infections and intoxications. In relation to compound animal feeds, this Directive at the moment only covers salmonella in poultry feeds. It calls for the authorities to sample compound feed at the final production stage, if that feed is suspected of involvement in a *Salmonella* outbreak on a poultry farm.

4.8 The regulation on the Circulation and Use of Genetically Modified Feed Materials

currently under debate will not only restrict the use of novel feed materials to those specifically authorised but will also introduce the specific requirement to label the compound feed to indicate the presence of each authorised novel feed ingredient.

4.9 The Controls Directive 95/53 (EC, 1995 a)

fixes the principles which govern the organisation of official inspection in the field of animal nutrition. It is under this Directive that EU officials can monitor what controls member states have on the production of animal feed.

This EU legislation has a common objective – that is the production of safe animal feed that will not be harmful to the health of animals or humans consuming the food from these animals, nor to the environment in which these animals are kept, or on which their manure is spread.

5. FEFAC Guidelines for the Implementation of a Code of Practice for the Manufacture of Animal Feedingstuffs

The various Directives and the codes of good manufacturing practices lay down manufacturing conditions which must be fulfilled. For instance, in the Establishments Directive 95/69 (EC, 1995 b) there are specific requirements for facilities and equipment, personnel, production, quality control, traceability and record keeping. The "FEFAC Guidelines for the implementation of a Code of Practice for the manufacture of animal feedingstuffs" (FEFAC, 1999) aims to incorporate the requirements of the various directives, whilst taking into account the practical problems of compound animal feed production. In simple terms, these requirements are as follows:

5.1 Facilities and Equipment

".. must be located, designed, constructed and maintained to suit the manufacture of compound feedingstuffs containing additives"

Similar requirements are of course needed when medicines are being added to animal feeds. In order to comply with this requirement, close attention must be given to the design, installation and maintenance of equipment. Two specific areas to which attention must be given are the reduction or elimination of cross contamination in order to avoid additives or medicines appearing in the wrong feed, and the installation of some means of ensuring death of *Salmonella* during the production process. This might be attained by heat treatment or by chemical treatment. It is always going to be the case that raw materials entering the feed mill will, on occasions, be contaminated with *Salmonella* and other harmful bacteria while the feed compounder's customers will demand to receive a product that is free of *Salmonella*.

5.2 Personnel

" The manufacturer must have sufficient staff possessing the skills and qualifications necessary for the manufacture of the products concerned"

To meet this requirement, proper training must be given to all staff involved in the production of compound feeds. The level of this training will depend upon the skills required for a particular function to be carried out. The training will therefore range from senior employees giving on-the-job training to new recruits working in the production process, through to high level management training for the mill manager and degree level qualifications for the nutritionists within a company.

5.3 Production

" A qualified person responsible for production must be designated. The manufacturer must ensure that the different stages of production are carried out in such a way as to obtain the desired quality of the products concerned"

To meet this requirement, different procedures are required for each stage of the production process. These procedures, whilst not only specifying how each stage of the process is controlled, must give particular attention to the avoidance of cross-contamination and to the control of *Salmonella*. Staff should be trained to ensure that they have adequate knowledge to carry out their functions and to ensure that they are aware of the risks not only to animal health but also potentially human health if there are failures in the manufacturing process. Very careful attention needs to be given to the subject of production planning i.e. the scheduling of one batch of feed after another, and the need for flushing the system under some circumstances, in order for the production of safe feed to be maintained.

5.4 Quality Control

"A qualified person responsible for quality control must be designated. The manufacturer must draw up and implement a quality control plan to guarantee and check that the products in question comply with the specifications defined by the manufacturer and that they comply with the law."

Again, defined protocols are required to cover sampling, analysis, reporting of results, action following these reports and, if necessary, recall. Each batch of feed should be sampled and a defined number submitted for analysis. This analysis needs to:
1. cover nutritional quality,
2. check the level of additives present in the feed,
3. checks occasionally to ensure that there are not excessive quantities of undesirable substances such as Aflatoxin, heavy metals, pesticide residues, etc. and finally
4. involve tests for bacteriological quality in order to ensure that whatever *Salmonella* controls are present in the mill they are working effectively.

5.5 Traceability

"Samples must be taken and kept from each batch of the product or proportion of production. They should be kept for the period of the use of the product.

Again, defined practices are necessary if traceability is to be guaranteed. It is recommended that all suppliers are authorised and undertake to keep adequate records to ensure that each batch of raw material can be traced as far back as possible in the supply chain. The practice should ensure that, at intake, contamination does not occur between batches of raw materials and that the specific batch of raw material is always delivered to the correct storage bin. Production and finished product records should identify what goes into each batch of feed and which finished product storage area it is sent to. Finally, delivery records of the batch of feed to farm are aspects of ensuring traceability.

5.6 Records

The manufacturer must record:

(i) the name and address of the manufacturer and batch number, and quantity of the premixture or additive (or medicine) used
(ii) the nature and quantity of the feedingstuff manufactured."

Good manufacturing practice requires that all parts of the raw material supply chain raw material intake and storage, the manufacturing process, finished product storage and finished product supply must be recorded. All records should be signed by the person responsible for each stage of the process. Records should also be maintained with regard to plant design and, specifically, any changes in the plant layout or equipment. These records should also cover maintenance of the plant and staff training. Any complaints from customers should be recorded fully, investigated and action taken (again recorded). Should a recall of feed become necessary, this should once more be fully recorded, including information on what corrective action was taken to avoid a repeat of the problem. Records must be retained for a suitable period. In the case of the manufacture of medicated animal feeds, this is specified in legislation as a minimum of 3 years. It is suggested that a suitable period for the maintenance of all the production records, including formulations, etc is for a period such that the customer, if dissatisfied, could take legal action against the feed supplier. In the United Kingdom this is 6 years and 4 months after the date of the feed delivery.

6. Hazard Analysis and Critical Control Points

Many of the GMP codes of practice, as well as more recent legislation, refer to the technique of HACCP. The application of this technique calls for an assessment to be made of what hazards exist, which are the critical points within the process that this hazard could occur, and what control systems could be put in place to avoid a hazard occurring. The avoidance of cross contamination of additives and medicines from one batch of feed to another and the control of Salmonella within the production process both lend themselves to the application of this technique. Examples of the approach that might be taken in these two areas are as follows:

6.1 The control of cross contamination in the feed mill

An example of the application of the HACCP technique is the investigation into the problem of Nicarbazin and Lasolocid residues in eggs. Statutory surveillance (Tables 1 and 2) shows that in 1998 over 3% of UK battery eggs (nearly 7% in 1997) were contaminated with Nicarbazin, and over 2% with lasolocid (VMD, 1999). Clearly some of this contamination is due to on-farm problems such as the recycling of the coccidiostat via the faeces and back to the bird and, on some occasions, undoubtedly bad feed handling. Nonetheless, in response to supermarkets, the publication of these results by the authorities has led the egg packers to demand that egg producers obtain guarantees from their feed compounder that the feed supplied is free of any product which would cause undesirable residues in eggs. A severe problem that the feed compounder has in investigating this type of problem is that the level of detection of Nicarbazin in feed is a hundred times higher than the level of detection in eggs. Thus a negative result may be obtained in the feed, yet a positive

Table 1. Some Results of UK statutory surveillance for 1998 on free-range eggs (VMD, 1999).

Veterinary Medicine	No. of Samples:	MRL mg/kg	No. of samples above action level : i.e. limit of quantification	1997 for comparison
Dimetridazole	22	Not set	2 (8.0%)	5.94%
Nicarbazin	25	Not set	3 (12.0%)	6.85%
Lasolocid	23	Not set	0 (0.0%)	2.29%

Table 2. Some result of UK statutory surveillance for 1998 on battery eggs (VMD, 1999).

Veterinary Medicine	No. of Samples:	MRL mg/kg	No. of samples above action level : i.e. limit of quantification	1997 for comparison
Dimetridazole	172	Not set	2 (1.16%)	3.70%
Nicarbazin	193	Not set	6 (3.11%)	6.76%
Lasolocid	193	Not set	5 (2.59%)	1.45%

result may be obtained in the egg produced by the bird receiving the feed. The use of micro tracers to mimic the action of medicines or other additives is a technique which could be incorporated into HACCP studies in order to obtain better quantification of cross contamination in the mill. Results of studies of this nature carried out in feed mills indicate that material coming off the dust extraction unit at the supplement/medicine tipping point can contain a medicine some 20 mixes after the medicine has been tipped. Screw conveyors are particularly likely to lead to carry-over. For instance, a 7 metre screw conveyor can lead to a carry-over of 0.36% of the previous mix to the next mix. In the case of Nicarbazin this alone would be almost enough to give a measurable residue in the egg. Even with fairly well designed equipment, the first two mixes after the addition of a particular additive are likely to contain 1% of the added level of that additive. Traces of the additive, as indicated by the presence of micro tracers, may be present up to 10 batches after the time of addition. The conclusion from this type of HACCP approach is that levels as low as 0.5mg of Nicarbazin in a kilogram of feed can give significant measurable effects in eggs and that to avoid problems the ideal is to split mills or production systems into those handling medicines and Zootechnical products as opposed to those making plain feeds such as layer mashes. Obviously careful production planning to ensure that no sensitive feed is made for at least 2 batches after the addition of problematical additives possibly allied with flushing can significantly reduce the risk. However, this approach needs to be accompanied by the replacement of pieces of equipment proved by the HACCP Study to cause contamination. These studies also demonstrate the importance of plant maintenance.

6.2 Ensuring *Salmonella* kill

Whilst either heat or chemical treatment can be used within the production process to kill off *Salmonella*, it is very difficult due to the low number of organisms present in any batch of feed, to prove the effectiveness of the treatment. It is therefore necessary if one is going to demonstrate i) which are the parts of the manufacturing process which reduce or multiply the hazard, and
(ii) that whatever procedures are implemented at these critical control points is effective.

The Campden and Chorleywood Food Research Association in the United Kingdom has developed a procedure whereby a marker organism is used to replicate the effect of heat on *Salmonella* (report in press). The organism *Bacillus brevis* is a non pathogenic spore forming Bacillus which responds to heat treatment in a similar manner to *Salmonella*. Being non pathogenic, it can be safely added to a feed in the feed mill at high levels in the range of 10^6. Sampling at different points in the production process can then enable measurements to be taken of the number or log reductions that occur in the organisms present after different treatments. Given the highest levels of *Salmonella* that are found in raw materials, three log reductions can be taken as evidence of the virtual elimination of Salmonella. Table 3 below shows that if *Salmonella senftenberg* is eliminated, then *Salmonella enteritidis, S. typhimurium* DT 104, *Listeria monocytogenes, Yersinia pseudotuberculosis* and *Escherichia coli* 0157:H7 will also be eliminated. One UK company has applied this technique in a co-operative project, supported by MAFF, in a number of mills and obtained the log reductions in organisms present, as illustrated in Tables 4, 5 and 6.

Whilst this work is still to be fully reported, a number of interim conclusions have already been reached, these being that all systems involving heat give a reasonable kill stage in the production process. This kill is most likely to occur with steam injection to the conditioner. The preliminary results suggest that steam quality in terms of temperature and moisture is probably more important to the kill process than absolute temperature or the period of time that the feed is in the conditioner, expander or press.

Table 3. D-values for various micro-organisms.

Micro-organism	D-value at 80°C at 0.8 Aw (mins)
Salmonella anatum	12.3
Salmonella enteritidis	6.9
Salmonella senftenberg	11.8
Salmonella mbandaka	9..2
Salmonella typhimurium DT104	7.7
Salmonella derby	2.2
Listeria monocytogenes	5.4
Yersinia pseudotuberculosis	<0.3
Eschericia coli 0157:H7	6.3

Table 4. Log reduction of Bacillus brevis *in Mill 1.*

Long term conditioner/pelleter	
Log reduction post conditioner	6.3
Log reduction post press	6.5
Log reduction post cooler	6.5
Short term conditioner (80°C) pelleter	
Log reduction post conditioner	6.9
Log reduction post press	8.6
Log reduction post cooler	8.1
Short term conditioner (64°C) pelleter	
Log reduction post conditioner	2.3
Log reduction post press	5.5
Log reduction post cooler	5.3

Table 5. Log reduction in Bacillus brevis *in Mill 2.*

Log reductions	Post conditioner	Post expander	Post pelleter	Post cooler
Conditioner/expander/pelleter	15.4	11.9	12.2	10.0
Conditioner/pelleter	15.1	NA	13.5	11.0
Conditioner/expander	27.8	31	NA	31.0
Conditioner	25.0	NA	NA	24.4

Table 6. Log reduction in Bacillus brevis *in Mill 3.*

Double conditioner/double pelleter	
Log reduction post conditioner	7.5
Log reduction post press	7.5
Log reduction post cooler	7.5

The adoption of the HACCP technique is now becoming a common practice in the industrial production of safe animal feeds in Europe. However, one has to question whether it is yet applied when farmers produce their own compound feed on farm.

7. Consumer demands

In addition to the foregoing legislative requirement, and the adoption of industry Codes of Practice, there has been extensive media interest in many countries of the EU in human disease problems which are, or are reputed to be, foodborne. In recent years the media has given wide attention specifically to:

(a) BSE and what is claimed to be a corresponding disease in man nvCJD
(b) *Salmonella* in eggs and meat
(c) Bacterial drug resistance in humans which they claim may well be linked to the use of antibiotics, both as therapeutic drugs and as growth promoters in animal production.

There have also been other health scares related to animal production, but not necessarily involving animals feeds. The most significant of these is probably *E. coli* 0157:H7. As a result of this media attention the consumer is increasingly demanding safe animal products as food. This in turn means that buyers of farmers' products, be they slaughter houses buying meat, packers buying eggs or dairies buying milk, are increasingly demanding tighter controls on production at the farm level, and in turn on the producer of the feed compounds used to feed the animals producing the meat, milk or eggs. Similar (and maybe more stringent) demands are coming from the large supermarket groups in many countries within the EU. In the United Kingdom this is leading supermarkets to develop their own feed quality schemes, all of which are increasingly demanding action by the compound feed producer to ensure the safety of the feed at the time it goes to the farm.

This pressure, both from legislation and consumers etc. has led to the compound feed industry associations in a number of EU countries to enhance the FEFAC Code of Practice by the development of more detailed codes of good manufacturing practices in order to aid their members in meeting the increasingly stringent requirements of the market place. Today's and tomorrow's objective must be 'safe feed = safe food'.

8. Conclusions

The feed compounders' objective is to ensure the production of safe animal feed in order to contribute to the production of safe food. Whilst compliance with the requirements of EU legislation should take the compounder well towards this objective, further input is undoubtedly necessary if the full objective is to be attained. It is recommended that all compound feed producers should follow a system of good manufacturing practice. The FEFAC Code of Practice for the Production of Safe Animal Feeds is a good starting point, but each mill requires specific procedures to be laid down. The application of the HACCP technique to assess where hazards and risk occur, and to eliminate the risk is an essential step in ensuring the production of safe animal feed. FEFAC hopes that Codex, with whom we are working, will adopt a similar Code of Practice so that the production of safe animal feed will become universal thus ensuring that animal products, wherever they are produced, will be as safe as possible for the human consumer.

References

EEC, 1970. Council Directive 70/524/EEC concerning additives in feedingstuffs. OJ No.L270/1 of 14.12.1970

EEC, 1979. Council Directive 79/373/EEC on the marketing of compound feedingstuffs. OJ No.L86/30 of 6.4.1979

EEC, 1982. Council Directive 82/471/EEC concerning certain products used in animal nutrition. OJ No.213/8 of 21.7.1982

EEC, 1986. Council Directive 86/362/EEC on the fixing of maximum levels for pesticide residues in and on cereals. OJ No.L221/37 of 7.8.1986

EEC, 1990(a). Council Directive 90/167/EEC laying down the conditions governing the preparation, placing on the market and use of medicated feedingstuffs in the Community. OJ No.L92/42 of 7.4.90

EEC, 1990(b). Council Directive 90/220/EEC on the deliberate release into the environment of genetically modified organisms. OJ No.L117/15 of 8.5.90

EEC, 1990(c). Council Directive 90/667/EEC laying down the veterinary rules for the disposal and processing of animal waste, for its placing on the market and for the prevention of pathogens in feedstuffs of animal or fish origin. OJ No.L363/51 of 27.12.90

EEC, 1991, Commission Decision 91/516/EEC establishing a list of ingredients whose use is prohibited in compound feedingstuffs. OJ No.L281/23 of 9.10.91

EEC, 1992, Council Directive 92/117/EEC concerning measures for protection against specific zoonoses and specific zoonotic agents in animals and products of animal origin in order to prevent outbreaks of food-borne infections and intoxications. OJ No.L62/38 of 15.3.93

EEC, 1993. Council Directive 93/74/EEC on feedingstuffs intended for particular nutritional purposes. OJ No.L237/12 of 22.9.93

EC, 1994. Commission Decision 94/381/EC concerning certain protection measures with regard to bovine spongiform encephalopathy and the feeding of mammalian derived protein. OJ No.L172/23 of 7.7.94

EC, 1995(a). Council Directive 95/53/EC fixing the principles governing the organisation of official inspections in the field of animal nutrition. OJ No.L265/17 of 8.11.95

EC, 1995(b). Council Directive 95/69/EC, laying down conditions and arrangements for approving and registering certain establishments and intermediaries operating in the animal feed sector. OJ No.L332/15 of 30.12.95

EC, 1996. Council Directive 96/26/EC on the circulation of feed materials. OJ No.L125/35 of 23.5.96

EC, 1999, Council Directive 1999/29/EC on the undesirable substances and products in animal nutrition. OJ No.L115/32 of 4.5.99

FEFAC, 1999. FEFAC Guidelines for the implementation of a Code of Practice for the manufacture of animal feedingstuffs. Fédération Européenne des Fabricants d'Aliments Composés, rue de la Loi 233, Bte 3, B-1040, Brussels, Belgium.

VMD, 1999. The Veterinary Medicines Directorate Annual Report on Surveillance for Veterinary Residues in 1998, Veterinary Medicines Directorate, Woodham Lane, New Haw, Addlestone, Surrey, KT15 3LS, UK

Modern methodologies for isolation/typing of animal-borne human pathogens: A research update on epidemiological tools
I. Recombinant antigens for antibody-based diagnosis

Christine Citti*, Marion Brank and Renate Rosengarten
Institute of Bacteriology, Mycology and Hygiene, University of Veterinary Medicine, Veterinärplatz 1, A-1210 Vienna, Austria, (Corresponding author: E-mail: Christine.Citti@vu-wien.ac.at)

Summary

Control of foodborne pathogens at the pre-harvest level necessitates the early identification of subclinically infected or nonshedding carriers to enable their segregation and minimizing the risk of spreading and enhancing food safety. Assays that assess the presence of circulating serum antibodies offer a suitable and practical alternative to direct detection of the pathogen or of its components, as they can identify infected animals within a large herd even in the absence of shedding organisms. For the development of sensitive and specific serological tests as control tools for epidemiological monitoring and forecasting of infections in the pre-harvest phase the identification of highly immunogenic antigens that are specific for a particular agent is required. Gene expression technology offers new approaches to produce in large quantities pure antigens for the improvement of serological assays. This approach as well as its benefits and limitations are described using as example the non-human pathogen *Mycoplasma bovis*. This agent is frequently involved in mastitis as well as enzootic pneumonia of cattle causing substantial economic losses to the world beef and dairy industry due to the lowered feed efficiency and milk production. One of the major advantages of using recombinant proteins in diagnostic test systems is that they limit problems due to cross- or non-specific reactivity and allow the establishment of well-defined and highly standardized serodiagnostic assays that can be designed in a high-throughput ELISA format for application in epidemiological surveillance and control programmes.

Keywords: gene expression technology, recombinant antigens, serological assays, epidemiological tools, *Mycoplasma bovis*

1. Introduction

An important strategy to control foodborne pathogens at the pre-harvest level is the early identification of subclinically infected or nonshedding carriers to enable their segregation and thereby minimizing the risk of spreading and enhancing food safety.
In principle, the specific detection of any particular bacterial pathogen can be achieved in two different ways. Directly by detection of the pathogen itself using classic procedures such as cultivation followed by microscopic observations and biochemical or serological confirmation or by the detection of selected components specific of the pathogen, such as nucleic acids or proteins. One drawback of these methods is that they rely on the presence of organisms in the samples at a detectable concentration that depends on the sensitivity of the test which is performed.

An alternative to the direct detection of the pathogen is the so-called indirect approach which consists of the detection of specific antibodies generated by the host post infection. The drawback of this approach is that it cannot discriminate between hosts that are currently infected and those that are in the post-infection state, as circulating antibodies can persist in the host up to several months even in the absence of the pathogen. On the other hand, it allows the identification of asymptomatic, infected animals within a large herd even in the absence of shedding organisms. This can be performed by simple test systems, such as ELISA, Western blot or agglutination assays. In the field of clinical microbiology, the determination of specific antibodies directed towards structural or functional antigenic proteins of a given pathogen by ELISA is certainly the most commonly used serological tool for epidemiological studies because of the ease of this technique and because of its ability to assess a large number of samples at once in a short period of time.

The development of sensitive and specific serological tests as control tools for epidemiological monitoring and forecasting of infections in the pre-harvest phase requires the identification of highly immunogenic components that are specific for a particular agent. Gene expression technology offers new approaches to produce in large quantities pure antigens as recombinant proteins for the improvement of serological assays. Apart from limiting the problems due to cross- or non-specific reactivity, one of the advantages of using recombinant proteins in diagnostic test systems is that they allow the establishment of well-defined and highly standardized serological assays that can be designed in a high-throughput ELISA format for application in epidemiological surveillance and control programmes. Alternative uses of recombinant antigens include the production of specific reagents such as monoclonal or monospecific polyclonal antibodies for the identification or typing of a particular pathogen, the development of functional assays, and the production of subunit vaccines.

2. Classic Immunoassays Using Native Antigen Preparation

Despite their widespread use, immunoassays are compromised by a number of significant problems that may affect the reliability and accuracy of their outcome (Fierz, 1998). In serological tests designed to assess the presence of specific host antibodies directed to components of a given pathogen, the nature of the target antigen, as well as the criteria and methods chosen for its preparation, purification and testing are crucial. The better an antigenic preparation is defined and controlled over its production, the more the assay is likely to be accurate. In any case, the performance of the assay in terms of sensitivity and specificity, relies on the selection of the target antigen that has to fulfill several fundamental criteria: the target antigen must be (i) specific for a given pathogen, (ii) immunogenic in the host at an early stage of the infection, preferably before apparition of the clinical signs, and (iii) expressed by all pathogenic strains and isolates.

A large number of the serological assays currently available is based on the use of complex material such as whole cell lysates or partially purified fractions, such as membrane preparations, as target antigens. While most problems due to non-specific reactions result from the presence of impurities in the antigen preparation and can be solved by improving the quality of the reagents, false positive reactions due to true cross-reacting epitopes can only be eliminated by removing the interfering antigenic determinant from the preparation. This might not always be possible, specially when the presence of cross-reacting epitopes in a particular host is not always known and may vary depending on environmental factors or on the occurrence of concomitant infections.

3. Advantages in Using Recombinant Proteins as Target Antigens

In recent years, a new generation of target antigens for antibody-based serological assays has emerged which is based on less complex antigenic material and uses recombinant antigens generated by genetic engineering (see below for example). Overall these new reagents offer multiple advantages (Table 1) as they can be produced in a large amount with relative ease and allow the development of well-standardised serodiagnostic assays at low cost.

Table 1. Advantages of Recombinant Antigens in Serological.

Parameter	Recombinant antigens	Whole-cell extract or purified fractions
Antigen quality	Well-defined	Not well-defined
Antigen quantity	Easily adjustable	Not easily adjustable
Batch-to-batch consistency	High	Low
Cross-reactivity	Low	High
Safety	High	Low
Large-scale preparation	Easy	Not always feasible
Cost	Low	High

3.1 Large Scale Production and Safety Assurance

A major limitation in using whole cell proteins or purified fractions is the need to maintain large volumes of cultures which might be technically demanding and costly for some pathogens due to their fastidious nutritional requirements (Shenai et al., 1996). As well, the production of large cultures of highly pathogenic bacteria, viruses or parasites constitute a potential threat for the environment and for the technical personnel handling them. In this respect, the production of recombinant proteins using for instance *E. coli* or yeast cells as expression hosts, provides both a safe and economically suitable alternative. In the line of safety, genetic engineering allows the design of recombinant proteins that have lost their virulence attributes but have retained their antigenic properties. Furthermore, the tremendous advances in molecular biology techniques offer the prospect of producing recombinant proteins from bacteria for which there may be no known permissive cell line or other biological system for their propagation.

3.2 Controlled Antigen Quality and Standardisation

Another problem which may be encountered in preparing crude extracts as antigenic target is the presence of high levels of proteolytic activities (Ostoa-Saloma et al., 1989) that may contribute to a reduction in the antigenicity of the extracts upon storage over a period of time. As well, the extent to which a particular impurity in a particular antigenic extract is recognized by host sera is not always known, therefore optimizing the purity of the preparation is always an advantage with consequential outcomes, as long as the quality of the epitope is not compromised. Because the content of an antigen preparation based on recombinant proteins is better defined than that of whole cell lysates or purified fractions, potential contaminants are better known and procedures for their elimination can be set up and easily standardized (Babbit et al., 1990). Finally, the use of recombinant antigens is highly suitable for a better standardisation and reproducibility of serological assays, specially in (i) improving the batch-to-batch consistency of the antigen preparation which may be affected in whole-cell lysates by the intraclonal or intrastrain antigenic variation of a given pathogen and (ii) allowing a more accurate measurement of the amount of antigen to be used.

3.3 Immunoassay Control

A critical point in establishing detection systems for antibodies is the use of the right control preparation that does not contain the target antigen but is otherwise identical to the preparation with the antigen. Such control can be easily obtained in the case of recombinant proteins since they are genetically engineered and produced in heterologous systems which do not naturally express the foreign product.

4. Generation of Recombinant Antigens

4.1 Basic Principles

The basis for the production of a recombinant protein is the expression of an open reading frame into prokaryotic or eukaryotic heterologous systems. Using standard molecular procedures, heterologous genes are introduced into an expression vector under the transcriptional control of a promoter. Because the overexpression of foreign proteins might be detrimental to the host cells at an early stage of the multiplication, most systems use promoters that are activated on demand by addition of an inducer (see for illustration Fig. 3 B). This is for instance the case of the bacterial promoters P_{tac} or T7 that respectively require a source of isopropyl β-D-thiogalactopyranoside (IPTG) or T7 RNA polymerase to be functional. In addition to an inducible promoter, these expression vectors also provide a ribosome binding site and a start codon which are localized downstream of the promoter and upstream of a multiple cloning site (MCS) for the in-frame insertion of the sequence to be expressed. A number of vectors additionally offers a variety of tags adjacent to the MCS which purpose is to facilitate the purification and the detection of the expressed protein as illustrated in Fig. 1.

Most tags are linked to the N-terminal portion of the protein of interest, but some vectors offer the alternative of fusing the tag to the C-terminal part of the protein. Furthermore these tags contain a region overlapping the MCS and encoding a cleavage motif for releasing the product of interest. Finally, some expression vectors are equipped with a sequence signal located upstream of the tag for periplasmic localization.

4.2 Prokaryotic versus Eukaryotic Expression Systems

In the past few years, the number of expression systems commercially available for the production of recombinant proteins suitable for use in bacteria or in various eukaryotic cells has constantly increased. The most documented are (i) recombinant *E. coli* cells, (ii) recombinant yeast cells, (iii) a baculovirus-driven insect cell expression system (Kost and Condreay, 1999), (iv) recombinant vaccinia viruses for the transient expression of foreign proteins into various mammalian cells (Fuerst et al., 1986) and (v) mammalian chinese hamster ovary cells for continuous expression of recombinant proteins (Collen et al., 1984). The problem in using bacterial systems is that they do not support most of the post-translational modifications achieved by eukaryotic systems and are therefore not suitable for the expression of eukaryotic or viral proteins that may require specific glycosylation, acylation or phosphorylation for displaying their antigenic properties. Overall, the choice of the expression system mainly depends on the nature of the protein to be expressed, and to whether modifications or secondary structures are of antigenic relevance for the assay. The pro's and con's of these systems and detailed protocols for expression of foreign antigens along with tips for problem solving have been extensively reviewed (Reschl, 1998).

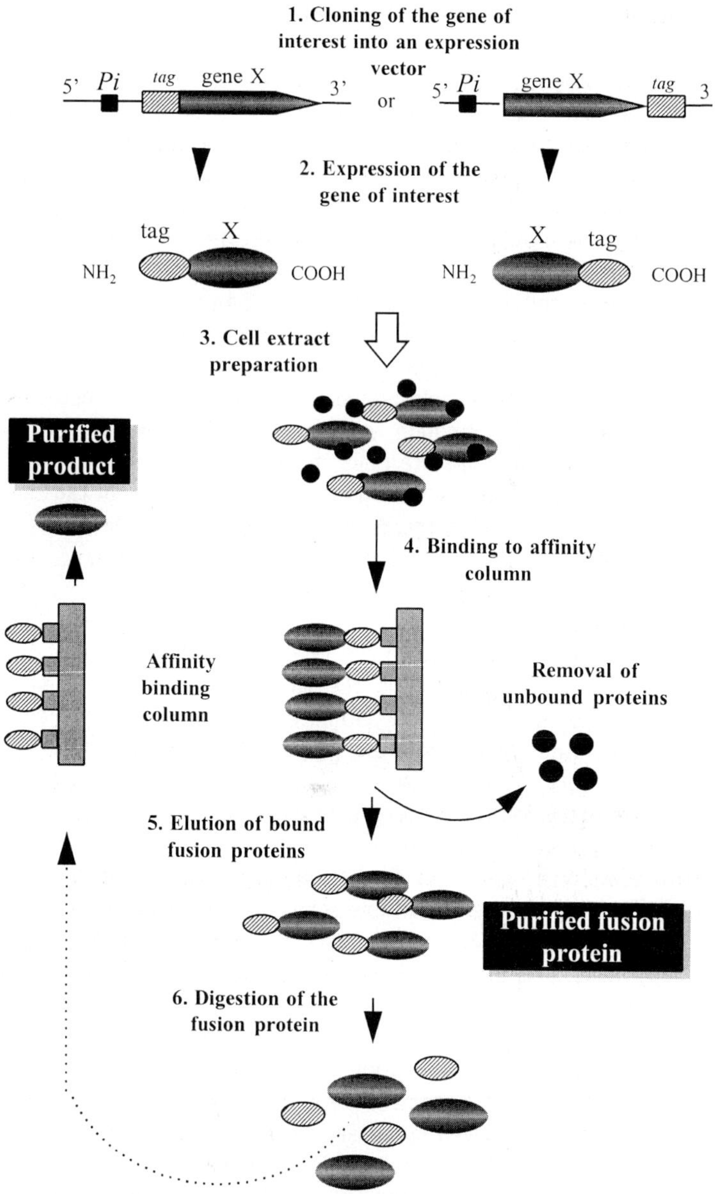

Figure 1. Outline illustrating the successive steps to be performed for the purification of a recombinant fusion protein. The gene of interest (shaded open arrow) is cloned into an expression vector, in frame with a sequence encoding a tag (hatched open arrow). Depending whether the gene is inserted downstream or upstream of the tag-encoding sequence, the recombinant product will be fused to the tag at its N-terminal or C-terminal end. Following transformation of the host cells, the expression of the fusion protein will be induced by adding a factor that will activate the transcription of the inducible promoter (Pi). Shaded circles: protein of interest; hatched circles: tag; black circles: proteins from the host cells; vertical shaded rectangle: affinity column.

4.3 Production of Recombinant Fusion Proteins

As mentioned above, a wide range of expression vectors are commercially available for the production of recombinant proteins. Most bacterial expression vectors either use the T7 promoter or the P_{tac} in conjunction with a polyhistidine tag which can bind to nickle-nitrilotriacetic acid (Ni-NTA) columns, or with naturally occurring proteins which include (i) the maltose binding protein (MBP), a protein which has an affinity to amylose, (ii) the gluthatione S-transferase (GST) which has an affinity to gluthatione, and (iii) protein A which has an affinity to immunoglobulin. The eukaryotic expression system based on the pFASTBACHT series of bacculovirus shuttle vectors (GIBCO BRL) also express polyhistidine-tagged proteins.

Since the factors leading to high-level expression of the protein of interest in foreign hosts are usually not known, the choice of which affinity tags to use for the expression of a particular protein is empirical. Following purification, the fusion protein can be directly used, but in most of the cases the removal of the affinity tag is necessary either for proper folding of the protein of interest or for avoiding false positive serological reactions due to the presence of antibodies directed to the tag in the sera to be tested. Because removal of the tag using chemical procedures is often difficult to control, it is mainly performed by enzymatic cleavage using proteases, such as the enterokinase, the thrombin or the factor Xa. Table 2 presents few systems commercially available for the expression of recombinant proteins in *E. coli* or in mammalian cells.

As mentioned above, the factors influencing the level of expression of a particular recombinant protein in a prokaryotic or eukaryotic host system are usually unknown. However, a number of relevant problems often encountered with the expression of recombinant proteins can be overcome. These include solubilization, toxicity and proper refolding of the recombinant proteins. As well, the poor expression of some mycobacterial genes in *E. coli* partly result from their content in low-usage *E. coli* codons. Codon replacement by site-directed mutagenesis has been recently shown to enhance the production of recombinant *Mycobacterium tuberculosis* antigen 85A and superoxide

Table 2. Most Commonly Used and Commercially Available Expression Systems.

Affinity tag	Affinity column	Elution ligand	Supplier	Vector[1] designation
Glutathione S-transferase	Gluthatione sepharose	Reduced gluthatione	Pharmacia	pGEX
6Xhis	Ni-NTA[2]	Imidazole	Quiagen,	pQE
			Novagen	pET
			GibcoBRL	pFASTBACHT
Maltose Binding Protein	Amylose resine	Maltose	New England Biolabs	pMAL
Protein A IgG Binding Domain	IgG sepharose	IgG	Pharmacia	pEZZ 18
				pRIT2T

[1]and derivates [2]Nickel-nitrilotriacetic acid

dismutase in *E. coli* indicating that for some bacterial genes this strategy might be applicable for obtaining overexpression in *E. coli* (Lakey et al., 2000). In the case of the recombinant antigen 85A, an urea-based resolubilisation protocol similar to methods used in refolding of other recombinant proteins was also necessary to obtain the soluble recombinant antigen.

5. Selection of a Target Antigen

The general approach to undertake for the identification of potential target antigens to be further used as recombinant products in serological assays is illustrated in Fig. 2.

Preliminary studies are commonly performed by Western blot analysis using whole-cell extract or purified fractions of the type strain and/or well-characterised isolates of a given

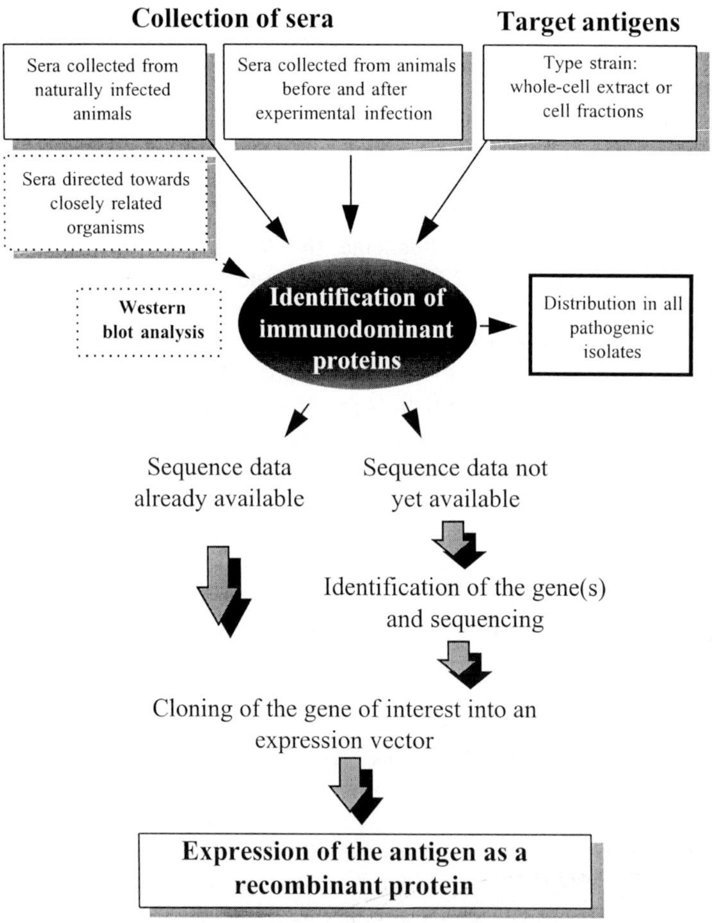

Figure 2. Outline illustrating the overall strategy to be undertaken for the production of a recombinant antigen to be further used as a target antigen in immunoassays.

pathogen and a collection of sera. The selection of the sera that compose the collection is crucial and ideally include (i) sera collected from natural outbreaks, (ii) reference sera collected from animals which have never been in contact with the pathogen, (iii) reference sera collected from experimentally infected animals, before and at different time points after the begin of infection, and (iv) sera raised to other microorganisms that are likely to induce cross-reacting antibodies. In most of the cases, the results allow the investigator to identify potential antigens that are specific of a particular pathogen and are recognized by all sera collected from infected animals. Additional valuable information can be obtained such as whether a particular epitope (i) stimulate an early and lasting humoral immune response, (ii) is linear or conformational, (iii) belongs to a single product of a defined molecular size. Once the potential target antigen has been identified and its presence confirmed in all field isolates and strains (or at least those which have been proven pathogenic), the number of steps necessary for ultimately produce the antigen as a recombinant protein varies, depending whether the corresponding gene sequence is already available. Although the identification of the gene of interest may be time-consuming, the progress made over the last 10 years offers several well- documented strategies to perform this task.

6. Recombinant Antigens in Serodiagnostic Applications: *Mycoplasma bovis* Serology as an Example

6.1 *Mycoplasma bovis* Infection and Serological Response

Mycoplasma bovis is one of the most pathogenic mycoplasma species of cattle in which it induces various clinical manifestations including pneumonia, mastitis and arthritis (Kirk and Lauermann, 1994; Pfützner et al., 1996; Stipkovits et al., 1993). These are globally prevalent chronic diseases of low mortality but high morbidity and are economically damaging due to their devastating effect on feed efficiency and milk production (Trevor, 1997). Culling of diseased animals due to *M. bovis* mastitis or respiratory disease is the usual consequence, due to the drastic reduction of milk yield, loss of weight gain, and loss of carcass and carcass condemnation. In absence of an efficient antibiotic therapy or vaccination, the control of *M. bovis* diseases relies on the identification and the removal of infected animals which requires early disease-stage diagnostic tests to enable the segregation of subclinically infected animals or non-shedding carriers. Until now the management of this problem was based on the identification of the organism either by laborious and expensive classic isolation and cultivation procedures followed by colony immunobinding assays or directly in clinical samples by PCR (Ayling et al., 1997). These techniques, however, require the presence of organisms at a concentration that is sufficient enough to allow their detection depending on the sensitivity of the test. To enable a reliable identification of infected animals even in the absence of shedding organisms, the most suitable and practical approach currently available is the monitoring of specific anti-*M. bovis* serum antibodies. However, a major limitation of the few ELISAs which are currently in use is that they are based on complex whole-cell extract preparations that do not take into account the high variation of *M. bovis* membrane components occurring while the organism is grown in large culture. Antigenic variation can severely complicate the final outcome of routine identification of mycoplasmas isolated from clinical material based on colony immunoblot analysis using monoclonal or polyclonal antibodies (Rosengarten and Yogev, 1996). For instance, heterogeneous colony blot patterns can be observed when *M. bovis* PG45 colonies are immunostained with conventional diagnostic hyperimmune antisera. These heterogeneous patterns reveal populations displaying weak or strong staining as the result of antigenic variation of surface components and do not indicate the presence of another cross-reacting mycoplasma species. In terms of serodiagnostic assays, antigenic variation introduces a major drawback by compromising the content of the antigen preparation which may vary from batch to batch, or from one laboratory to another one. Therefore, the reproducibility and sensitivity of these assays cannot be ensured. To overcome this problem, the identification and isolation of immunogenic *M. bovis* surface components that can be produced reproducibly in large quantities as recombinant products generated by gene expression technologies and used as highly sensitive and specific antigen targets has been an ongoing effort of the authors of this review article, and part of these efforts are summarized below.

6.2 Candidates for the Design of a Recombinant Antigen

In *M. bovis*, a gene family has been identified and its members, designated *vsps*, are subjected to high frequency oscillation in expression resulting in emergence of subpopulations within a single strain exhibiting phenotypic differences in their surface composition (Behrens et al., 1994; Rosengarten et al., 1994). This phenomenon is driven by a mutational process that spontaneously generates DNA rearrangements within the *vsp* locus and governs the ON and OFF switching of Vsp expression (Lysnyansky et al., 1996). In the type strain PG45, 13 *vsp* genes have been characterized and shown to be clustered within a 14 kpb DNA sequence (Lysnyanski et al., 1999), three of which encode for products that have been identified at the protein level, namely VspA, VspB and VspC. These three products are antigenically related as they share several epitopes, most of which are among the major immunodominant components that strongly stimulate the host humoral immune response (Rosengarten et al. 1994, Brank et al., 1999). In a recent study (Brank et al., 1999), it was shown by Western blot analysis that sera collected from experimentally or naturally infected animals specifically recognise predominant epitopes that are carried by the VspA product, independently of (i) the clinical signs, (ii) the geographical regions where the *M. bovis* outbreaks occurred, (iii) the animal history, and (iv) the origin of the *M. bovis* strains or isolates responsible for the infection. In addition, it was shown that (i) *M. bovis* epitopes cross-reacting with a serum directed to *Mycoplasma arginini*, a mycoplasma also found in cattle, belong to products other than the VspA protein, and (ii) all the field isolates tested so far contain *vspA*-related DNA sequences (Poumarat et al., 1999). In light of these data, it was concluded that VspA would provide a good candidate for the design of a recombinant antigen for antibody-based diagnosis of *M. bovis* infection.

6.3 Production of Recombinant VspA Proteins

The overall strategy used for the production of a recombinant VspA protein (FP-VspA-I) in *E. coli* using the pMAL system (New) is depicted in Fig. 3.

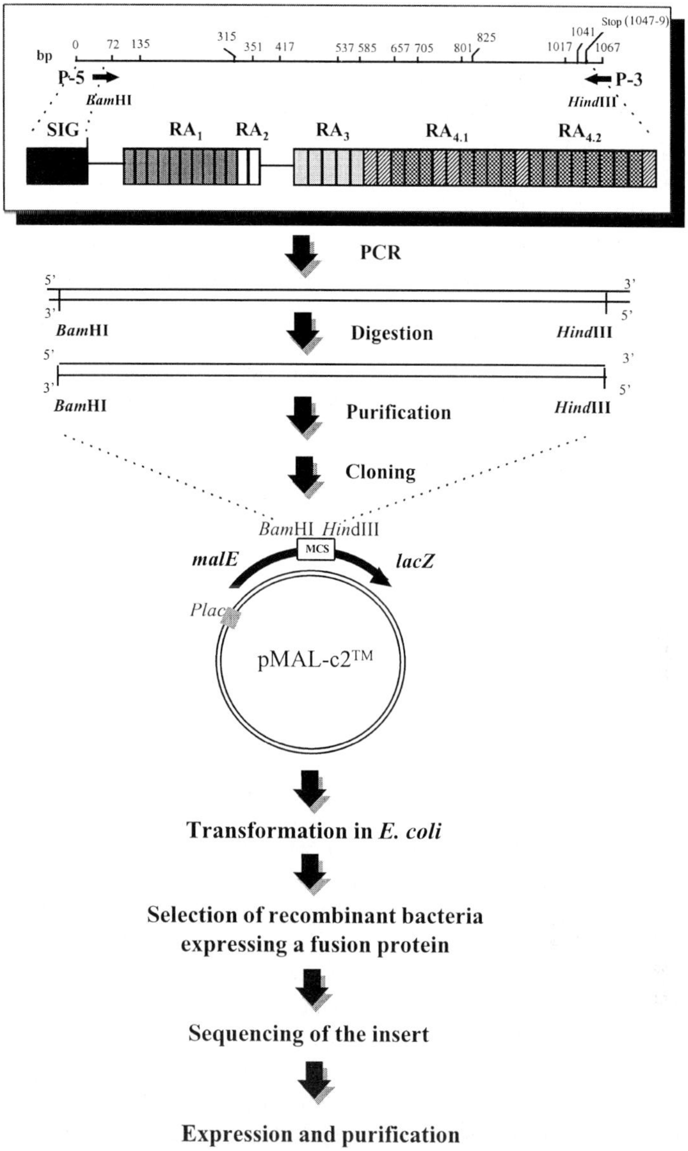

Figure 3. Outline illustrating the cloning strategy used to insert the coding region of the mature VspA protein into the pMAL-c2^{TM} expression vector for production of a recombinant VspA fusion protein, FP-VspA-I. The VspA ORF is shown as a rectangle with the solid black box representing the N-terminal signal peptide and the different hatched or shaded blocks designated R_A1, R_A2, R_A3, $R_A4.1$, and $R_A4.2$ representing short repeated amino acid motifs. The line above the VspA ORF represents the corresponding DNA fragment from which the vspA PCR product was obtained using two flanking primers P-5 and P-3. Numbers above the line indicate the number of nucleotides, with 0 corresponding to the first base of the GTG initiation codon. The two restriction sites BamHI and HindIII were engineered in the primer P-5 and P-3, respectively, for the in-frame cloning of the vspA gene sequence downstream of the malE gene of the pMAL-c2^{TM} vector. Included in the P-3 primer is a stop codon for termination of the protein.

The vector pMAL-c2 (New England Biolabs; Maina et al., 1988) was used to create a VspA fusion protein with the MBP. Primers flanking the DNA sequence encoding the mature VspA (located downstream of the peptide signal), P-5 and P-3, were designed according to the sequence previously published (Lysnansky et al., 1996). In order to allow the directional cloning of the *vspA* DNA insert into the pMAL-c2 vector, in frame with the *malE* gene encoding the MBP, *BamH*I and *Hind*III restriction sites were engineered into P-5 and P-3, respectively. Using these primers, PCR-DNA fragments containing the *vspA* coding sequence (without the signal sequence) were generated, digested by the appropriate restriction enzymes and ligated into the dephosphorylated *Bam*HI-*Hin*dIII pMAL-c2 vector after gel-purification. Following transformation of competent *E. coli* cells, recombinant bacteria containing the construct of interest were selected on the basis of their ability to express upon IPTG induction a fusion protein displaying in SDS-PAGE a molecular weight higher than that of the MBP-β-galactosidase. To ensure that during cloning into the bacterial host no mutation had occurred that would alter the composition of the protein of interest, the DNA insert of a selected clone was sequenced and confirmed to have 100 % identity with that of the original *vspA* gene. Using a similar approach, MBP fusion proteins that contain selected portions of the VspA product were constructed. The fusion protein carrying the full-length VspA was designated FP-VspA-I, while fusion proteins containing VspA truncated versions were designated FP-VspA-II to IV (Fig. 4A).

6.4 Testing of Recombinant VspA Fusion Proteins

Bacteria containing a recombinant plasmid with an appropriate insert size were selected, and production of the respective four fusion proteins was obtained after addition of IPTG (see for illustration Fig. 4B). The migration in SDS-PAGE of the fusion proteins is shown in panel C of Fig. 4. Subsequent Western blot analysis revealed that they all react with Vsp-specific monoclonal antibodies (MAbs) (Fig. 4D). While FP-VspA-I, -II and -III bind the MAb 1A1, FP-VspA-IV only reacts with MAb 1E5. This indicates that the first 106 amino acid of the N-terminal portion contains the 1A1 epitope while the 1E5 epitope is localised at the C-terminal end between amino acid 155 and 325. The ability of the VspA fusion proteins to detect *M. bovis* infection was assessed by Western blot analysis using a collection of approximately 100 serum samples collected from (i) healthy cattle, (ii) natural outbreaks of *M. bovis* infection and disease, (iii) calves experimentally infected with *M. bovis*, and (iii) rabbits immunized with closely related mycoplasma species (Brank et al., 1999; see also Fig. 4E and 4F). Because a few serum samples were cross-reacting with the MBP part of the fusion protein (see Fig. 4F, lanes 2 to 4), the FP-VspA-I was digested by the factor Xa to avoid false positive reactions. The results showed that the FP-VspA-I reacts specifically with sera collected from cattle naturally infected with *M. bovis* and is recognized by serum antibodies as early as six days following experimental infection (Fig. 4F, lanes 2 and 5), i.e., before apparition of any clinical symptoms. The use of the recombinant FP-VspA-I as target antigen in ELISA is currently being tested. Of the truncated VspA fusion proteins, FP-VspA-IV (designated FP-VspA-II in Brank et al., 1999) was similarly recognized by sera collected at day 6 after experimental infection, while FP-VspA-II (designated FP-VspA-IV in Brank et al., 1999) and FP-VspA-III were only reacting with sera collected 21 days post infection.

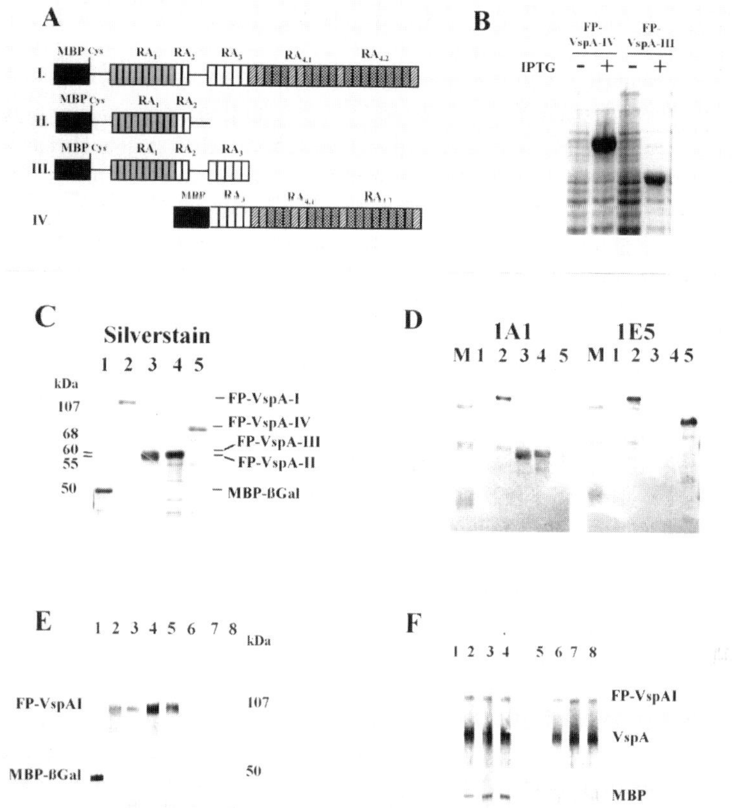

Figure 4. Features of FP-VspA-I to -IV and seroreactivity of the VspA immunogenic domains with Vsp-specific MAbs and sera from cattle. (A) Schematic representation of the different domains that compose FP-VspA-I to IV. (B) Electrophoretic migration in SDS-PAGE of the total proteins of two recombinant E. coli clones expressing FP-VspA-IV or -III, in presence (+) or absence (-) of IPTG after Coomassie staining. (C) Electrophoretic migration in SDS-PAGE of FP-VspA-I to -IV and of the MBP fused to the β-galactosidase (MBP-βgal) after silver staining. Lanes 1 to 5 represent the total proteins of E. coli recombinant clones expressing the MBP-βgal, FP-VspA-I, FP-VspA-II, FP-VspA-III and FP-VspA-IV respectively. (D) Identical immunoblots immunostained with Vsp-specific monoclonal antibodies MAb 1A1 or MAb 1E5. Lanes 1 to 5 correspond to those shown in panel C, M corresponds to the size marker. (E) The FP-VspA-I and the MBP-β-galactosidase products were separated by SDS-PAGE, transferred onto nitrocellulose membrane and immunostained with anti-MBP serum (lane 1), MAb 1A1 (lane 2), MAb 1E5 (lane 3), serum collected from a calf experimentally infected with M. bovis (lane 3), serum collected from a natural M. bovis outbreak (lane 5), serum from sheep infected with M. agalactiae, a mycoplasma closely related to M. bovis (lane 6), serum from an M. bovis-free animal (lane 7), anti-M. gallisepticum rabbit hyperimmune, a mycoplasma which shares antigenic epitopes with M. bovis (lane 8). (F) The FP-VspA-I product digested with the factor Xa was separated by SDS-PAGE, transferred onto nitrocellulose membrane and immunostained with sera from two individual calves experimentally infected with an identical inoculum of M. bovis (lanes 1 to 4 and 5 to 8 respectively). Sera were collected after experimental infection at day 3 (lane 1 and 5), 6 (lanes 2 and 6), 9 (lanes 3 and 7) and 12 (lanes 4 and 8), respectively.

7. Drawbacks in Using Recombinant Antigens

As with any technological advances, there is a number of drawbacks in using recombinant proteins in diagnostic assays for the detection of infectious diseases or asymptomatic carrier hosts. Consequences linked to the specificity of the target antigen can have dramatic outcomes such as misleading or erroneous test results. The selection of an appropriate antigen to be expressed and the acquisition of preliminary knowledge regarding its immunogenicity in the host is therefore the highest priority. The host humoral response to a pathogen may vary from one individual to another one, and immunoassays that are based on a single antigen might not detect all infected individuals (Livingston et al., 1999). However, the use of a cocktail of specific antigens can alternatively overcome this problem. Another drawback is that the recombinant product might not display non-linear antigenic epitopes after expression in foreign host systems due to improper folding or post-translational modification. For this reason, careful pretesting of the recombinant antigens following their production is required prior to their incorporation in diagnostic assays. Finally, the pressing need of serodiagnostic tests for newly emerging pathogens often favors the development of immunoassays based on whole-cell extracts rather than on recombinant proteins, as the prerequired knowledge concerning the nature of the antigenic products to be expressed and the identification of their corresponding DNA sequences may require several months up to several years.

8. Use of Recombinant Antigens in Alternative Applications

One of the major success of modern molecular biology is the production of recombinant gene products that provide reagents for assessment of fundamental issues of cell biology, such as gene expression and function, and the nature of products involved in gene regulation or enzymatic activities of a particular protein. In clinical microbiology, recombinant proteins have emerged as valuable tools not only in replacing complex target antigens in ELISA or Western blot analysis, but also in generating specific reagents, such as monoclonal or polyclonal antibodies for bacterial typing or subunit vaccines for disease control. The use of recombinant proteins as subunit vaccines present the advantages of (i) being safe for the environment (no risk of reversion to virulence), (ii) being precisely controlled, and (iii) providing immunity without the risk of causing significant disease symptoms that can be induced with vaccines based on attenuated pathogens. An additional advantage is that the use of recombinant proteins as vaccines allows the design of serological assays that can discriminate between vaccinated and naturally infected hosts. For instance, bacculoviruses have been engineered to express hemagglutinine (HA) genes of recent avian influenza virus types (AI). Vaccine protocols employing these vaccines will not elicit an immune response against internal AI proteins and therefore will not interfere with epidemiological surveys of natural influenza infection in the field that are based on the detection of host antibodies directed towards internal AI proteins (Crawford et al., 1999).

9. Conclusions

9.1 What has been achieved

Over the past 10 years the tremendous progress that has been made in the field of molecular biology have provided microbiologists with an increasing panel of techniques that have allowed the design of more refined and more effective diagnostic assays. Recombinant proteins were soon adopted as key antigens in serological assays but their potential in routine diagnosis was compromised by the high background interference with host cell proteins. This was later overcome by the introduction of recombinant fusion proteins which facilitated the purification of the antigen of interest from the host cell proteins. Because viruses possess a rather simple and small genome, recombinant proteins have been widely used by virologists for instance to discriminate between closely related viruses, e.g., between HTLV-I and HTLV-II (Chen et al., 1990; Miyagawa et al., 1998) or to detect specific infections via the identification of circulating antibodies, e.g., to HIV (commercial recombinant HIV antigens are manufactured by several companies). Concerning food-borne pathogens, several recombinant proteins have been produced for the improvement of diagnostic assays either to detect circulating host antibodies to a given pathogen or to detect antigens by immuno-capture assays. Some examples include the production of truncated forms of Listeriolysin O as a specific and sensitive antigen for serodiagnosis of listeriosis (Gholizadeh et al., 1996), the production of recombinant antigens encoded by the LEE locus of *E. coli* O157 for the detection of patients infected with verocytotoxin-producing *E. coli* (Jenkins et al., 2000), or the production of recombinant fimbrial proteins for the specific detection of *Salmonella enteritidis* infection in poultry (Rajashekara et al., 1998, 1999). Within a period of six months, two independent reports have described the potential of different recombinant antigens for the detection of *Toxoplasma gondii* infection (Lecordier et al. 2000; Li et al. 2000).

9.2 What has been neglected

The production of recombinant antigens in clinical microbiology has rapidly increased over the past few years and their use as diagnostic or epidemiological tools in ELISA is more and more accepted as long as validation tests are progressing along to ascertain their performance in the field. However, the production of specific MAbs or monospecific polyclonal sera directed towards an amino acid sequence specific of a given pathogen or serovar using recombinant fusion antigens has been neglected. This is rather surprising, specially when sequence data are already available. Advantages in using recombinant proteins for this purpose include the facts that they (i) display a limited repertoire of epitopes compared to more complex preparations, (ii) can be easily purified avoiding for instance difficulties in generating enough material from columns or gels for immunization, and (iii) facilitate the screening of the appropriate MAb. For instance, using a recombinant construct of the VlpA product of the swine pathogen *Mycoplasma hyorhinis*, a specific rabbit antiserum has been produced that can distinguish this particular product from the seven other members of the Vlp family with which VlpA shares several antigenic motifs (Citti and Wise, 1995). Overall, this strategy could certainly benefit to the development of specific reagents for bacterial typing.

9.3 What needs to be done

The ultimate goal of a given diagnostic assay is to detect a particular infection before apparition of the clinical symptoms and before it spreads to other healthy animals. Consequently it has to be specific and sensitive. As mentioned above, new reagents for the improvement of these assays are currently emerging or are already available for a large number of pathogens. The critical question which in most cases remains to be answered is how do they compare and compete with existing serological tests already in use in diagnostic laboratories. Validation in the field of these new assays with reference material followed by standardization among laboratories as well as the definition of which type of assay will answer accurately and reproducibly a particular diagnostic question are immediate and future challenges.

References

Ayling, R.D., Nicholas, R.A., and Johansson, K.E., 1997. Application of the polymerase chain reaction for the routine identification of *Mycoplasma bovis*. Vet. Rec. 141, 307-308.

Barbitt, P.C., West, B.L., Buechter, D.D., Kuntz, I.D., and Kenyon, G.L., 1990. Removal of proteolytic activity associated with aggregates formed from expression of creatine kinase in *E. coli* leads to improved recovery of active enzyme. Biotechnology 8, 945-949.

Behrens, A., Heller, M., Kirchhoff, H., Yogev, D., and Rosengarten, R., 1994. A family of phase- and size-variant membrane surface lipoprotein antigens (Vsps) of *Mycoplasma bovis*. Infect. Immun. 62, 5075-5084.

Brank, M., Le Grand, D., Poumarat, F., Bezille, P., Rosengarten, R., and Citti, C., 1999. Development of a recombinant antigen for antibody-based diagnosis of *Mycoplasma bovis* infection in cattle. Clin. Diagn. Lab. Immunol. 6, 861-867.

Citti C., and Wise K.S., 1995. *Mycoplasma hyorhinis vlp* gene transcription: critical role in phase variation and expression of surface lipoproteins. Mol. Microbiol. 18, 649-660.

Chen,Y.M. Lee, T.H., Wiktor, S.Z., Shaw, G.M., Murphy, E.L., Blattner, W.A., Essex, M., 1990. Type-specific antigens for serological discrimination of HTLV-I and HTLV-II infection. Lancet 336, 1153-1155.

Collen D., Stassen, J.M., Marafino, B.J. Jr, Builder, S., De Cock, F., Ogez, J., Tajiri, D., Pennica, D., Bennett, W.F., Salwa, J., and Hoyng, C.F., 1984. Biological properties of human tissue-type plasminogen activator obtained by expression of recombinant DNA in mammalian cells. J. Pharmacol. Exp. Ther. 231, 146-152.

Crawford, J., Wilkinson, B., Vosnesensky, A. Smith, G., Garcia, M., Stone, H., and Perdue, M.L., 1999. Baculovirus-derived hemagglutinin vaccines protect against lethal influenza infections by avian H5 and H7 subtypes. Vaccine 17, 2265-2274.

Fierz, W., 1998. Basic problems of serological laboratory diagnosis. In: (Ed.) Reschl, U., Methods in Molecular Medicine, Vol. 13: Molecular Diagnostic of Infectious Diseases, Human Press Inc., Totowa, NJ, pp 443-471.

Fuerst, T.R., Niles, E.G., Studier, F.W., and Moss, B., 1986. Eukaryotic transient-expression system based on recombinant vaccinia virus that synthesizes bacteriophage T7 RNA polymerase. Proc. Natl. Acad. Sci. USA. 83, 8122-8126.

Gholizadeh, Y., Poyart, C., Juvin, M., Beretti, J.L., Croizé, Berche, P., and Gaillard, J.L., 1996. Serodiagnosis of listeriosis based upon detection of antibodies against recombinant truncated forms of Listeriolysin O. J. Clin. Microbiol. 34, 1391-1395.

Jenkins, C., Chart, H., Smith, H.R., Hartland, E.L., Batchelor, M., Delahay, R.M., Dougan, G., and Frankel, G., 2000. Antibody response of patients infected with verocytotoxin-producing *Escherichia coli* to protein antigens encoded on the LEE locus. J. Med. Microbiol. 49, 97-101.

Kirk, J.H., and Lauermann, L.H., 1994. Mycoplasma mastitis in dairy cows. The compendium 16, 541-551.

Kost, A.T., and Condreay, J.P., 1999. Recombinant baculoviruses as expression vectors for insect and mammalian cells. Curr. Opin. Biotechnol.10, 428-433.

Lakey, D.L., Voladri, R.K., Hager, C., Samten, B., Wallis, R.S., Barnes, P. F., and Kernodle, D.S., 2000. Enhanced production of recombinant *Mycobacterium tuberculosis* antigens in *Escherichia coli* by replacement of low-usage codons. Infect.Immun. 68, 233-238.

Lecordier, L., Fourmaux, M.P., Mercier, C., Dehecq, E., Masy, E., and Cesbron-Delauw, M.F., 2000. Enzyme-linked immunosorbent assays using the recombinant dense granule antigens GRA6 and GRA1 of *Toxoplasma gondii* for detection of immunoglobulin G antibodies. Clin. Diagn. Lab. Immunol. 7, 607-611.

Li, S., Maine, G., Suzuki, Y., Araujo, F.G., Galvan, G., Remington, J.S., and Parmley, S., 2000. Serodiagnosis of recently acquired *Toxoplasma gondii* infection with a recombinant antigen. J. Clin. Microbiol. 38, 179-184.

Livingston, R.S., Riley, L.K, Hook, R.R Jr, Besch-Williford, C.L, and Franklin C.L., 1999. Cloning and expression of an immunogenic membrane-associated protein of *Helicobacter hepaticus* for use in an enzyme-linked immunosorbent assay. Clin. Diagn. Lab. Immunol. 6, 745-750

Lysnyansky, I., Rosengarten, R. and Yogev, D., 1996. Phenotypic switching of variable surface lipoproteins in *Mycoplasma bovis* involves high-frequency chromosomal rearrangements. J. Bacteriol. 178: 5395-5401.

Lysnyansky, I., Sachse, K., Rosenbusch, R., Levisohn, S., Yogev, D., 1999. The *vsp* locus of *Mycoplasma bovis*: gene organization and structural features. J Bacteriol. 181, 5734-5741.

Maina, C.V., Riggs, P.D., Grandea, A.G., Slatko, B.E., Moran, L.S., Tagliamonte, J.A., McReynolds, L. A., Guan, C. D., 1988. An *Escherichia coli* vector to express and purify foreign proteins by fusion to and separation from maltose-binding protein. Gene. 74, 365-373.

Miyagawa, E., Yoshiki, A., Fujii, N., Honda, H., Ueno, E., Kurano, Y., and Ito, S., 1998. Establishment of monoclonal antibody, gp21-34, against HTLV-II envelope protein (p20E). Segregation of human T cell lymphotropic virus type I and II infections by antibody reactivity to unique viral epitopes. Hybridoma 17, 347-354.

Ostoa-Saloma, P., Cabrera N., Becker I., and Perez-Munfort R., 1989. Proteinases of *Entamoeba histolytica* associated with different subcellular fractions. Mol. Biochem. Parasitol. 32, 133-144.

Pfützner, H., and Sachse, K., 1996. *Mycoplasma bovis* as an agent of mastitis, pneumonia, arthritis and genital disorders in cattle. Rev. Sci. Tech. 15, 1477-1494.

Poumarat, F., Le Grand, D., Solsona, M., Rosengarten, R. and Citti C., 1999. Vsp antigen and *vsp*-related sequences in field isolates of *M. bovis*. FEMS Microbiol. 173:103-110

Rajashekara, G., Wanduragala, D., Halvorson, D.A., Nagaraja, K.V., 1999. A rapid strip immunoblot assay for the specific detection of *Salmonella enteritidis* infection in chickens. Int. J. Food Microbiol. 53, 53-60.

Rajashekara, G., Munir, S., Lamichhane, C.M., Back, A., Kapur, V., Halvorson, D.A., Nagaraja, K. V., 1998. Application of recombinant fimbrial protein for the specific detection of *Salmonella enteritidis* infection in poultry. Diagn. Microbiol. Infect. Dis. 32, 147-157.

Reschl, U., 1998. (Ed.) Methods in Molecular Medicine, Vol. 13: Molecular Diagnostic of Infectious Diseases, Human Press Inc., Totowa, NJ.

Rosengarten, R., Behrens, A., Stetefeld, A., Heller, M., Ahrens, M., Sachse, K., Yogev, D., and Kirchhoff H., 1994. Antigen heterogeneity among isolates of *Mycoplasma bovis* is generated by high-frequency variation of diverse membrane surface proteins. Infect. Immun. 62, 5066-5074.

Rosengarten, R., and Yogev, D., 1996. Variant colony surface antigenic phenotypes within mycoplasma strain populations: implications for species identification and strain standardization. J. Clin. Microbiol. 34, 149-158.

Shenai, B.R., Komalam, B.L., Arvind, A.S., Krishnaswamy, P.R., and Rao, P.V., 1996. Recombinant antigen-based avidin-biotin microtiter enzyme-linked immunosorbent assay for serodiagnosis of invasive amebiasis. J. Clin. Microbiol. 34, 828-833.

Stipkovits, L., Rady, M. and Glavits R., 1993. Mycoplasmal arthritis and meningitis in calves. Acta Vet. Hung. 41, 73-88.

Trevor, R., 1997. Dairy calf pneumonia, the disease and its impacts. Vet. Clin. N. Am.-Food A. 13, 379-391.

Modern methodologies for isolation/typing of animal-borne human pathogens: A research update on epidemiological tools
II. Quantitative real-time PCR

Dieter Klein and Walter H. Günzburg
Institute of Virology, University of Veterinary Sciences, Veterinärplatz 1, A-1210 Vienna, Austria, (Corresponding author: E-mail: Dieter.Klein@vu-wien.ac.at)

Summary

As a result of increased sensitivity of molecular methods in detecting microbiological agents, one aspect in the diagnosis of human pathogens has become more and more important: the quantification of infectious agents. In the recent past, real-time PCR-based assays have gained favour for this purpose, since they are characterized by a wide range of quantification (7-8 logarithmic decades), a high technical sensitivity (< 5 copies) and a high precision (< 2% standard deviation). Time-consuming post-PCR steps are not necessary and have decreased the risk of cross-contamination. The possibility of automated sample preparation has improved turn-around time and allows the development of high throughput screening assays. Introduction of multiplex assays for either subtyping of infectious agents or the simultaneous calculation of pathogen relative to the food stuff has further increased the accuracy of these assays. However, the higher risk of false negative results due to mutations in the primer or probe binding region necessitates to carefully investigate the molecular epidemiology of the pathogen of interest before the design of reliable real-time PCR based assays.

Keywords: TaqMan, real-time PCR, quantification, pathogens, food safety

1. Introduction

In the last decade the development of molecular based methods for the detection of pathogens has grown exponentially. Due to the higher technical sensitivity of these methods two aspects have become more and more important: i. the subtyping of pathogens to distinguish pathogenic and less pathogenic strains and ii. the quantification of the microorganism of interest.
In the recent past real-time PCR based assays (Heid et al., 1996; Lie et al., 1998; Jordan, 2000) have gained favour for both purposes especially the quantification of infectious agents (Surynarayana et al., 1998; Klein et al., 1999; Leutenegger et al., 1999; Hofmann-Lehmann et al., 2000; Klein et al., 2001), including food pathogens (Oberst et al., 1998; Chen et al., 2000; Kimuraet al., 1999; Hein et al., 2001). The experimental set-up of this method is based on PCR containing a fluorogenic probe binding to a target sequence within

the amplified PCR product. Three kinds of probe chemistry can be used in these assays: i. hydrolysis probes, ii. hybridization probes and iii. molecular beacons or hairpin probes. All three chemistries are based on the increase of a reporter fluorescence if the target nucleic acid is present. This chapter will focus mainly on hydrolysis probes as an example, since they have been the first on the market and thus most information about this system is available, but the topics discussed can be transferred to the other chemistries as well.

2. Principle of real-time PCR

The experimental principle of hydrolysis probe assays is based on the 5′ nuclease activity of *Taq* DNA polymerase (Holland et al., 1991) resulting in the cleavage of a probe, which is labeled at both ends with two different fluorogenic dyes: the reporter and the quencher (Figure 1). The probe is chosen to hybridize to an internal sequence of the PCR target sequence. After excitation with an argon laser, the energy from the reporter fluorochrome is transferred to the quencher fluorochrome and only the quencher emits quantifiable amounts of light. The quenching effect occurs only as long as both fluorochromes are present as part of the intact probe. During the elongation of the complementary strands, the probe is displaced and cleaved by the 5′ - 3′-nuclease activity of *Taq* DNA polymerase, thus eliminating the quenching by the separation of the two fluorochromes. The resulting increase in the reporter fluorescence is proportional to the amount of amplified product during the PCR and can be measured every 7 seconds in the 96 well format, which is the basis for the real-time measurement of the amount of the PCR product.

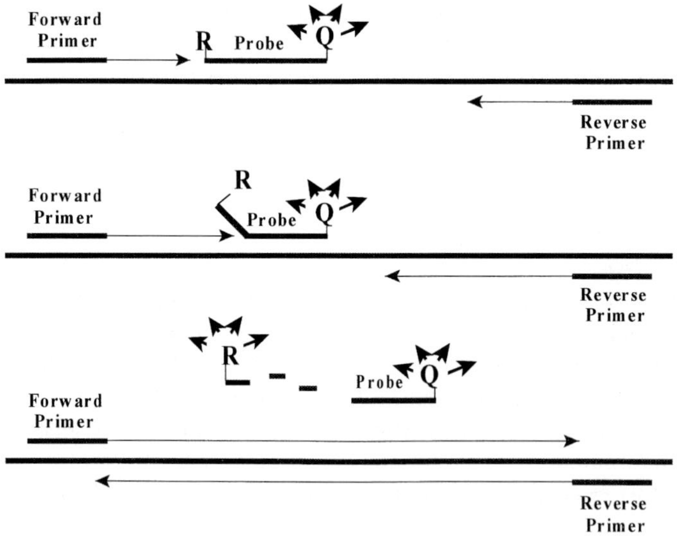

Figure 1. Principle of the real-time PCR.
The principle of the real- time PCR is based on the 5′ nuclease activity of Taq DNA polymerases, which is used to cleave an internal probe, resulting in the increase of reporter fluorescence. The increase in fluorescence is proportional to the amount of PCR product and can be measured every 7 seconds in the 96-well format.

The experimental set-up enables the PCR to be monitored during all phases of the reaction, which is the basis for the determination of the amount of input target sequence in the exponential phase of the reaction (Figure 2).

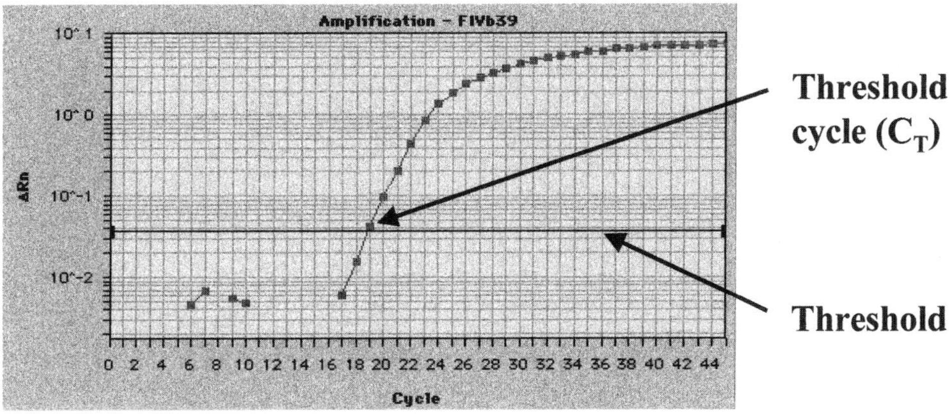

Figure 2. Amplification plot.
The normalized amount of the reporter fluorescence (dR_n), which is proportional to the amount of PCR product is shown in relation to the PCR cycle number. The threshold used to estimate the threshold cycle number (C_T) is indicated This on-line measurement of PCR product accumulation enables the utilization of the exponential phase of the PCR reaction for quantification.

The first cycle which is above a defined threshold (which is either chosen or calculated automatically as 10 times the standard deviation of the background fluorescence) is defined as the threshold cycle (C_T). The C_T reflects the earliest point in the amplification curve which can be used for quantification of the amplified product, leading to a minimized variation in the estimation of template input (Figure 3).

3. Fidelity of real-time PCR assays

The detection of fluorescence during the exponential phase of the PCR avoids time-consuming post-PCR steps and thus decreases turn-around time and costs. Another advantage of the measurement during the exponential phase of the reaction is that in contrast to the high variation at the end of the reaction there is only very little variation at the point were the C_T value is estimated, which explains the very low coefficients of variations, which have been found in intra-assay and inter-assay experiments (Leutenegger et al., 1999; Klein et al., 2000a; Hein et al., 2001; Klein et al., 2001).

In addition, this experimental set-up enables the calculation of initial target copies over a wide range (Figure 4).

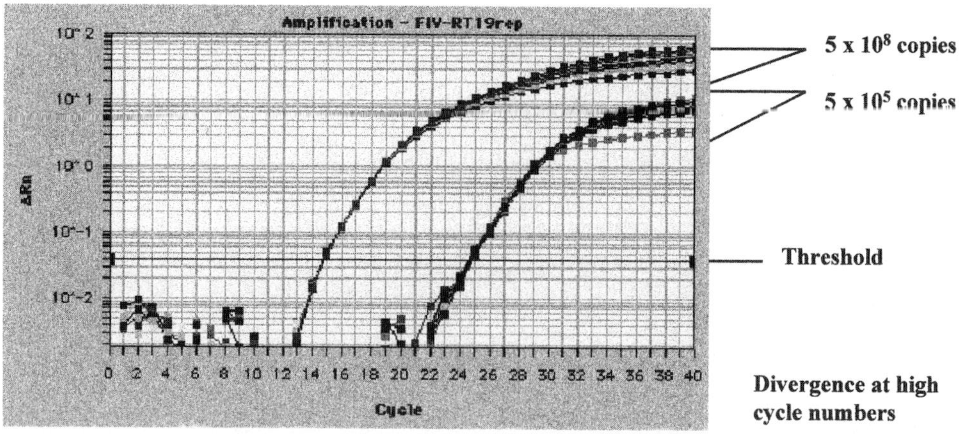

Figure 3. Reproducibility at different phases of the real-time RT-PCR.
Two different amounts of in vitro transcribed RNA were measured twelve times in one real-time RT-PCR. The calculated normalized reporter signal ()R_n) is plotted on a logarithmic scale in correlation to the PCR cycle. The threshold used to estimate the threshold cycle numbers (C_T) of each sample is indicated and demonstrates the high precision of the measurement during the early phase of the amplification compared to the high variation at the end of the amplification (40 cycles).

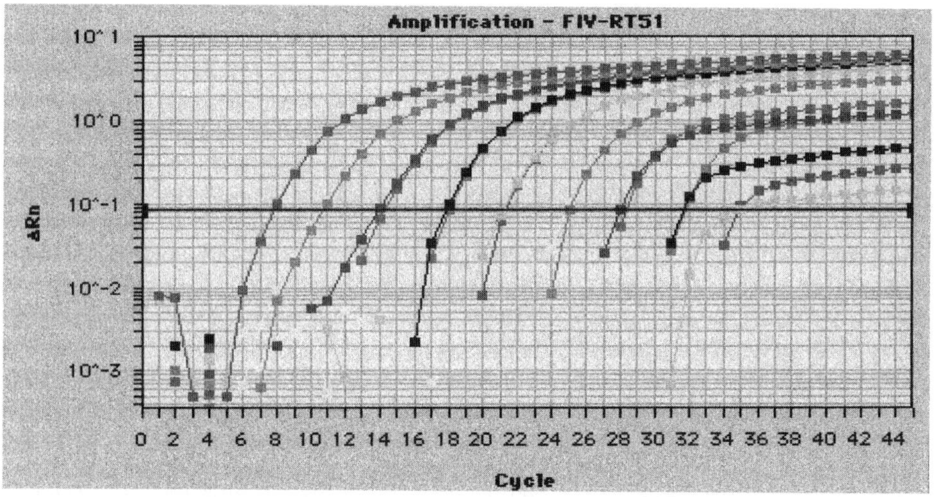

Figure 4. Amplification plot of a RNA standard dilution series.
The dynamic range of quantification of a real-time RT-PCR system is demonstrated using a dilution series of in vitro transcribed RNA standard. Tenfold dilution series were made in PCR grade water containing 3 µg t-RNA per ml as carrier RNA and measured in duplicate.

The reporter fluorescence at the end of the reaction, which corresponds to the thickness of the band on an agarose gel, is similar for a wide range of input target copies. In contrast, clear difference in C_T values (PCR cycle number where the PCR curve meets the threshold) can be observed over a wide input range. These C_T values are used to calculate a standard curve, which is linear over 7-8 logarithmic decades (Figure 5) (Klein et al., 1999; Leutenegger et al., 1999, Klein et al., 2000a; Hein et al., 2001; Klein et al., 2001)

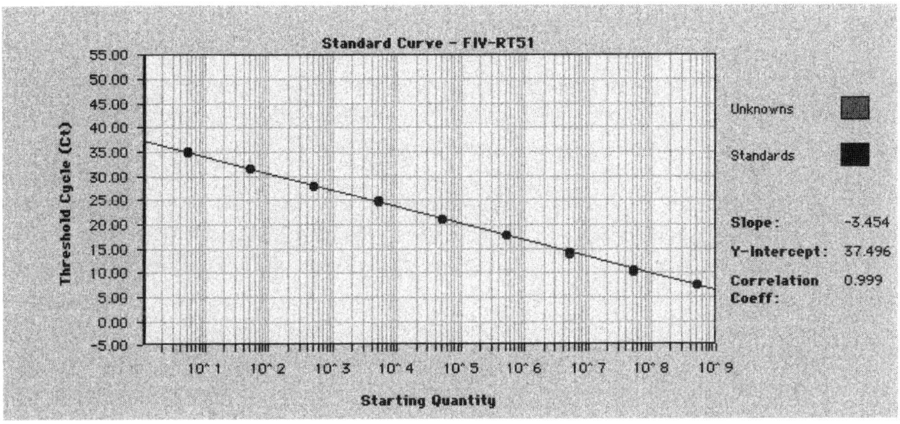

Figure 5. Standard curve.
The resulting standard curve of Figure 4 is illustrated. The coefficient of correlation and the slope are indicated.

Beside the wide range of quantification this method is characterized by a high technical sensitivity (in terms of how many copies can be quantified) (Gut et al., 1999; Leutenegger et al., 1999; Klein et al., 2000 Hofmann-Lehmann et al., 2000; Klein et al., 2001).

In summary, real-time PCR assays are characterized by a wide dynamic range of quantification (7-8 logarithmic decades), a high technical sensitivity (< 5 copies) and a high precision (< 2% standard deviation). Similar results have been obtained from real-time RT-PCR assays (Klein et al., 2001). Another advantage of this fully closed test-system, which decreases the risk of cross-contamination by PCR products and avoids time-consuming post-PCR steps, is the possibility to develop high-throughput assays.

4. Influence of Mutations on Sensitivity

In the light of the use of PCR based assays for diagnostic purposes the number of false-negative results is very important. This is especially true for microorganisms with high mutation rates, e.g. retroviruses. If such mutations are located in regions where the primers or the probe would bind, it will influence the specificity of the assay. These effects represent the basis for the development of allele specific real-time PCR assays (Steinborn et al., 2000), but are a major drawback for the development of diagnostic real-time PCR assays. For example: major sequence variations of two or more nucleotides in the probe binding region prevent detection and thus quantification (Klein et al., 1999; Klein et al.,

2000b). However, two mutations in the primer binding region can be tolerated, depending on the resulting mismatch and the position of the mismatch, while three inhibit binding of the primers (Klein et al., 2000b). In contrast, minor mismatches in this region reduced the reaction efficiency (Klein et al., 1999), and had a significant effect (up to four logarithmic decades) upon quantification (Klein et al., 2001).

Despite testing of a variety of mismatches in different assays no general rules regarding their effect on quantification can be drawn (Klein et al., 2001). The exact effect of the mismatch has to be calculated for each assay/isolate combination. This result highlights the importance of the knowledge of the molecular epidemiology of the microorganism of interest, which is a prerequisite for reliable real-time PCR assay development.

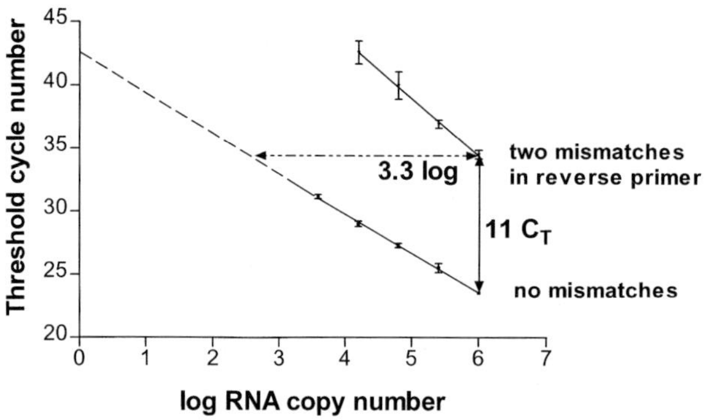

Figure 6. Impact of sequence variations on viral load estimation (adapted from Klein et al., 2001). A fourfold dilution series of a FIV isolate was measured in triplicate in two real-time RT-PCR assays. One assay has no mismatches while the other displayed two mismatches in the reverse primer. The difference in the measured C_T values at the highest point between the two standard curves are indicated by arrows. In order to illustrate the resulting difference in viral load (log RNA copy number) the standard curve from the assay without mismatches was extrapolated to higher C_T values and the C_T values of the same samples measured in the assay with two mismatches were drawn in this standard curve as well.

5. Multiplex Real-Time PCR Assays

One possibility to overcome the risk of false negative results is the development of real-time PCR assays, which target highly conserved regions of the microorganism (McGoldrick et al., 1998; Gut et al., 1999, Hein et al., 2001). The sensitivity of the assay can be increased by these kind of approaches, but it increases as well the risk of false-positive results due to the detection of related microorganisms (Gut et al., 1999).
Another possibility to increase the sensitivity of the assay is the development of assays, which contain more than one probe. If these probes are labeled with different reporter fluorochromes they can be detected simultanously. These kind of multiplex real-time PCR assays provide an elegant tool to distinguish certain subtypes of microorganisms. Additionally, the different subtypes can be quantified within one assay, which is of special interest, if these subtypes reflect different levels of virulence.

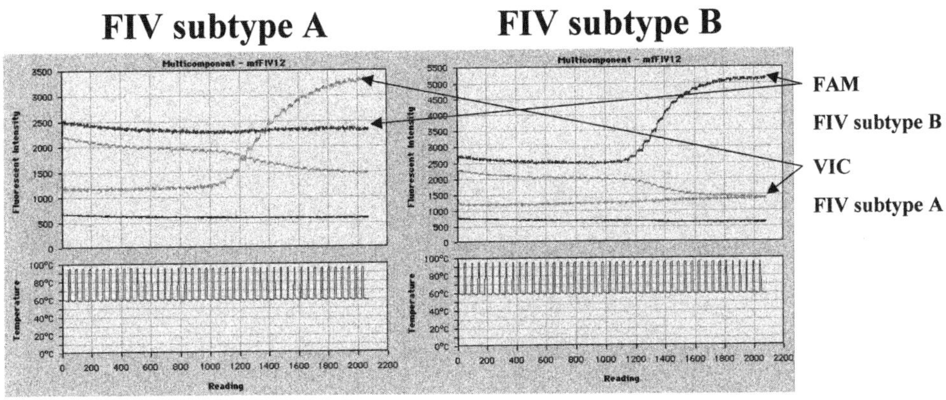

Figure 7. Subtype specific real-time PCR.
The fluorescence of the three measured fluorochromes (FAM as reporter for FIV subtype B; VIC as reporter for FIV subtype A and TAMRA as quencher for both) is plotted against PCR cycle.

Real-time PCR assays can be applied as well in thoses cases, where the amount of the pathogen has to be quantified relative to the number of host cells or volume (e.g. 'proviral' and 'viral load') or to the amount of a particular food stuff. Multiplex real-time PCR assays, which allow the simultaneous calculation of the pathogen and the host/food stuff, have been developed for these purposes. They allow to minimize both the inter-assay variation as well as the accumulation of errors (Klein et al., 2000a; Klein et al., 2001).
These multiplex assays need even more optimization than monoplex assays, since the components of two assays, each containing two primers and one probe, compete for common reagents. If the two target genes are found in unequal amounts this can result in lower sensitivity of the less abundant target, because the more abundant target assay has used up all common reagents. Therefore the primer concentration of the more abundant target has to be limited to increase sensitivity of the less abundant target.
To estimate to which extent the concentration of the primers of the more abundant target sequence can be reduced without affecting the quantification, a matrix of reactions, each with different concentrations of the more abundant forward and reverse primer has to be performed.

In these experiments the influence of the reduced primer concentration on the C_T value (as a parameter of the exponential phase of the reaction) can be investigated. While for higher primer concentrations (up to 60 nM each) the C_T value is similar, it increases dramatically at primer concentrations below 60 nM. Therefore 60 nM reflects the lowest primer concentration, which can be used without effecting the exponential phase of the reaction. However, to investigate the effect of the reduced primer concentration in a multiplex reaction in comparison to a monoplex reaction, the obtained standard curves have to be compared. This comparison is only reliable if the the obtained standard curves display high coefficients of correlation ($r^2 > 0.995$) over the range of quantification of interest. Figure 9 demonstrates that decreased primer concentration allows efficient quantification in the relevant range, although the reaction efficiency is reduced by 10%.

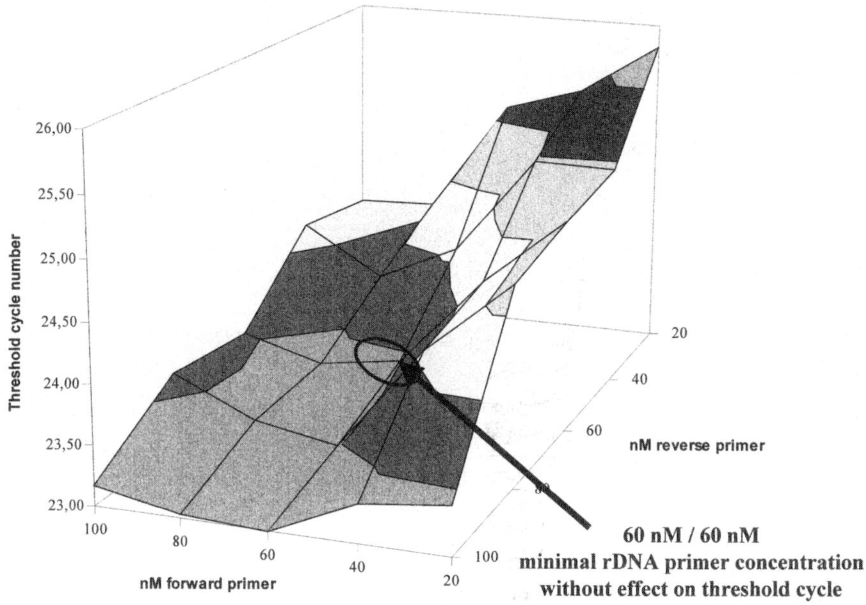

Figure 8. Primer matrix experiment.
Genomic DNA was measured in a real-time PCR targeting the rDNA genes. The primer concentration of the forward and the reverse primer was varied in 20 nM steps in the range from 20 to 100 nM. The obtained threshold cycle for each combination is illustrated (mean of duplicate measurements).

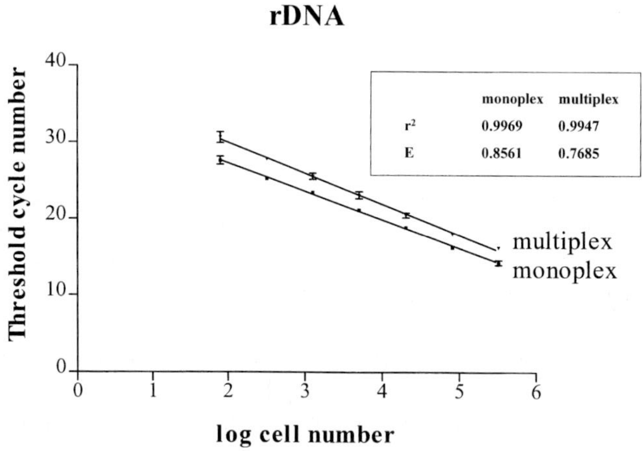

Figure 9. Influence on reaction efficiency (adapted from Klein et al., 2000a).
A fourfold dilution series of genomic DNA was measured in duplicate with a rDNA monoplex real-time PCR and a rDNA/EGFP multiplex real-time PCR with reduced primer concentration. The resulting standard curves are illustrated using the mean and the standard deviation for each dilution step. The corresponding coefficients of correlation (r^2) and the calculated reaction efficiencies (E) are illustrated in the box.

6. Applications in Food Hygiene

Foodborne diseases are a major public health concern. Since food contamination can occur at different phases of the production, rapid testing of samples at the different stages during the process is necessary to increase food safety. Conventional ways of identification and quantification of foodborne pathogens are generally laborious and time-consuming. Hence, a growing number of rapid real-time PCR based assays for foodborne pathogens have been developed in the recent past (Bassler et al., 1995; Chen et al., 1997; Kimura et al., 1999; Oberst et al., 1999; Chen et al., 2000; Hein et al., 2001).
The advantage of real-time PCR based assays is the low turn-around time, which is an important factor in modern food production. Although PCR based methods detect only the genetic material of the microorganism and do not allow any conclusions about the infectivity of the detected pathogen, clear correlations between colony forming units (CFU) and copy numbers have been demonstrated (Hein et al., 2001).

7. Conclusions

Real-time PCR based assays have a number of advantages compared to commonly used methods for the detection and quantification of infectious agents. The avoidance of any time-consuming post-PCR steps and the possibility of automated sample preparation has improved turn-around time and allows the development of high throughput screening assays, which is the basis for rapid testing of samples at different stages during food production. Future implementation of these assays will help to avoid foodborne diseases and improve food safety. However, a critical survey of the molecular epidemiological situation of the pathogens of interest is a prerequisite for reliable assay development and improved food safety.

References

Bassler, H.A., Flood, S.J., Livak, K.J., Marmaro, J., Knorr, R., Batt, C.A., 1995. Use of a fluorogenic probe in a PCR-based assay for the detection of Listeria monocytogenes. Appl. Environ. Microbiol. 61, 3724-3728.

Chen, S., Yee, A., Griffiths, M., Larkin, C., Yamashiro, C.T., Behari, R., Paszko-Kolva, C., Rahn, K., De Grandis, S.A., 1997. The evaluation of a fluorogenic polymerase chain reaction assay for the detection of Salmonella species in food commodities. Int. J. Food Microbiol. 35, 239-250.

Chen, W., Martinez, G., Mulchandani, A., 2000. Molecular beacons: a real-time polymerase chain reaction assay for Salmonella. Anal. Biochem. 280, 166-172.

Gut, M., Leutenegger, C.M., Huder, J.B., Pedersen, N.C., Lutz, H., 1999. One-tube fluorogenic reverse transcription-polymerase chain reaction for quantitation of feline coronaviruses. J. Virol. Methods 77, 37-46.

Heid, C.A., Stevens, J., Livak, K.J., Williams, P.M., 1996. Real time quantitative PCR. Genome Res.6, 986-994.

Hein, I., Klein, D., Lehner, A., Bubert, A., Brandl, E., Wagner, M., 2001. Detection and quantification of Listeria monocytogenes and Listeria innocua by a new real-time quantitative PCR assay. Res Microbiol., in press.

Holland, P.M., Abramson, R.D., Watson, R., Gelfand, D.H., 1991. Detection of specific polymerase chain reaction product by utilizing the 5′ to 3′ exonuclease activity of Thermus aquaticus DNA polymerase. Proc. Natl. Acad. Sci. U S A 88, 7276-7280.

Hofmann-Lehmann, R., Swenerton, R.K., Liska, V., Leutenegger, C.M., Lutz, H., McClure, H.M., Ruprecht, R.M., 2000. Sensitive and robust one-tube real-time reverse transcriptase-polymerase chain reaction to quantify SIV RNA load: Comparison of one- versus two-enzyme systems. AIDS Res. Hum. Retroviruses 16, 1247-1257.

Jordan, J.A., 2000. Real-time detection of PCR products and microbiology. New technologies for life sciences: a Trends guide. 61-66.

Kimura, B., Kawasaki, S., Fujii, T., Kusunoki, J., Itoh, T., Flood, S.J., 1999. Evaluation of TaqMan PCR assay for detecting Salmonella in raw meat and shrimp. J. Food Prot. 62, 329-335.

Klein, D., Janda, P., Steinborn, R., Muller, M., Salmons, B., Gunzburg, W.H., 1999. Proviral load determination of different feline immunodeficiency virus isolates using real-time polymerase chain reaction: Influence of mismatches on quantification. Electrophoresis 20, 291-299.

Klein, D., Bugl, B., Günzburg, W.H., Salmons, B., 2000a. Accurate estimation of transduction efficiency necessitates a multiplex real-time PCR. Gene Ther. 7, 458-463.

Klein, D., Musil, C., Hirt, R., Gold, P., Thalhammer, J.G., Günzburg, W.H., 2000b. (Possibilities and limitations of molecular diagnostic methods in clinical microbiology: the Feline Immunodeficiency Virus as an example). Wien. Tierärztl. Mschr. 87, 269-277.

Klein, D., Leutenegger, C.M., Bahula, B., Gold, P., Hofmann-Lehmann, R., Salmons, B., Lutz, H., Gunzburg, W.H., 2001. Influence of preassay and sequence variations on viral load determination by a multiplex real-time reverse transcriptase-polymerase chain reaction for feline immunodeficiency virus. JAIDS: in press.

Leutenegger, C.M., Klein, D., Hofmann-Lehmann, R., Mislin, C., Hummel, U., Böni, J., Boretti, F., Guenzburg, W.H., Lutz, H., 1999. Rapid feline immunodeficiency virus provirus quantitation by polymerase chain reaction using the TaqMan fluorogenic real-time detection system. J. Virol. Methods 78, 105-116.

Lie, Y.S., Petropoulos, C.J., 1998. Advances in quantitative PCR technology: 5′ nuclease assays. Curr. Opin. Biotechnol. 9, 43-48

McGoldrick, A., Lowings, J.P., Ibata, G., Sands, J.J., Belak, S., Paton, D.J., 1998. A novel approach to the detection of classical swine fever virus by RT-PCR with a fluorogenic probe (TaqMan). J. Virol. Methods 1998 72, 125-135.

Oberst, R.D., Hays, M.P., Bohra, L.K., Phebus, R.K., Yamashiro, Paszko-Kolva, C., Flood, S.J., Sargeant, J.M., Gillespie, J.R., 1998. PCR-based DNA amplification and presumptive detection of Escherichia coli O157:H7 with an internal fluorogenic probe and the 5′ nuclease (TaqMan) assay. Appl. Environ. Microbiol. 64, 3389-96.

Steinborn, R., Schinogl, P., Zakhartchenko, V., Achmann, R., Schernthaner, W., Stojkovic, M., Wolf, E., Muller, M., Brem, G., 2000. Mitochondrial DNA heteroplasmy in cloned cattle produced by fetal and adult cell cloning. Nat. Genet. 25, 255-257.

Suryanarayana, K., Wiltrout, T.A., Vasquez, G.M., Hirsch, V.M., Lifson, J.D., 1998. Plasma SIV RNA viral load determination by real-time quantification of product generation in reverse transcriptase-polymerase chain reaction. AIDS Res. Hum. Retroviruses 14, 183-189.

Pre-harvest health and quality monitoring, risk assessment and their relevance to the food chain

Jos P. Noordhuizen[1] and John D. Collins[2]
[1]Department of Farm Animal Health, Faculty of Veterinary Medicine, Utrecht University, The Netherlands; E-mail: j.p.t.m.noordhuizen@vet.uu.nl [2]Faculty of Veterinary Medicine, University College Dublin, Ireland; E-mail:jcollins@vetmed.ucd.ie

Summary

The integrated nature of the food chain creates an inter-dependence between the producer in the pre-harvest phase of production and the processor and distributor in the later stages in order to ensure both the quality and safety of the final product for the consumer. That inter-dependence is acknowledged in the requirements of the *corpus* of legislation that controls international trade, namely, the requirement to ensure that such trade in animals and foods of animal and vegetable origin conforms with sanitary and phytosanitary rules and is in accordance with the so-called "precautionary principle". Risk assessment is an integral component of this approach and while the disciplines associated with its implementation are demanding, they are nevertheless based on practices that are well known to veterinarians and food hygienists alike. The challenge facing the herd clinician is to ensure conformity by his client with the requirements of the market both in terms of quality and safety of the product purveyed.
The application of longitudinally integrated safety and quality assurance schemes within a commercially viable framework is now to be regarded as a indispensable component of the modern food animal practice. This, together with the delivery of a focussed programme of veterinary food hygiene and inspection service, can form the basis for a sustainable veterinary contribution to public health by providing a reliable basis for consumer confidence in the safety and quality of the food supply and at the same time contribute to the future viability of agricultural production.

Keywords: herd health, livestock trade, risk assessment, food safety

1. Introduction

Reliable pre-harvest monitoring of quality-related aspects of the production of food of animal origin has evolved over the past twenty years. The application of quality and safety control procedures is being most prominently applied in each step of the food chain involving the dairy industry. Quality control of dairy products both on the farm and throughout all stages of primary production are now well established in most developed countries with the result that dairy products are regarded as being among the most frequently tested food products on the market to-day. Despite this, dairy products do not always meet the demands of the market, or of the individual consumer, not least because of some uncertainties, as perceived in the marketplace, regarding the safety of such foods, and often without foundation. The reasons for this are not fully understood but may be attributable in part at least to a lack of communication between the primary producer and the consumer, hence, leading to a certain perception. This leads to a lack of understanding, or to mis-understanding, of the nature of food animal production, at a time when there is a growing demand for so-called natural products of known origin.

2. Herd Health Programmes: the Need for Animal and Farm Monitoring

Food animal production methods have undergone substantial changes over the past fifty years. These changes are mirrored in the shift in general veterinary practice, from the earlier classical, consultative practice that was focussed upon the treatment of the individual sick animal, to a more population-oriented advisory practice in the 1980's and 1990's. The rate of change has varied from country to country and, worldwide, from region to region, with consequential differences in the nature and quality of the products produced under the different regimes in operation in each region. Veterinary herd health and production management programmes offered by today's veterinary practices are examples of this development. The main focus of these programmes is on-farm economics (productivity; costs; income) and especially operational management, through the application of veterinary-zootechnical skills and a wider knowledge of the commercial and societal consequences of disease in the food-producing animal. Key elements in such a programme include a strengths & weaknesses assessment and priority setting at the start, clinical animal and farm inspection, data monitoring, herd problem analysis and prevention, most of which procedures can be carried out during regular and frequent farm visits. The execution of such programmes, for example on dairy farms, follows pre-set operational protocols (see example in Figure 1, Brand et al., 1996).

These protocols define the farm objectives, identification of methods and materials required in each case, followed by analysis of production and other data, the interpretation of collected information including putative risk factors for disease occurrence, and the development and application of preventive procedures and the critical monitoring of the effectiveness, or otherwise, of their implementation. Overall, product definition, structure of the programme, planning together with farm specificity, are paramount keys.

The monitoring of both animals (clinical inspections), farm conditions (e.g. housing, feeding, risk factors) and available data sources (e.g. milk production, somatic cell counts and bacteria counts) is the basis of such herd health and production management programmes. Through this monitoring one can obtain signals of deviating farm performance in the areas under investigation, being significant as an early warning system and as starting point for discussions between farmer and veterinarian about how to tackle the -impending -problem (Zaaijer and Noordhuizen, 2001).

In recent years there has been a growing tendency to merge dairy herd health programmes with the evidence of quantitative veterinary epidemiological studies in order to provide both the farmer and the veterinarian with science-based, quantitative risk profiles of population

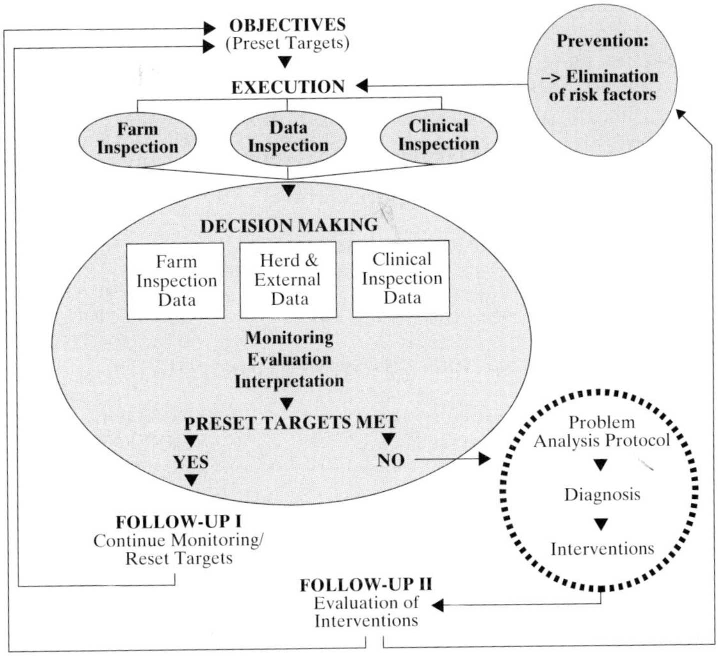

Figure 1. Example of a protocol for dairy herd health and production management. This protocol comprises 3 areas: routine monitoring, problem analysis and prevention.

events. Furthermore, these risk profiles provide an economic justification for veterinary advice leading to prospective investment in healthy animal production systems that is individualised for specified farms according to the particular requirements of each case.

Further advantages of such risk profiles is the setting of priorities of veterinary-zootechnical actions to be taken. In some cases this would mean that action is directed towards the most prominent risk condition(s), while in other cases it could economically be more appropriate to focus on a few, less prominent risk conditions.

3. Risks Associated with Trade in Livestock

The role of the veterinarian and other scientists in the assessment of risk requires to be tempered by the observation, by Krebs and Kacelnik (1997), and others, that risk is the currency that has replaced wealth production as a means of measuring decisions of industry, politicians and others. The systematic analysis of risk in relation to animal health risk is complex. It is comprised of three procedures, *viz.* hazard analysis, risk management and risk communication. While each of these procedures has a number of elements, and in most cases are data-dependent, nevertheless, the principles involved are not unfamiliar to veterinarians, and have provided a basis for decision making in this and other areas, including food hygiene and veterinary public health, for many years (Figure 2).

	RISK ANALYSIS	
CLINICAL PRACTICE		FOOD HYGIENE
	RISK ASSESSMENT	
• DIAGNOSIS	• HAZARD IDENTIFICATION • HAZARD CHARACTERISATION • EXPOSURE ASSESSMENT • RISK CHARACTERISATION	• DIAGNOSIS • DISPOSITION • RECOGNITION
	RISK MANAGEMENT	
• TREATMENT • INTERVENTION • AUDITS	• MONITORING • SURVERILLANCE	• CERTIFICATION • DETENTION • DESTRUCTION • AUDITS
	RISK COMMUNICATION	
• CONTROL • PREVENTION • ADVICE	• ADVICE, INSTRUCTION • TRANSPARENCY	• CONTROL • PREVENTION • ADVICE
	COST BENEFIT OF RISK REDUCTION	

Figure 2. Correlation of the elements of risk analysis in clinical practice and food hygiene practice.

In 1999, the Office International des Epizooties defined the objective of risk analysis as the identification of potential hazards associated with the animal or animal product, assessment of the likelihood of introduction, establishment and spread within the importing country of such a hazard, and estimation of the potential consequences relating to its introduction as summarised in Figure 3 (OIE, 1999).

Risk Analysis in Relation to International Trade

"... Animal health risk analysis is a process comprised of hazard analysis, risk management and risk communication".
"... Its objective is to

- *identify the potential hazards associated with a commodity,*
- *assess the likelihood of its introduction, establishment in the importing country, and*
- *estimate the potential consequences relating to its introduction".*

Source: Office International Epizooties (1999)

Figure 3. Risk analysis in relation to international trade in animals and products of animal origin.

Quantitative risk analysis extends such investigations further by the development of strategic models based on real life observations and data relating to confirmed cases or outbreaks. The statistical analysis of risk regarding the likelihood of entry and spread of such diseases as classical swine fever and foot and mouth disease have long been the subject of decision making at national government level (Acree and Ahl, 1991)

Quantitative risk analysis of the recent outbreak of classical swine fever in the Netherlands, by means of simulation models, has provided further evidence of the need for vigilance in respect of importation and internal movement of pigs and their products, particularly in densely populated livestock areas ((DPLA's) of the European Community (de Vos and Saatkamp, 2000).

A similar approach has been adopted by Wilesmith *et al.* (2000) in the risk assessment of bovine spongiform encephalopathy in Great Britain, with a view to defining geographical risk and the level of propagation risk to livestock populations and the effectiveness of the removal for slaughter of entire herds in which cases have occurred, as compared to removal of cohorts of the case (Figure 4).

Regional risk factors in the BSE epidemic in Great Britain

Rate of progression of the regional epidemic and the ultimate proportion of holdings* affected in a region was likely to have been determined by three factors, namely

1. The relative contribution of cattle to meat and bone meal and absolute amount of such meal consumed by cattle
2. Effectiveness of rendering in reducing the infectivity of of the meat and bone meal
3. The degree of effectiveness of the control measures

(* unit of interest: holdings (n=118,057); no. of BSE +ve holdings: 30,042)

Figure 4. Identification of risk factors affecting the progression and extent of the bovine spongiform encephalopathy *epidemic in Great Britain (After Wilesmith et al., 2000).*

This approach, which provides a scientific basis for decision making in animal disease management at the national level, requires a complete understanding of the interactions of all the factors involved, both from an economic and veterinary perspective, if it is to succeed.

4. Animal Health and International Trade

Recent developments such as the introduction of the Sanitary and Phytosanitary (SPS) Measures implicit in the World Trade Organisation agreement, and the European Union Commission's White Paper on Food Safety (CEC, 2000), both of which advocate adoption of the so-called Precautionary Principle, are of considerable relevance in this regard (Figure 5).

Risk analysis relates to:

- World Trade Organisation and its Agreement on the Application of Sanitary and Phyto-Sanitary Measures (WTO, 1995)
- CEC's White Paper on Food Safety (2000)
- Policy Formulation on Animal Health and Environmental Issues at National and Regional Level, *e.g. movement controls, quarantine, vaccine use*
- Decision Making in Veterinary Practice, *e.g. effluent management, infection/disease control*
- Decision Making in Food Hygiene and Safety in Plants, *e.g. control of animal disease, foodborne agents*

Figure 5. Role of risk analysis in relation to food animal production and food safety at the regional and international level.

Not in the least because trade implies risk, and the security considered to be offered by the enforcement of a zero-tolerance approach to disease exclusion in international trading is untenable, more particularly because disease agents do not respect political boundaries, with or without the assistance of illegal operators engaged in inter-state smuggling. In this context, regionalisation, i.e., the establishment of zones or regions, in the international context, of comparable animal health status and comparable (disease) risk management levels, has much to recommend it, provided the means for defining and establishing such zones is scientifically sound and is, at all times, transparent (Zepeda et al., 2000). Against the background of serious outbreaks of animal diseases that may or may not have any public health association, of consumer concerns in general (ethics, society acceptance, welfare, environmental issues), together with increased pressure from the markets, it is logical and with good reason, that structural changes in the primary production sector must now take place.

Animal health economics is an emerging new branch in the field of economics and more specifically agro-economics (McInerney, 1988, in Brand *et al.*, 1996). It addresses decision questions at the farm level, the sector and national level and the international level. Livestock disease is part of an economic process. Disease affects the production process by hampering productivity, increasing production costs and sub-utilisation of resources on the one hand. On the other hand, further disease losses occur because the supply of products can be affected, the value of products diminished and the resources partly lost. The discipline of animal health economics has recently contributed to decision problems such as regarding the risks of introduction and spread of foot-and-mouth disease or regarding the occurrence of classical swine fever in Europe (de Vos and Saatkamp, 2000). It also has contributed to the decision about yes/no vaccination against Aujeszky disease at national level, and if yes, how frequently this should be done. It has also contributed to the decision-making process at farm level focusing, for example, on the economically optimal replacement policy on dairy farms or sow farms. Different methodologies are used in this decision-making process, e.g. cost-benefit analysis, decision tree analysis, simulation models, decision support models with or without geographical information systems, risk assessment. The last few years there is a tendency of merging animal health economics' principles and methods with veterinary epidemiological principles and methods with the purpose to further optimise decision-making under risk conditions such as occurring in animal disease control.

5. Food Chain Issues

The World Trade Organisation's agreement on the Application of Sanitary and Phytosanitary (PSP) Measures requires that sanitary and phytosanitary measures are scientifically based (WTO, 1995). This requirement highlights the role of epidemiologists and veterinary food hygienists in supporting sustainable agriculture and in providing the assurance of food safety and quality that is a *sine qua non* of international trade to-day. The key elements of the SPS agreement are risk analysis, regionalisation, harmonisation, equivalence and transparency. Two of these elements, namely risk analysis and regionalisation, are dependent on information derived from epidemiological surveillance. This approach provides a basis for the practical application of the so-called "precautionary principle" in regard to importation control of animals and animal products.

The European Commission's White Paper on Food Safety (CEC, 2000) acknowledged the need to ensure that all the elements of risk analysis had to be seen to apply to each phase of the food chain, and to the pre-harvest phase in the first instance. The latter means that on-farm quality risk management programmes have to be developed and implemented (Noordhuizen and Welpelo, 1996). The safety of food of animal origin in respect of the potential risk posed by the TSEs of animals has again defined the importance of sound data on the quality and safety of animal feedstuffs, traceability of food animals and their products and an appreciation of the dynamics of internal trade of animals in each country and region. The recent move to active surveillance of TSEs will provide further data on the regional and national incidence of these diseases, for use in the strategic design of control and preventive measures based on quantitative risk analysis.

This pro-active approach has an immediate application in the assessment of the risk posed by foodborne agents such as the *Campylobacter* spp., *Toxoplasma* spp. and *Listeria* spp. and the Norwalk-like agents and other pathogens associated with fish and fish products. In each case, there is a need for direct involvement of epidemiologists, veterinary food hygienists and other scientists in the elucidation and control of the risk factors that contribute to the exposure of the consumer to these agents at all levels of production and processing. Furthermore, occupational exposure to these and other agents is often understated. Yet this aspect is of veterinary public health concern, in so far as it relates to the primary producer and processor in particular, receives comparatively little attention. There is an urgent need for a risk assessment in these groups in order to properly address the economic losses that accrue from exposure to these and other foodborne hazards, at the employee or consumer level. In particular, detailed risk analysis of potential for spread of the causal agent to the human population from this source has been a focus of concern for all (Attenborough and Matthews, 2000). Also, while not usually regarded as a cause of disease in animals *per se*, the enterocytotoxigenic strains of *Escherichia coli*, including *E. coli* 0157:H7, and their distribution and potential for spread to consumers, have been the subject of recent studies, the final outcome of which is as yet, unclear. This particular pathogen represents a real challenge for the primary producer, as well as for the processor, since on-farm conditions and the carriage and pre-slaughter conditions of food animals as well as the contamination of carcasses at slaughterhouses are likely to be the critical factors in determining the level of risk for the ultimate consumer.

6. Other Examples of Risk Assessment in Livestock Production in Relation to Human Health

There are other food safety-related challenges facing the primary producer that can best be resolved through the efficient implementation of herd health and safety programmes in which the quantitative analysis of risk is likely to be a key factor. Decision making in regard to the safety of foods of animal origin from (i) dairy herds exposed to paratuberculosis, and (ii) cattle herds exposed to *Mycobacterium bovis,* are prime examples. In the case of the latter disease, there have been recent developments in diagnosis, based on the in-vitro assessment of cell-mediated immune responses in exposed populations, which can more clearly define the level of risk posed by infected but non-tuberculin test responsive animals (Collins, 2000). The interferon-γ assay is one such example (Table 1).

Table 1. Risk assessment of the disclosure of further tuberculin reactor cattle in herds containing animals previously deemed positive to the interferon γ assay for tuberculosis. After Collins (2000).

Longitudinal study on 26 tuberculosis-restricted herds: disclosure of reactors

	tbt+ve	tbt-ve	odds ratio
At first reactor tuberculin test (tbt)			
77 cattle IFN-γ +ve at index test	21	56*	
			8.90
371 cattle IFN-γ -ve at index test	15	356	
At second reactor tuberculin test (+56 days)			
54 cattle IFN-γ +ve at index test	6	48	
			7.16
233 cattle IFN-γ-ve at index test	4	229	

* two of these cattle slaughtered 10 days after the index test were lesion + ve

7. Conclusions

The relevance of on-farm factors including spatial and temporal relationships, herd or flock size, purchasing policy, animal movements and of farm practices including effluent management and the need for concise and accurate information on each of these and other aspects of animal production become apparent in the process and, once again, illustrate the need for astuteness on the part of the clinician, epidemiologist or veterinary food hygienist and other advisers in terms of the quality and consistency of their monitoring observations and other data, as submitted for spatial and statistical analysis.

Environmental concerns already apply to the continued use of antimicrobial agents in animal production and these concerns, along with those relating to the risk to public health of such usage require a scientific approach to the estimation of risk along the lines of quantitative risk analysis. Likewise, the genetic component of host susceptibility to diseases ranging from brucellosis, salmonellosis, cryptosporidiosis, mycobacterial diseases, along with scrapie and other transmissible spongiform encephalopathies (TSEs), as examples, is already under investigation by this and other means. Recently, the European Commission has published a report on Harmonization of Risk Assessment procedures in Europe (DG Health & Consumer Protection). This represents a further step on the road to formal risk assessment applications at farm, sector, memberstate and EU level.
The principles of risk analysis have much in common with those that apply both to clinical practice and to food hygiene. The veterinary specialist in each of these disciplines is well equipped to meet the challenges posed by the new order of world trade and thereby contribute to the advancement of the veterinary sciences and the eventual attainment of a sustainable supply of healthy food for all.

References

Acree, J.A. and Ahl, A.S. (Eds.), 1991. Animal Import Risk Analysis, *Proceedings of a USDA/APHIS/PPD International Seminar*, 64 pp. Ottawa, Canada, August 1991.

Attenborough, M. and Matthews, K.R., 2000. Food safety through the meat supply chain. Journal of Applied Microbiology Symposium Supplement 2000, 88:14S-148S.

Brand, A., Noordhuizen, J.P.T.M. and Schukken, Y.H., 1996. Herd health and production management in dairy practice. Wageningen Pers Publ., Wageningen, The Netherlands, 543 pp.

CEC, 2000. White Paper on Food Safety. COM (1999) 719 final. Brussels: Commission of the European Communities.

Collins, J.D., Monaghan, M., McGill, K., Kelly, A. and Fitzsimons, T., 2000. A longitudinal study of cattle found positive to the interferon - assay for *Mycobacterium bovis* infection. In *Proceedings of the Ninth International Symposium of the International Society for Veterinary Epidemiology and Economics*, 1263-1265. Breckenridge, Colorado, August, 2000.

De Vos, C.J. and Saatkamp, H.W., 2000. An approach for quantitative risk analysis of contagious animal disease introduction into densely populated livestock areas of the European Union. In *Proceedings of the Ninth International Symposium of the International Society for Veterinary Epidemiology and Economics*, 1220-1222. Breckenridge, Colorado, August, 2000.

Krebs J.R. and Kacelnik A., 1997. Risk: a scientific view. Proceedings of The Royal Society 1997:31-43.

Noordhuizen, J.P.T.M. and Welpelo, H.J., 1996. Sustainable improvement of animal health care by systematic quality risk management according to the HACCP concept. The Veterinary Quarterly 18:121-126.

OIE, 1999. In: *75 Years, 1924-1999*. Office International des Epizooties. Paris

Wilesmith, J.W., Ryan, J.B.M., Stevenson, M.A., Morris, R.S., Pfeiffer, D.U., Lin, D., Jackson, R. and Sanson, R.L., 2000. Temporal aspects of the epidemic of bovine spongiform encephalopathy in Great Britain: holding-associated risk factors for the diseases. Veterinary Record 147: 319-325.

WTO, 1995. Agreement on the application of Sanitary and Phytosanitary Measures. World Trade Organisation, wwwwto.org

Zaaijer, D. and Noordhuizen, J.P.T.M., 2001. Monitoring rumen fill, faeces consistency and undigested particles in faeces in order to evaluate reproductive performance in dairy cows. The Veterinary Quarterly (in press)

Zepeda, C., Salman, M. and Ruppanner, R., 2001. International trade, animal health and veterinary epidemiology: challenges and opportunities.Preventive Veterinary Medicine, 48: 261-271.

Human pathogens associated with on-farm practices - Implications for control and surveillance strategies

Andreas Hensel[1] and Heinrich Neubauer[2]
[1]Institute of Animal Hygiene and Public Veterinary Health, An den Tierkliniken 17, D-04103 Leipzig, Germany; Corresponding author, [2]Institut für Mikrobiologie, Sanitätsakademie der Bundeswehr, Neuherbergstr. 11, D-80937 München, Germany

Summary

A number of communicable diseases is transmitted from animals to humans - known as zoonoses. These diseases transmitted either by animal products like foodstuff or direct contact can cause substantial health problems and severe economic loss in developing and in industrialized countries. Various zoonotic pathogens are considered to be emerging or re-emerging. Examples for emerging pathogens are *Yersinia enterocolitica*, *Salmonella enteritidis*, *Escherichia coli* O157:H7. "Classical" zoonotic diseases such as anthrax, brucellosis, leptospirosis, rabies, or bovine tuberculosis have been controlled or eliminated in many industrialized countries, but remain still endemic in developing regions. New zoonotic agents such as the Nipah virus (Malaysia, 1999), the "bird flu" influenza virus A (H5N1) (Hong Kong, 1997) and with special respect to Europe prions causing bovine spongiform encephalopathy (United Kingdom, 1985) can be recognized. Further threats to human health are animal-associated opportunistic parasites or bacteria (e.g. *Toxoplasma gondii*, *Listeria monocytogenes*) when they infect immunocompromised patients e.g. HIV positive patients.

The upward trend in the number of zoonotic diseases can be traced to environmental, technical, medical, agricultural and industrial changes. The WHO fact sheets list many reasons: the alteration of the environment affecting the size and distribution of animal species and vectors, an increasing human intrusion into previously un-populated areas (e.g. tropical forests) favouring the first contact of people and 'new' agents; the deterioration of public health and veterinary infrastructures particularly in developing regions (e.g. illegal slaughtering of animals and inappropriate waste disposal practices); misuse of antibiotics and antimicrobial drugs in humans and animals, which can hasten the evolution of resistant microbes; deforestation, changes in climate and weather; the continuing evolution of pathogenic microorganisms; new medical techniques like xenotransplantation; industrialization and intensification of the animal production (crowding); changes in food processing, food distribution and the nutritional habits of consumers (consumption of 'raw' animal products such as non-pasteurized milk); increase of international movement of people and of international trade of animals and animal products and finally the increased numbers of immunocompromised persons. Water, air, soil, feed - contaminated by animal waste -, inadequate waste management practices, uncontrolled animal-to-animal contact or contact to wildlife play an important role in the epidemiology of zoonotic diseases at the pre-harvest level. The food-processing industry

can also have a significant influence on the epidemiology of zoonotic diseases. Effluents which contaminate the environment may originate from slaughterhouses e.g. from sewage contaminated with faeces. Carcasses may serve as carriers for infectious agents to the food-processing plants where they eventually contaminate the processed food. Rodents and pests may also be vectors.
Hygienic animal keeping guarantees that the spread of pathogens in the animal population of a farm, to other farms, to food-processing plants and to the consumer can be minimised. Hence, an effective surveillance and control of zoonoses is required aided by strict control strategies and legislation.

Keywords: zoonoses, spread and transmission, control and countermeasures

1. Introduction

Looking at the pathogens causing human disease and being associated with on-farm practices different groups of agents based on their ecology and their mode of transmission can be identified. The handling of farm animals including breeding, keeping and production of food and goods can lead to direct transmission of agents from animals to the animal keepers and their helpers. Examples for such agents are the dermatophytes *Trichophyton* and *Microsporus* or the re-emerging bacterium *Burkholderia mallei* causing chronic glanders in solipeds and a fatal syndrome in humans if misdiagnosed or being untreated. Typical human pathogens like hepatitis virus A or human enteroviruses can contaminate food and goods if the hygienic status on the farm is low or guidelines are purposely ignored for instance to save production costs by the manager in charge or by the employees to 'make life easier'. On the other hand human nutrition is one of the main sources for human infections and infestations and it is also an important vehicle for spreading the causative agents all over the world. These agents can either be classical causative agents of zoonoses like *Bacillus anthracis*. They can also be introduced into the food chain as contaminants after having occupied the environment of the food or product manufacturing plants, like *Salmonella* or *Listeria* species. Moreover, the importance of the agents is varying according to the geographic region of concern. *Brucella* and *Leptospira* species cause regularly outbreaks in eastern Europe, *Bacillus anthracis* is of major concern in less developed countries of Asia and Africa and western Europe is facing a growing number of foodborne diseases caused by *Campylobacter* species. Besides *Salmonella* and *Campylobacter, Trichinella*, *Shigella*, *Staphylococcus aureus*, *Clostridia,* and *Listeria* are important food borne disease causing agents in European countries (WHO Surveillance Programme for Control of Foodborne Infections and Intoxications in Europe, 7th Report, 2000; www.bgvv.de/publik/who/7threport/7threp_fr.htm).

This review will illustrate causative chains, countermeasures, and surveillance strategies for zoonotic agents and contaminants of environmental origin.

2. Zoonoses - Towards a Broadened View

Many of the infections which are associated with on-farm practices are classical zoonoses. The WHO (1959) defined zoonoses as 'Diseases and infections which are naturally transmitted between vertebrate animals and man'. A guideline on zoonoses (EU 92/117 EWG) followed this definition and specified the agents which can cause a zoonosis: bacteria, viruses, fungi, and parasites. The lack of clearance in the term 'naturally transmitted' and ignoring the actual immune status of the host in this definition have caused a lot of confusion on the entity of 'zoonosis'. The direction of transmission was also used as a criteria for the grading of zoonoses: anthropozoonosis (from man to animal), zooanthroponosis (from animal to man) and amphixenosis (both directions can occur). Another approach to classify zoonoses made use of the type of life cycle of the agent causing disease: direct zoonosis (transmission without or little changes of the agent), cyclo-zoonosis (more than one species or vertebrate host required), meta-zoonosis (a stage of propagation in an invertebrate vector and an extrinsic incubation period required) and sapro-zoonosis (a vertebrate host and a non-animal site or reservoir required) (Schwabe, 1964). Sapronosis or geonosis is sometimes used for diseases caused by agents which have a non-living habitat, infect animals and humans on occassion but cannot be transmitted to another host. Transmission can be horizontal (inter- and intraspecific) or vertical (from one generation to the other in a single species). Examples of agents for which classical 'definitions' do not match satisfactorily are the algae *Prototheca*, the fungus *Cryptococcus*, or the bacteria *Listeria* or *Salmonella*. In general, we consider every agent which can cause a disease or infection in both humans and animals, as a zoonotic agent independent from its life cycle or natural habitat. These agents are of special interest in food hygiene and have to be controlled to protect the consumer but also the environment, wildlife, lifestock and the producer of animals or animal products.

3. Selected Human Pathogens - Spread and Transmission

Approximately 150 diseases are considered to have a zoonotic character (Teufel, 1998). Table 1 displays a selection of those agents we consider to be of main interest for on-farm practices including reservoirs and the frequency they are reported. It is obvious that this chapter cannot describe all zoonotic agents exhaustively so we will focus on some diseases - emerging and re-emerging, classical and 'new' - to show current trends.

Bacillus anthracis is a very good example for a zoonotic agent which has a reservoir in soil. Herbivores get infected when taking up the spores with their feed. Humans usually get infected directly when the bacteria or spores are incorporated via contaminated food or when handling contaminated material during processing. Anthrax is therefore a direct and indirect (via vector or vehicle) zoonosis (Table 2).

In developed countries anthrax in animals and man is rare. Effective veterinary countermeasures like notification, culling and destruction of infected animals, treatment of suspected cases, and vaccination of livestock in endemic areas reduced the incidence of the disease. Industrial cutaneous and pulmonary anthrax in man was controlled by an

Table 1. Important human pathogens of animal or environmental origin associated with on farm practices.

AGENT	RESERVOIRS	FREQUENCY
Bacteria		
Mycobacterium spec.*	Wildlife, livestock, man	Frequent, world wide
Brucella spec. *	Domestic and wildlife	Regional, world wide
Campylobacter spec.+	Domestic and wildlife, fowl	Frequent, world wide
Clostridium perfringens+	Soil, water, ingesta of animals	Sporadic to frequent, world wide
Eschericha coli (EHEC/STEC)+	Man, animals, (cattle)	Frequent, world wide
*Erysipelothrix rhusiopathiae**	Livestock	Sporadic or group wise, world wide
Leptospira interrogans subsp.*	Domestic animals and pets, rodents	Regional, world wide
Listeria monocytogenes+	Soil, feed, milk products	Sporadic, group wise, world wide
*Bacillus anthracis**	Soil (abbatoirs), ruminants, pigs	Sporadic, world wide, regional
Staphylococcus aureus / toxins+	Man, pets	Sporadic frequent, world wide
Salmonella-serovars+, #	Domestic mammals, fowl, reptiles, water, waste water	Frequent, group wise to endemic, world wide
Shigella spec. +	Man, feces, waste water	Frequent world wide
Clostridium botulinum #	Environment, animals	Sporadic, world wide
Clostridium tetani #	Soil, waste water, man, animals	Sporadic, world wide
Yersinia spec. +	Livestock, rodents (?)	Sporadic, world wide
Pasteurella spec.* +	Livestock, rodents	Sporadic
*Chlamydia psittaci**	Fowl	Sporadic to group wise
Viruses		
Influenza virus*	Fowl, pigs	Sporadic to endemic, world wide
West-Nile fever virus*	Fowl, horses, cattle, sheep, man, biting midges	Sporadic, world wide
Rift-Valley fever virus*	Domestic and wildlife, man, biting midges	Sporadic to endemic
Orthopox viruses*	Livestock, cats, monkeys, wildlife	Sporadic, world wide
Equine encephalitis viruses* (Eastern, Western, Venezuelan)	Rodents, fowl, horses, snakes, biting midges	Sporadic
Rabies virus*	Domestic and wildlife, bats	Sporadic, world wide
Fungi or Algae		
Microsporum spec. *	Livestock	?, worldwide
Trichophyton spec. *	Livestock	?, worldwide
Aspergillus spec. *, #	Livestock, soil	?, worldwide
Candida spec. *	Man, livestock	?, worldwide
Prototheca spec. +, #	Soil, water	?, worldwide

Table 1. continued

AGENT		RESERVOIRS	FREQUENCY
Parasites			
Protocoa:	*Amoebia* spec.$^+$		
	Toxoplasma	Lifestock, cats, rodents	
	Giradia spec.$^+$		
Trematodes:	*Fasciola*$^+$		
Cestodes:	*Taenia*$^+$		
	Echinococcus$^+$	Dogs, cats, foxes	Sporadic, world wide
Nematodes:	*Ascaris*$^+$		
	Trichinella$^+$	Lifestock, rodents	Sporadic to group wise, world wide
	Anisarkis$^+$	Fresh water fish	
Unknown origin: prions (?)$^+$		Cattle, sheep, food	Sporadic to endemic

*: zoonosis mainly transmitted by contact
$^+$: foodborne zoonosis
#: environmental contamination

Table 2. Transmission of various zoonoses.

Horizontal direct transmission of zoonoses			
Disease	Sheddding by	Entry	Mode of transmission
Anthrax	Carcass, blood	Skin	Contact
Brucellosis	Genital tract	Skin (lesions)	Contact
Orthopox viruses	Crusts	Skin (lesions)	Contact
Mycoses	Skin	Skin surface	Contact
Helminthiases	Faeces	Mouth	Contact

Horizontal indirect transmission of zoonoses				
Disease	Shedding by	Entry	Mode of transmission	Vehicle
Anthrax	Meat	Mouth	Alimentary	Food
Brucellosis	Milk	Mouth	Alimentary	Food
Tuberculosis	Milk	Mouth	Alimentary	Food
Psittacosis	Faeces	Respiratory tract	Aerogenic	Air, dust
West Nile fever	Blood	Skin	Via vectors	Insects
Helminthiases	Faeces	Mouth	Alimentary	Food
Prion-induced disease	'Risk material' ?	Mouth	Alimentary	Food

improved factory hygiene when producing wool, hides, meat, and bones. Quality requirements forced the industry to start up processing already in the country of origin of animal products reducing the import of raw materials which could be contaminated with spores. On the other hand, the prosperity of many industrial countries lead to a rise in the import of bone meal for fertilizers or feed. Raw meat of exotic animals, exotic fruits, spices, and vegetables are easily available and are consumed increasingly not only from people who are allergic to domestic food. This is a possible reason why anthrax cases occur in non-endemic areas. Re-introduction with products which are spread on fields and meadows might even lead to the establishment of new local foci with an increased risk for livestock (Turnbull, 1998). Anthrax highlights also the problems of pest control in developing or undeveloped countries of endemic areas in Africa and Asia. Public health and veterinary regulations are often not existing or are ignored due to poverty or customs and traditions. A short review on the PROMED mails (a program for monitoring emerging diseases of the International Society for Infectious Diseases) of the last year can teach us the shortcomings hygienists face at the pre-harvest stage in such countries. In Kazakhstan (August 2000) various human anthrax cases were reported. Governmental countermeasures like police controls and sanitary checkpoints were of little value as cattle breeders and sellers could avoid these controls by using hidden paths. Criminal machinations can also be assumed when animals with a certificate of vaccination get ill with *B. anthracis*. Hence, inadequate handling of the vaccine by a veterinarian might also be possible. A lack of infrastructure or of interest from governmental personnel involved can cause life threatening situations. Although a case of animal anthrax was suspected correctly a time delay of several days in the final diagnosis occurred. Meanwhile the meat was processed to food. In this outbreak a case of cutaneous anthrax was reported which turned into a fatal septic form when treated unsufficiently by the patient on its own. A dramatic rise in number of human cases in an outbreak can be observed in very poor countries like Zimbabwe (January 2001) when people are forced to beef cadavers. In this outbreak at least 980 patients were involved. As it is not possible to control anthrax in wildlife outbreaks can be observed in countries with a high number of free-living herbivors like South Africa involving man and livestock (Januar 2001). In India where rendering plants for carcass disposal are rare and vultures still 'dispose' most of the carcasses of animals, a die-off of the vulture population raised the fear that anthrax can emerge in future (February 2001). In South Africa most of the infected animals are game which cannot be vaccinated. Infections in man are again due to eating the carcasses of died cattle (February 2001). It is clear that many of these mails which are reported on a private basis do not provide the exact facts but they can be used to show actual trends in zoonoses (http://www.promedmail.org).

An interesting distribution of brucellosis outbreaks caused by *Brucella* species is reported for Europe (WHO, 2000). In Mediterranean countries like Italy, Portugal and Spain as well as in the eastern European countries of the former Soviet Union brucellosis is still an important cause for human disease with more than 1000 outbreaks every year. In contrast, Germany like most of the northern European countries is considered to be free of brucellosis. However, in 1999 20 cases of human brucellosis have been notified. About 15 of these cases have been imported, 9 from Turkey. Sources of infections were the contact to sheep and goats and the consumption of raw milk or products made from raw milk

(Anonymous, 2000). An unintentional import of *Brucella* via the living animals but also through imigrants should always be considered (Rasch et al., 1997). Further efforts have to be made in the future to control this zoonosis already at the farm level with strengthened veterinary countermeasures like vaccination, serological survey, and reimbursement of the animal owner whose life stock has been destroyed to get acceptance from the farmers.

Glanders caused by the bacterium *Burkholderia* (*Pseudomonas*) *mallei* is a chronic disease in solipeds. It leads to disastrous losses in horse populations in the past and is a rare but often fatal infection in humans (Neubauer et al., 1997). It can be transmitted to humans by contact or by the consumption of contaminated meat. Glanders is one of the first zoonoses which was distributed all over the world with its host, the war animal horse. With the decline of the horse population due to motorization and strict controls in trade, the number of glanders infections in man and horses has dropped. For many years it was not reported to the OIE and was considered to be eradicated. However, in a survey in the European part of Turkey evidence was found that it is still widely distributed in the domestic horse population but being ignored by the veterinary administration (Arun et al., 1999). The infection was distributed by commonly used bridles, drinking troughs, and mangers. Transport, crowding and bad keeping conditions maintained the infection. In the north-east of Brazil, namely in the States of Alagoas, Ceara, Maranhao, Pernambuco, Piaui and Sergipe it was recognized by the local authorities in 1999 and 2000 again and notified to the OIE. Glanders is still a threat to solipeds and humans demanding the need for keeping up the strict, expensive and 'old-fashioned' countermeasures which are in action in many countries for more than a century.

Since the eradication of variola virus - the causative agent of smallpox in humans - new threats have emerged from two closely related orthopox virus species. In Africa, a growing number of smallpox like infections has been recognized in humans. Monkey pox virus was isolated from infected humans and monkeys during various outbreaks (Jezek and Fenner, 1988). The declining immunity in the human population after cessation of smallpox vaccination, hunting and eating of monkeys, deforestation due to overpopulation and therefore closer contact to monkeys or reservoir animals like squirrels were identified to be the main reason for infections in humans (Mukinda et al, 1997; Neubauer et al., 1998). Due to their impact on human health pox viruses were always under special investigation including sophisticated epidemiological studies. Monkey pox virus infection and the reasons for its emergence could be considered as the 'index case' for a long list of 'new' zoonoses like Nippah virus, Meanangle virus or even Ebola virus infections namely the increasing human intrusion into previously un-populated areas. Although the animal ancestor of vaccinia virus is unknown it was commonly supposed that the vaccinia virus strains which have been used in the campaign against smallpox would not be able to establish new natural reservoirs. However, Buffalopox virus in India and Cantagalo virus in Brazilia may have derived from the locally used smallpox vaccine strains infecting nowadays cattle and humans. Cantalago virus infection is even considered to be a emerging disease (Damaso et al., 2000). Therefore, these two agents have an extraordinary position in the class of zoonoses; they are indeed man made.

A special problem in controlling human pathogens on the farm level is the spreading of an agent by wildlife vectors which can cover far distances in short times. Migrating birds can carry biological agents along their travelling routes but also in areas far away from their normal range by accident for example when displaced by storms. The outbreaks of West Nile fever in Europe and Northern America in the last five years are well documented examples for this phenomenon. In the Old World the virus is introduced by viraemic migratory birds into European countries like Italy, France or Romania from the African continent. Ornithophilic mosquitoes like *Culex pipiens* are then the local vectors and birds are again the amplifying hosts. Humans and horses get diseased mainly in or near wetlands where the birds have contact to large numbers of biting midges. The disease can take a severe course in humans with fatal outcome. In contrast to Europe, West Nile fever is a new zoonosis for the United States starting with a die-off of birds and several human cases in New York City. Normal migration between Europe and America, displacement by storms or legal/illegal imports have been discussed as the source of the outbreaks. Whether West Nile fever virus can establish a steady reservoir in the temperate areas of the Northern hemisphere in the light of a steadily global warming up will be experienced in the future (Work et al., 1955; Hubálek and Halouza, 1999; Rappole et al., 2000). In the year 2000 another emerging arthropod-borne virus, Rift Valley fever virus, spread into Saudi Arabia and Yemen on the Arabian peninsula. But not only viruses can be transported by migrating animals over thousands of miles. The appearance of *Mycobacterium africanum* in cattle in northern Bavaria, Germany (Weber et al., 1998) gave rise to the assumption that this bacterium was imported into a tuberculosis free country by a swallow.

Intoxications caused by toxins of *Staphylococcus aureus, Clostridium perfringens, or Bacillus cereus* have to be considered as important sources of human and animal disease. Helminth infections or infestations with protozoa play an important role as human pathogens in Asian countries. Dermatomycoses are a threat to animal keepers all over the world. Although these diseases cause enormous economic losses they will not be discussed as this review can focus only on some outstanding highlights.

4. Factors Influencing Human Pathogens at the Farm Level

The interaction of the pathogen with the individual animal and the herd on the farm and with the human host plays a crucial role in the maintenance of the infectious circle. The factors influencing this circle are numerous and cover all fields of environmental, animal, and human hygiene so that only some extraordinary items will be discussed. This review does not and can not claim to be complete.

4.1 The Agent

We have to look at the agents itself and its characteristics. Bacterial pathogens can change their resistance to antibiotics and disinfectants rapidly due to their short replication time and the selection of naturally occurring mutants. Besides naturally occurring resistance acquired resistance by transformation, transduction and conjugation may result in multi-resistant strains. Due to the problem of antibiotic resistance but also the costs or impracticality of improvements in hygiene and management vaccination e.g. in poultry may become more important. (Zhang-Baber et al., 1999) New variants may even develop by means of interactions between wild and vaccine strains displaying 'new' syndromes or vaccine strains may revert to the wild type, then. The status of non-culturable but viable pathogenic bacteria has to be assessed in the future for its importance in the transmission of pathogens at the farm level. (Pommepuy et al., 1996) In viruses the genetic re-assortment driven by antigenetic drift and antigenetic shift is a well-known source for new RNA viruses, e.g., influenza viruses in pigs, fowl, horses, and man. Virulence and contagiousness can vary in between the same species tremendously. *Yersinia enterocolitica*, *Salmonella* species and *Campylobacter* species produce asymptomatic diseased shedders in animals. The infective dose - even for agents which are adopted only to humans - is high (Table 3).

Table 3. Oral infectious doses of various pathogens (colony forming units for experimentally infected adult human volunteers).

	Percentage of clinically diseased patients			
Bacterium	0 - 25 %	26 - 50 %	51 - 75 %	76 - 100 %
Shigella dysenteriae	10^1-10^4	10^2-10^4	10^3	10^4
flexneri	10^2-10^4		10^3-10^9	10^6-10^8
Vibrio cholerae	(10^3)-10^3	10^8-10^{11}	10^4-10^6	
Salmonella typhi	10^5-10^8	10^5-10^7	10^4	10^8-10^9
S. derby		10^7		
S. pullorum		10^9		
Escherichia coli	10^6-10^8	10^8	10^6-10^{10}	10^{10}
Streptococcus faecalis	10^9	10^{10}		
Clostridium perfringens		10^8	10^9	10^9

SCS engineers (1979) modified]

4.2 The animal host

The immunity of the individual animal and the herd decides about the persistence of an agent in the population. Maternal antibodies may protect the newborn animal from infection. Immunity can be only serotype specific or even cross protective. The immunity can be humoral, T-cell mediated, or associated to the mucous membrane. Keeping conditions like crowding, hygiene measures, nutrition, stable construction, air filtration but also breeding hygiene and veterinary control are prominent factors in animal health reflecting the status of the immune system. Even the management system can influence the excretion of pathogens (Wray et al., 1987). In an interesting study, Davies et al. (2000) described reasons for the dirtiness of slaughtering beef and therefore for a possible contamination with pathogens: age, feed type, coat length and clipping, journey distance (over 150 miles) and time, and the status of the abattoir. Three of these risk factors are clearly connected to the host. Young (age), 'shorthair' (breed), clipped cattle were cleaner than other cattle.

4.3 Humans - Hosts and Cause

Another important factor in the life cycle of a zoonotic agent is the human being itself. The deterioration of public health and veterinary infrastructures in developing regions by wars and civil wars, overpopulation, new medical techniques like xeno-transplantation, industrialization, changes in food processing, food distribution and the nutritional habits of consumers (consumption of 'raw' animal products such as non-pasteurised milk), increase of international movement of people and of international trade and finally the increased numbers of immunocompromised persons contribute to steadily changing human 'host population' (WHO Fact Sheet 185, 1997). Education in personnel hygiene and in the way to work with animals can reduce the risk of exposure to zoonotic pathogens. Campagnolo et al. (2000) demonstrated that smoking and drinking beverages while working with infected pigs was a risk for humans to get infected with leptospirosis. Consequently, washing hands after work reduced the risk of infection, too. This is also true in pathogenic strains of *E. coli*, like EHEC (Baljer and Wieler, 1999). Often daily work is not brought in connection to an increased risk for the livestock. So parking of feed trucks near the entrance of the change room was one of the risk factors for an infection of broilers with *Salmonella* (Rose et al., 1999). Dealers and their sub-contributors can also significantly amplify the dissemination and spread of *Salmonella* (Wray et al., 1990). The role of veterinarians and other officials like meat inspectors in the spreading of infectious diseases has to be evaluated in the future.

4.4 Environment, Water and Air

The environment in which animals are brought to produce animal goods is often already contaminated. Pastures can be already contaminated with zoonotic agents. This contamination can be caused by grazing livestock, wildlife, sewage water, fertilizer (manure, dung, liquid manure), or, nowadays, sludge. A good example for the contamination of pastures with a zoonotic agent by livestock is *B. anthracis* (see above). Faeces of wildlife birds contain often *Salmonella* which may contaminate the environment. Kohler (1993) found *S. typhimurium*, *S. saint paul* and *S. enteritidis* in depots and animal meal plants. *Salmonella* contaminated water of north German cost regions was discussed to be a potential source of infection for grazing cows (Reuss and Zur Horst, 1975). Fertilizers made from hen faeces may contain *Salmonella* as well as *E. coli*. A storage of the faeces at least twelve weeks outdoors can reduce the number of these bacteria and produce a hygienic product (Schleiff and Dorn, 1997). Sludge can be contaminated with all pathogenic agents which can be found in household, industrial plants, hospitals and farms. Table 4 lists various factors which influence the spread of agents in the soil.

But also livestock which is kept steadily in stables or houses can be exposed to infectious agents. Air conditioning can be connected with poor air quality. In the air of a pullet house Wathes et al. (1991) found bacterial commensals originated from the animal skin but also

Table 4. Parameters influencing the spreading of microorganisms in the soil.

Quality of soil	• Texture • Size and distribution of particles • Structure and proportion of organic material • pH • Size and distribution of pores • Soil density
Chemical-biological factors	• Ionic strength of the various solutions in the soil • pH of infiltrating water • Origin and distribution of the organic mass in biogenic residuals (concentration and amount) • Microorganisms • Tenacity and size of microorganisms
Environmental influences	• Temperature • Ability to absorb water • Water flow rate
Criteria for distribution	• Drying out between seasons • Times of distribution (seasons)

(Weigel, 1995)

Scopulariopsis and *Aspergillus* species. The concentrations of airborne fungi were positively correlated with ammonia concentrations in stables. Madec et al. (1999) proved that a residual contamination can be found on stable floors of pig houses if cleaning and disinfection is neglected. Removal of hidden slurry, in-time starting of damping, and thorough washing but also the correct time of disinfection and correct dosage of the disinfectant are crucial. Rough surfaces have higher contaminations than smooth ones. Rose et al. (1999) showed that the *Salmonella* contamination of the house in which broiler-chickens were kept was significantly related to *Salmonella* contamination of the flock at the end of the rearing period. The tenacity of bacteria in the environment or in food can then lead to infections of animals even after years. *S. enteritidis* can persist in empty poultry houses at least for one year. The bacteria were also found outdoors surviving in small pockets of litter and fan dust but in dust particles swept from the floor and in food troughs (Davies and Wray, 1996). Early work on the survival of *Salmonella* in soil was done by Platz (1981, 1980). Table 5 displays survival times of zoonotic agents which are discussed in this review in rooms, soil, and water. The emission of airborne pathogens from animal producing plants and their influence on the human health will be an interesting scientific challenge in the future.

Salmonella species can also survive in insect vectors if pest control is neglected (Table 6).

Table 5. Survival of zoonotic pathogens in environmental compartments.

Pathogen	Closed rooms	Soil	Water
Salmonella spp.		> 3 a (forest soil) > 50 a (farmland)	>14 d
Campylobacter spp.		> 30 d	
Bacillus anthracis	> 70 a	> 70 a	Iinfinite, half time at 30°C ~ 650 a
Listeria monocytogenes	> 2 a	Infinite	2 a
Brucella abortus	> 50 d	> 100 d	> 90 d

[Strauch (1996), modified]

Table 6. Survival of Salmonella *on or within insects.*

Serovar	Vector	Survival time
S. enteritidis	Cockroaches	17 d
S. typhimurium	Cockroaches	14-16 d
S. enteritidis	Fleas	till death
	on flies	> 24 < 30 h
	within flies	7 d
S. typhimurium	On flies	> 16 > 24 h
	within flies	5 d

[Mitscherlich and Marth (1984), modified]

4.5 Feed

Many feeds are contaminated with pathogens infected already during production or during further proceeding. So, feed meal instead of small pellets was connected to a higher risk for the infection of broilers with *Salmonella* (Rose et al., 1999). *Salmonella* are present on the environment of food meal producing plants and may contaminate sterile produced feed lots (Kohler, 1991). The most well-known example for a food borne disease is the infection with *Listeria monocytogenes* due to spoiled silage (Sanaa et al., 1993). Food from contaminated pastures can lead to new infections in livestock (see above). Drinking water proved to be one of the most important risk factors in the transmission of campylobacteriosis in chicken (see below).

4.6 *Campylobacter* spp. Infection - Risk Factors and Countermeasures

The number of infections with *Campylobacter* spp. has nearly reached the number of *Salmonella* infections in humans in many countries. The disease is mostly associated with poultry. Some recent studies investigated the reasons for the spread of the agent in the flocks. Evans and Sayers (2000) found that effective hygiene barriers (good state of repair of the houses, disinfectant boot dips, a high standard of cleansing and disinfection of the drinking water equipment) reduced the risk for the prevalence of *Campylobacter*. Adequate cleansing and disinfection of the houses prohibited survival of *Campylobacter* in the environment. Rodents were not identified as a source of infection. Despite, other farm animals like cattle, pigs, sheep, and fowl other than poultry, were identified as possible risk factors (Van de Giessen et al., 1998; 1996; Kapperud et al., 1993). In a Norwegian study also the geographic region and the season (autumn) were important risk factors (Kapperud et al., 1993). Spreading could be assumed to occur via the farmer's footwear or by his dirty hands. Countermeasures should therefore concentrate on hygiene measures, including thorough cleaning and disinfection procedures, change or at least dip of footwear at the entrance of each broiler house, control of vermin, and other hygienic precautions. Preventive measures should also include disinfection of drinking water (Van de Giessen et al., 1998; 1996; Kapperud et al., 1993).

References

Anonymous, 2000. Wichtige Infektionskrankheiten in Deutschland Jahresbericht 1999 Teil 3: Importierte Infektionskrankheiten: Brucellose. Epidem. Bul. 29: 233.

Arun, S., Neubauer H., Gürel, A., Ayyildiz, G., Kusu, B., Yesiledere, T., Meyer, H. and Hermans, W., 1999. Equine Glanders in Turkey. Vet. Rec., 144: 255-258

Baljer, G., and Wieler, L.H., 1999. Animals as a source of infections for humans—diseases caused by EHEC. Dtsch. Tierarztl. Wochenschr. 106:339-343.

Campagnolo, E.R., Warwick, M.C., Marx, H.L., Cowart, R.P., Donell, H.D., Bajani, M.D., Bragg, S.L., Esteban, J.E., Alt, D.P., Tappero, J.W., Bolin, C.A. and Ashford, D.A., 2000. Analysis of the 1998 outbreak of leptospirosis in Missouri in humans exposed to infected swine. J. Am. Vet. Med. Assoc. 1;216:676-682.

Damaso, C.R.A., Esposito, J.J., Condit, R. C. and Moussatche, N., 2000. An emergent poxvirus from humans and cattle in Rio de Janeiro State: Cantagalo virus may derive from Brazilian smallpox vaccine. Virology, 277: 439-449.

Davies, M.H., Hadley, P.J., Stosic, P.J. and Webster, S.D., 2000. Production factors that influence the hygienic condition of finished beef cattle. Vet. Rec. 146:179-183.

Davies, R.H. and Wray C., 1996. Persistence of *Salmonella enteritidis* in poultry units and poultry food. Br. Poult. Sci. 37:589-596.

Evans, S.J. and Sayers, A.R., 2000. A longitudinal study of *Campylobacter* infection of broiler flocks in Great Britain. Prev. Vet. Med.10:46(3):209-23.

Hubálek, Z. and Halouza, J., 1999. West Nile fever - a reemerging mosquito-borne viral disease in Europe. Emerg. Infect. Dis. 5:643-50.

Jezek, Z. and Fenner, F., 1988. Human Monkeypox. Karger, Basel, München & New York.

Kapperud, G., Skjerve E., Vik, L., Hauge, K., Lysker, A., Aalmen, I., Ostroff, S.M. and Potter, M., 1993. Epidemiological investigation of risk factors for *Campylobacter* colonization in Norwegian broiler flocks. Epidemiol. Infect. 111:245-255.

Kohler, B., 1993. Example of the concentration of salmonellae in the environment. Dtsch. Tierarztl. Wochenschr. 100:264-274.

Madec, F., Humbert, F., Salvat, G. and Maris, P., 1999. Measurement of the residual contamination of post-weaning facilities for pigs and related risk factors. Zentralbl. Veterinarmed. [B] 46:37-45.

Mitscherlich, E. and Marth, E.H., 1984. Microbial survival in the environment. Springer Berlin, New York & Tokio.

Mukinda, V.B., Mwema, G., Kilindu, M., Heymann, D.L., Khan, A. S. and Esposito, J.J., 1997. Re-emergence of human monkeypox in Zaire in 1996. Monkeypox Epidemiologic Working Group. Lancet 349:1449-50.

Neubauer, H., Finke, E.-J. and Meyer, H., 1997. Human Glanders. International Review of the Armed Forces Medical Services, LXX, 10/11/12: 258-265.

Neubauer, H., Reischl, U., Ropp, S., Wolf, H., Esposito, J.J. and Meyer, H., 1998. Specific detection of monkeypoxvirus by PCR. J. Virol. Methods, 74: 201-207

Philipp, W., 1996. Die Bedeutung kommunaler Rest- und Abfallstoffe bei der Verbreitung bakterieller Zoonoseerreger. In: Vorbeugemaßnahmen bei der Zoonosebekämpfung. DVG, Giessen (ISBN 3-930511-29-0) pp. 68-88.

Platz, S., 1981. Modellversuche zur Persistenz von Salmonellen auf landwirtschaftlichen Nutzflächen. Zbl. Bakt. Hyg.I. Abt. Org. B. 173:452-456.

Platz, S., 1980. Studies on survival of *Salmonella typhimurium* in different types of soils under outdoor climatic conditions. Zbl. Bakt. Hyg.I. Abt. Org. B.. 171:256-68.

Pommepuy, M., Butin, M., Derrien, A., Gourmelon, M., Colwell, R.R. and Cormier, M., 1996. Retention of enteropathogenicity by viable but nonculturable *Escherichia coli* exposed to seawater and sunlight. Appl. Environ. Microbiol. 62:4621-4626.

Rappole, J. H., Derrickson, S. R. and Hubálek, Z., 2000. Migratory birds and spread of West Nile virus in the Western Hemisphere. Emerg. Infect. Dis. 6:319-28.

Rasch, G., Schöneberg, I., Apitzsch, L. and Menzel, U., 1997. Brucellose-Erkrankungen in Deutschland. Bundesgesundheitsbl. 40 (2): 50-54.

Reuss, U. and Zu Horst, B. (1975) Mit Salmonellen kontaminierte Oberflächengewässer als potentielle Infektionsquellen für Weiderinder im nordwestdeutschen Küstengebiet. Forum Umwelthygiene 26, 285-290.

Rose, N., Beaudeau, F., Drouin, P., Toux, J.Y., Rose, V. and Colin, P., 1999. Risk factors for *Salmonella enterica* subsp. *enterica* contamination in French broiler-chicken flocks at the end of the rearing period. Prev. Vet. Med. 39:265-277.

Sanaa, M., Poutrel, B., Menard, J.L. and Serieys, F.J., 1993. Risk factors associated with contamination of raw milk by Listeria monocytogenes in dairy farms. Dairy Sci. 76:2891-2898.

Schleiff, G. and Dorn, W., 1997. Hygienic-bacteriologic evaluation of methods for production of dry poultry feces manure. Zentralbl. Hyg. Umweltmed. 199:475-495.

Schwabe, C.N., 1960. Veterinary medicine and human health. William and Wilkens, Baltimore

SCS Engineers, 1979. Health effects associated with wastewater treatment and disposal systems - state of the art review. US-EPA, Office of Research and Development, Research Friangle Park, N. C. 27711/USA/EPA-600/1 - 79-016a.

Strauch, D., 1996. Hygieneproblematik bei der biologischen Abfallbehandlungen - Einführung und Überblick - Problembereiche aus veterinärmedizinischer Sicht - 53. Informationsgespräch in Delmenhorst/Ganderkesee "Hygieneaspekte bei der biologischen Abfallbehandlung" Schriftenreihe des Arbeitskreises für die Nutzbarmachung von Siedlungsabfällen (ANS) e. V. (ISBN 3-924618-31-3) 32, pp. 25-60,

Teufel, P. and Hammer, Ph., 1999. Welche Zoonosen gibt es? Dtsch. tierärztl. Wschr. 106:311-318.

Turnbull, P.C.B., 1998. Anthrax. In: Zoonoses, Oxford University Press, Oxford, pp. 1-16.

Van de Giessen, A.W., Tilburg, J.J., Ritmeester, W.S. and Van der Plas, J., 1998. Reduction of *Campylobacter* infections in broiler flocks by application of hygiene measures. Epidemiol. Infect. 121:57-66.

Van de Giessen, A.W., Bloemberg, B.P., Ritmeester, W.S. and Tilburg, J.J., 1996. Epidemiological study on risk factors and risk reducing measures for *Campylobacter* infections in Dutch broiler flocks. Epidemiol. Infect. 117:245-250.

Wathes, C.M., Johnson, H.E. and Carpenter, G.A., 1991. Air hygiene in a pullet house: effects of air filtration on aerial pollutants measured in vivo and in vitro. Br. Poult. Sci. 32:31-46.

Weber, A., Reischl, U. and Naumann, 1998. Demonstration of *Mycobacterium africanum* in a bull from North Bavaria. Berl. Munch. Tierarztl. Wochenschr. 111:6-8.
Weigel, T., 1995. Hygienische Untersuchungen zum Verhalten von Bakterien und Viren im Boden. Hohenheimer Umwelttagung, 27 (ISBN 3-9803862-2-8) pp. 195-198.
Work, T.H., Hurlbut, H.S. and Taylor, R.M., 1955. Indigenous wild birds of the Nile Delta as potential West Nile virus circulating reservoirs. Am. J. Trop. Med. Hyg. 4:872-8.
World Health Organization, 1959. Zoonoses: Second report of the Joint WHO/FAO Expert Commitee.
World Health Organization, 2000. Surveillance Programme for Control of Foodborne Infections and Intoxications in Europe, 7th Report.
Wray, C., Todd, N., McLaren, I., Beedell, Y. and Rowe, B., 1990. The epidemiology of *Salmonella* infection of calves: the role of dealers. Epidemiol. Infect. 105:295-305.
Wray, C., Todd, J.N. and Hinton, M., 1987. Epidemiology of *Salmonella typhimurium* infection in calves: excretion of *S. typhimurium* in the faeces of calves in different management systems. Vet. Rec. 121:293-296.
Zhang-Barber, L., Turner, A.K. and Barrow, P.A., 1999. Vaccination for control of *Salmonella* in poultry. Vaccine 17:2538-2545.

Emerging pathogenic bacteria of special interest; epidemiological concerns

Eystein Skjerve
Norwegian School of Veterinary Science, P.O.Box 8146 dep., N-0033 Oslo, Norway; E-mail: Eystein.Skjerve@veths.no

Summary

An evolutionary perspective on foodborne infections is presented. Whilst traditional food supply relied upon traditional effective technologies such as as drying and heat treatment, the modern fresh-food market comprises new ecological niches for foodborne agents. Classic infections in milk and meat have been put under control, but the emergence of new and re-emergence of old enteric bacteria in the food chain represents a major problem for the modern food industry. Major changes in the animal industry linked to a steady increase of biological products represent the evolutionary driving force behind the increasing problems. This is exemplified by the emergence of psychrotrophic *Bacillus cereus*, shiga-toxin producing *E. coli* and new varieties of *Salmonella* spp. Recent problems associated with *M. paratuberculosis* present a dramatic new challenge for animal industry and consumer safety. To face the challenges ahead, a major change in the approach to epidemiology and veterinary public health strategies is called for.

Keywords: emerging diseases, food-borne agents, infections, evolution

1. Introduction

Consulting the literature on this topic is an interesting experience. The term 'emerging' has been extremely popular since it was more or less introduced in the late 1980's. Over the past 15 years a substantial number of scientific and popularised books, scientific papers and other articles have dealt with the problem of (re-) emerging infectious diseases, and a new journal called 'Emerging Infectious Diseases' was introduced. In august 2001 a Medline search yielded 2747 papers including the term "emerging infection(s)". From 47 hits on 'emerging infection' in 1990, the number increases to 301 in 2000. Presently, classic references to the issue are found in Emerging Infectious Diseases, e.g. papers by Morse (Morse, 1995), not to mention the various available web resources from amongst others WHO/FAO, CDC, NIH the EU.

The renewed interest in infectious diseases was largely kindled by the HIV-pandemic, but is also due to the recurrence of tuberculosis in western countries. In addition, as clarified in the following, a range of foodborne diseases are associated with the increasing number of immunocompromised individuals in both affluent and poor societies.

The causal factors behind the (re)emergence of infections include biological, sociological and political factors. Factors often listed [e.g. Morse, 1995; National Center for Infectious Diseases (U.S.), 1994; National Institute of Allergy and Infectious Diseases (U.S.), 1992] include: i) intrusion into tropical forests, ii) inadequate public health infrastructure; iii) ecological changes, iv) climate changes, v) changes in human behaviour, vi) demographic changes, vi) antibiotic misuse, and finally, vii) increased trade.

A number of (re-)emerging diseases of main concern are foodborne. As early as 1992 review papers (e.g. Notermans and Hoogenboom-Verdegaal, 1992) listed a number of emerging foodborne problems (*Campylobacter* spp., *S. enteritidis*, pathogenic *E. coli*, *Listeria monocytogenes*, *Aeromonas* spp., *Yersinia enterocolitica* and moulds) and assumed changes in food- production and consumption - in conjunction with increasing numbers of ageing and immunocompromised individuals - were reponsible. Altekruse et al. (1997) discuss the rapidly changing epidemiological pattern of foodborne infections, as suggested by well-recognized pathogens such as *Salmonella* producing distinct new varieties as *S. typhimurium* DT 104 and egg-borne *S. enteritidis*, and microorganisms hitherto considered harmless (e.g. *Campylobacter* spp. and *E. coli* 0157: H7) starting to cause major concern. These authors suggested a set of factors behind this change, i.e.: i) human demographics, ii) human behaviour, iii) changes in industry and technology, iv) changes in travel and commerce, v) microbial adaptation, vi) economic development and land use and, finally, vii) breakdown of public health infrastructure. These factors are not much different from those listed by Morse (1995), but obviously some of the foodborne diseases are special.

During 1990's *E. coli* 0157:H7 caused disquiet in the US cattle industry, whilst BSE caused next to a disaster for the European cattle industry. Yet, of even more public health concern was the steadily increasing incidence of enteric foodborne infections linked to *Salmonella, Campylobacter* and what one now refers to as verotoxin (VTEC) or shiga-toxin (STEC) producing *E. coli*.

A range of foodborne diseases are waterborne or associated to the consumption of vegetables or sea-foods. However, this contribution will mainly concentrate on the diseases of 'traditional' veterinary interest, i.e those transferred to man by meat, milk and their products. Numerous authors have listed the foodborne agents of concern in this context. Rather than reiterating the views expressed in such reviews, this contribution aims to approach this topic in a slightly different way.

Originally, descriptions of emerging infections were primarily inspired by ecological considerations, whereas recent years have seen the publication of a range of classic papers on specific micro-organisms and discussed from a less ecological perspective. To a certain extent, we should get away from focusing on specific pathogens, and try to get a broader view of the situation. Following this approach we might become more familiar with emerging disease patterns rather than be concerned with specific agents. Of course agents cause infectious diseases, but often diseases caused by different agents are the result of the same mechanisms of infection. I would suggest that public health scientists should have the courage to point out the obvious reasons for the increased incidence of foodborne infections we witness today.

2. An Evolutionary Process

The emerging pattern of foodborne disease in man reflects the situation in the complex microbial system represented by our food chain(s). In a very interesting paper by Lederberg (1997), the author takes an evolutionary perspective of emerging infections, which seems to be an appropriate approach towards understanding the processes behind emerging infections. As appropriate when discussing an evolutionary process, one must discuss the emerging *patterns* of infections than the infections *per se*. Of the factors discussed by Lederberg numerous relate to microbial behaviour and -interaction. The unique *intraclonal* variation among microbes mixed with *interclonal* exchange of genes give microbes a strategic upper hand if we allow them to take the lead. Adding to this, microbe/host co-evolution tends to stabilize the relationship over time.

We must, in discussing zoonotic transfer, also realise that the majority of the immense number of cross-species transfer simply does not function. Hundreds, perhaps thousands of genes have to act in concert to enable a pathogen to evolve into a new species. On the other hand - as pointed out in worst-case scenario's - the situation is different if agents are introduced through xenotransplants (Onions and Witt, 2000) rather than by 'natural' means.

Our food production systems are subject to significant changes. Major elements describing the emergence of new productions systems include: i) industrialization of agriculture with concurrently animal-/herd densities increasing, ii) peri-urban husbandry taking over in many poor countries, iii) volume of global (long distance) trade on foods and animal feeds increasing, iv) consumer changes creating a larger demand for 'fresh' foods and 'designer products', and, finally v) the adoption of GMO's introducing a new dimension to food safety.

Thus, in any food safety initiative one must start by considering some basic information about the food production chain itself. Most countries have in the past decades established an industrialized production systems, with large herds in high-density populations. These herds also tend to cluster in particular geographical regions. Moreover, animal feeds are increasingly produced and distributed on a large scale, and most countries rely on large quantities of feed antibiotics for limiting animal disease and promoting animal growth. *Per saldo* we have hence created a perfect environment for many infectious diseases. The seriousness of the outbreaks of Swine Fever in Europe exemplifies the ramifications of these developments, just as the BSE epidemic can be seen as a consequence of the production systems we have come to rely upon. In short, modern animal husbandry may have been able to control many serious contagious diseases of animals (often of viral nature), but in the process has created an ideal environment for propagation of enteric pathogens.

By and large, the microbial 'gene soup' contained in our foods has emerged as the result of an evolutionary process, the selective pressure we impose deciding the direction of development. Historically, one has focused on establishing cause-effect relationships between husbandry management practices and occurrence of pathogens or genes in our food. The ongoing discussion on the consequences of using feed antibiotics as growth promoters is a typical example. Within a simple cause-effect framework it is not easy to prove that there is a relationship between antibiotic resistance in animals and man. However, from an evolutionary perspective, the connection is obvious and therefore food microbiologists are well-advised to reconsider Darwin's theorema, with a view to better grasp the real biological processes behind emerging patterns of disease. Following this concept I will in the next section first discuss classic infections and subsequently address more recent ('emerging') problems. Within the framework of this contribution the focus will be exclusvely on bacteria.

A word of warning is timely here. In an evolutionary framework, individual determinants may be hard to identify. Instead, a matrix of factors pushing a microbial system in one ore another direction must be discussed. Also, according to the modern system theory approach (Bar-Yam, 1997) the possibilities of predicting long-term consequences may be very low. When causal relations turn non-linear, which is the case in complex biological systems, such complicates epidemiological research (Skjerve, 2000).

3. The Classics: Milk and Meat

The 'classic' veterinary public health approach was one towards controlling a number of parasitic meatborne infections. The next stage (started between the 1st and 2nd World war) included a major effort to control important meat- and milkborne zoonotic bacteria such as *Mycobacterium bovis* and *Brucella* spp. Although useful as a diagnostic station for purposes of identifying dangers for the consumer, the ultimate success was primarily related to an abrupt interruption of the bacterial cycle by introducing pasteurisation of milk. Besides water sanitation, pasteurisation remains one of the major public health improvements during the 20th Century. In this light, it remains a mystery why some people in developed countries seem to discard this knowledge and insist on the consumption of unpasteurised milk. As the prevalence of *M. bovis* and *Brucella* is very low, the risk associated with these agents is not very high. However, major outbreaks of group streptococci caused by unpasteurised milk should be a warning (Benenson, 1995).

At the same time we have observed other emerging problems linked to milk consumption. A range of outbreaks of STEC (Holsinger et al., 1997) caused by farm milk has regularly affected farm visitors, often children (Benenson, 1995). Thus, in most countries a farmer should be careful to invite visitors to drink fresh milk.

With the use of drastic measures as pasteurisation, most routes of infections from animals to man through milk and milk products were broken. Relying on pasteurisation, the milk delivery system has become a simple, linear system where effective actions may be taken, the remaining issues of concern being linked to consumption of unpasteurised milk. However, microbial adaptation to cold chains has caused *Bacillus cereus* and *Listeria monocytogenes* to emerge as microorganisms of concern (Andersson et al., 1995; Holsinger et al., 1997; Holsinger et al., 1997; Notermans and Hoogenboom-Verdegaal, 1992). Obviously, the use of cold storage has created ideal opportunities for such psychrotrophic agents to proliferate in milk and dairy products.

The meat industry has a less impressive history of success. After having kept some of the major parasitic zoonoses in check, one has witnessed a range of enteric bacteria proliferating in the meat production chain. Meatborne bacterial diseases have steadily increased over the last half of the 20th century (Foster, 1997; Notermans and Hoogenboom-Verdegaal, 1992). Through an increased trade in animal feed and an intensification of animal husbandry, use of antibiotic growth promoters, an ideal environment for enteric bacteria was created. In contrast to the milk industry, no effective 'safety valve' was identified. Unless we manage to turn the evolutionary tide, only draconic measures such as irradiation (Gould, 2000; Tauxe, 2001) may provide some sense of security.

4. Emerging Patterns of Disease

Traditionally, outbreaks of foodborne diseases have been associated with a lack of hygiene or improper kitchen practices, during such events as church suppers and dinner parties. Agents as *Bacillus cereus*, *Staphylococcus aureus* or *Clostridium perfringens* have often been involved. Often both the dose and incidence were high, the outbreak local and easy to trace. Foods incriminated were generally cooked or processed. Using previous terms, this would again be a simple system with easily traceable causal links (Skjerve, 2000).

While outbreaks of the traditional type still exist, many foodborne outbreaks have recently been reported, which were caused by small doses and often affecting special risk groups ('YOPI': young, elderly, pregnant, immuno-compromised). Further, today more outbreaks are caused by fresh products of animal or plant origin.

As long as currently practised animal husbandry systems prevail, it is likely that the main problems will remain associated with agents such as shigatoxin-producing *E. coli, Salmonella, Campylobacter* and antibiotic resistance factors. New agents of bacterial or other origin will emerge, but we will increasingly have to turn our focus on genes rather than on bacterial 'species'. Horizontal transfer of genes between closely or even distantly related microbial agents determines the establishment and spread of specific pathogenicity, virulence or antibiotic resistance patterns (de la Cruz and Davies, 2000). The disease pattern in man will be related to the soup of genes occurring in domestic animals and other sources. Some simple and more complex examples are dealt with in the following.

4.1 Long term selection - *Bacillus cereus*

The story behind the emergence of psychrotrophic *B. cereus* strains is an interesting one. First described in some detail more than 10 years ago (van Netten et al., 1990), the discussion still circles around origin and timepoint of emergence. Irrespective of this discussion, *B. cereus* is difficult to control in the dairy industry (Andersson et al., 1995). It seems impossible to completely avoid its presence in all milk samples, its spores being hydrophobic, and therefore primarily attaching to the surfaces of the pipeline system im dairy processing operations where they might multiply and re-sporulate (Andersson et al., 1995). Pasteurisation heating is insufficient to kill the spores, while competition from other vegetative bacteria is eliminated. It seems that several *B. cereus* strains have become psychrotrophic over the years, allowing growth at temperatures as low as 4-6 degrees C (Granum, 1994). None of the methods sofar used to control hygiene in the dairy industry are able to keep *B. cereus* in check.

Bacillus spp. may survive for long periods as spores. As such, the genes of such bacteria will evolve more slowly than e.g. enteric bacteria. Still, there are indications for some gene transfer between various *Bacillus* types. Most likely, the selection of psychrotrophic *B. cereus* has taken a long time to evolve. However, the large scale of dairy processing units over decades has allowed certain varieties of *B. cereus* (or related bacteria) to pick up genes from other bacteria or modify its own genes and turn into a major food-borne hazard. For *B. cereus* there are no strong indications for any clonal spread, and the psychrotrophic strains will have a range of different origins. From an evolutionary point-of-view, the emergence of the psychrotrophic strains is a long-term consequence of a massive selection pressure through a continuous cold chain.

More worrying is the recently found close relationship between *B.cereus, B. anthracis* and *B. thuringensis* (Henderson et al., 1995). If *B. thuringensis* sprayed on crops turns into enterotoxin producing *B. cereus* when heated trouble is ahead.

4.2 Clonal spread I: *E. coli* 0157:H7

While stable spore-forming organisms may need a more continuous selection pressure to change their genomes, enteric bacteria such as *Campylobacter, E. coli* and *Salmonella* could be described as having a 'plastic' genome, i.e. they may integrate genes they need from the 'gene soup' of ther environment. For extremely labile organisms like *Campylobacter*, the main pattern seems to be variation and clone diffusion. However, the whole genome has now been described (Wren et al., 2001), and some aspects seem to be constant also for these bacteria (Pawelec et al., 2000; Tu et al., 2001).

While hypervariable bacteria may tend to remain heterogeneous, the emergence of shigatoxin producing *E. coli* (STEC) as *E. coli* 0157:H7 seems to be examples of an agent having identified an evolutionary niche where one or a few clones will remain stable for some time.

The 0157:H7 clonal group is only distantly related to other STEC strains belonging to other serotypes of *E. coli* (Whittam et al., 1988). There are strong indications that isolates of *E. coli* 0157:H7 obtained from geographically separate outbreaks and sporadic cases of hemorrhagic colitis and haemolytic uremic syndrome belong to a pathogenic clone that occurs throughout North America. For Europe the picture is more complex, as a range of other STEC's are also important, and some areas still have a very low level of 0157: H7 in cattle (Vold et al., 1998) as well as very low incidence of human infections.

An explanation for the emergence of 0157:H7 has been a possible hypermutation in a part of the genome of *E. coli*, but it has been difficult to support this view (Whittam et al., 1998). What is clear is that *E. coli* 0157:H7 is found in most cattle populations around the world, at different prevalences. Reviews (Meyer-Broseta et al., 2001) describe a heterogeneous picture, but strongly supporting the importance of cattle as the main carrier of 0157: H7.

The US were the first to report major problems with 0157: H7, and the organism probably originates from the large cattle herds of the US, with a mixture of animal purchase, long-distance feed transport and intense faecal-oral contact within many herds. As such, the problem is a direct consequence of the way the cattle industry is managed. As pointed out by Armstrong et al. (1996), the determinants behind the emergence of 0157:H7 remains elusive, but points to the simple fact that large units represent better possibilities for emergence of new varieties.

Using an evolutionary approach, 0157:H7 may have reached a local optimum in the cattle population in the US, then possibly spread all around the world, mixed with other bacteria and spread some of the genes in the form of pathogenicity islands to other *E. coli* as well (Perna et al., 1998). This may explain the more heterogeneous picture on the European continent, while areas like Scotland seem to represent an US-like situation.

In summary, on a large scale, intensive cattle industry linked to large distance trade in live animals and beef seems to present a situation where the STEC problem will remain. IF the STEC group only consisted of 0157:H7 in a specific cattle niche, we might be able to control the spread, but with the tendency for gene diffusion, this option is no longer feasible.

4.3 Clonal Spread II: *S. enteritidis* PT4/PT

While 0157:H7 established in an ecological niche and tended to spread and distribute their genes, the establishment of certain varieties of *S. enteritidis* in egg production presented another picture. When reports came in on an increase in the number of egg-borne *Salmonella* infections after 1985 (Rodrique et al., 1990), it was soon clear that *S. enteritidis* was on the increase on at least the continents of North America, South America, and Europe, and possibly Africa. Investigations soon revealed eggs to be the source and specific types of *S. enteritidis* found (PT4). Later also another phage type (PT6) (Evans et al., 1998) was associated with eggs.

There is nothing new about *Salmonella* colonisation of eggs. However, PT4 has now established as a vertical infection in the major egg industries of the world. How could this happen?

First PT4 seems to be specially adapted to ovaries of hens and may remain there for a long time. There are no important competing bacteria around, and PT4 seems to thrive, is transferred to the inside of the egg, and will subsequently grow in the egg if the egg is stored for too long time. Molecular analysis (Stanley et al., 1992) soon showed that a range of U.K. isolates of the epidemic phage type PT4 all belonged to a single clonal line, and that evolutionary lines or races exist within pandemic isolates associated with human gastroenteritis. Recent analyses of PT1, PT4 and PT6 (Laconcha et al., 2000) showed that strains from the different countries shared the same PFGE combination profile, confirming the close relationship among these phage types and the protracted spread of a single clone over a large geographical area. An interesting theory is the suggestion that *S. enteritidis* got an environmental niche when *S. gallinarum* was eradicated from most poultry industries (Rabsch et al., 2000). Possibly there was an evolutionary optimum for *S. enteritidis* settling in ovaries of hens. The following explosive spread was of course linked to large-scale trade on live chicken and eggs.

As recent results suggest, also PT4 seems to have spread their genes into a number of other phage types, but the evolutionary fixed PT4 is still around presenting major health problems in many countries. As the spread of genes is less pronounced than for 0157:H7, it should however be easier to fight PT4. There are also some promising results in reducing the prevalence of infection in eggs. It remains to be seen if it is possible to control PT4 in an international egg/chicken market.

4.4 Multiple reservoirs: *Salmonella typhimurium*

While clonal selection and then clonal spread and diffusion were the case for 0157:H7 and PT4, certain salmonellas represent another picture. *Salmonella typhimurium*, found all around the globe, remains one of the most important *Salmonella* types in man and can be found in a range of reservoirs. *S. typhimurium* has been associated with a range of domestic animals, outbreaks has also been associated with specific types of *S. typhimurium* from environmental sources. Norway represents an interesting example, where the Copenhagen variety of *S. typhimurium* was established in wild birds around 1960. This variety remains the most important source of domestic *Salmonella* infection in Norway, with infections through contact with drinking water and contact with wild birds as the main risk factors (Kapperud et al., 1998). A major outbreak was caused by contaminated chocolate in 1987 (Kapperud et al., 1990), and recent findings of infections caused by the same variety in hedgehogs indicate a persistent occurrence of a specific *S. typhimurium* type.

S. typhimurium seems to be very heterogeneous, with a range of different types able to cross barriers from animals and environment to man. However, the persistent colonisation of specific ecological niches seems to follow the same pattern as for other bacteria – if a local evolutionary optimum is reached the genome may remain rather stable for some time. The occurrence of multi-resistant *S. typhimurium* DT 104 seems to confirm this.

4.5 Horizontal gene transfer - Antibiotic resistance

The epidemiology of antibiotic resistance genes in epidemic multi-resistant *S. typhimurium* DT 104 of human and animal origin seems to be spread by the same inserted gene cassettes, irrespective of source and country of origin, supporting the suggestion of the spread of an epidemic clone (Ridley and Threlfall, 1998; Threlfall et al., 1994). It is not surprising that this would happen, with the massive use of antibiotics in the animal industry. The discussion about the relative importance of antibiotic growth promoters in transferring resistant strains or genes to man still seems pending. However, from an evolutionary perspective, this is a non-existing discussion. The use of low-dosage antibiotics in animals represents a massive selection pressure for emergence antibiotic resistant bacteria. The vancomycin- resistant enterococci (van den Braak et al., 1998) is another example, flumequine-resistant *Campylobacter* yet another (Koenraad et al., 1995).

Without going into further discussion on details, the remedy against the problem is obvious: the amount of antibiotics used in the animal industry should be reduced to a minimum.

4.6 The new - *M. paratuberculosis* ??

While a range of emerging foodborne infections has been revealed the last years, some cases are still unclear. One of them is a possible connection between the occurrence of *Mycobacterium paratuberculosis* in foods and Crohn's disease in man. In a recent review Chamberlin et al. (2001) claims that the evidence supporting *M. paratuberculosis* infection as a cause of Crohn's disease is mounting rapidly. The data implicating Crohn's as an autoimmune disorder may be viewed in a manner that supports the mycobacterial theory. The mycobacterial theory and the autoimmune theory are complementary; the first deals with the aetiology of the disorder, the second deals with its pathogenesis. Still, the possible causal routes may be difficult to trace completely. The survival of *M. paratuberculosis* during pasteurisation is still not fully clear (Grant, 1998; Keswani and Frank, 1998).

We must be prepared for a full-scale crisis in our dairy and meat industries if (when?) the causal links have become more clear. This may be a crisis to which even the BSE crisis will compare as relatively "happy days". With the high prevalence of infection in domestic animals, we face a tremendous challenge if we have to establish means to prevent the robust Mycobacteria to reach man through milk or meats.

5. Conclusions

5.1 Major Challenges in Epidemiology

Many theories about the epidemiology of foodborne agents are dominated by case reports from outbreaks. Often, the more important sporadic cases follow other routes than more spectacular outbreaks do. A major shift from traditional epidemiological methods to infectious dynamics is needed to fully comprehend the complex pattern of many foodborne infections. Standard case-control studies have been valuable in describing outbreaks and also some sporadic infections. The major limitations for case-control studies are the simple fact that patterns of foodborne diseases are continuously changing. We have to take up the challenge of studying complex food supply systems, using models form infectious disease modelling and ecology. Further, we must be aware that the causal mechanisms indicated by epidemiological studies may be biased and only temporarily correct.

Scientists have always been trained in explaining possible causal mechanisms. In public health we are more interested in intervention than in explanation, and we have to substantiate the possible beneficial effect of intervention into the food production or market. It is just too easy to forget this major philosophical step from explaining to predicting. Explanatory theories are not necessarily valid for intervention. Many epidemiological studies have failed to forward an understanding that intervention on different levels may not be supported by the same kind of study.

5.2 WTO, Trade and Risk assessment

Risk assessment represents the most expressed jump from using explanatory studies to present mathematical calculations to describe risk of future events to happen. Risk assessment has, after the establishment of the WTO, emerged as the main tool to minimize health hazards linked to food trade and production. Sofar, almost all risk assessments in our sector has been under the ordinary 'HACCP umbrella', and we await its use in regulating trade also based upon risks for spreading microbial diseases between developed countries. Risk assessments can help us in observing and interpreting food production chains in a systematic way. However, in an evolutionary framework, the models for risk assessment used sofar may be inappropriate to analyse the systems themselves. My prediction is that risk assessment will be used in simple regulatory systems to assess the impact of trade, effect of importing systems, but that we have to establish a much more comprehensive theory to understand and explain our food production systems within an evolutionary framework. Modern system theory as used in studies of ecological systems will be a valuable tool.

Up to now we have concentrated on short term technological fixes for adjusting production systems in an effort to minimize the hazard of introducing pathogens or non-wanted genes into the food chain. Hope that "rapid methods" should solve the problem has evaporated, but other short-term remedies are popping up, the last of them being risk assessment. Based on an evolutionary framework, there are several approaches. We can decide to limit the introduction of pathogens and genes in new populations by restricting trade on live animals or feeds or imposing heat treatment on animal feed. This option seems not to be very popular today with free trade as the main aim of development. Further, we may adjust the structure of the animal production to limit the animal and herd densities. In short – limit the extent of faecal-oral contact between animals and between herds. Strict slaughter hygiene and an integrated chain control will supplement this work to minimize the faecal-oral contact between animals and man. Then the final challenge remains. There is a strong support for the view that biological production needs biological borders to be able to survive for longer periods. By shortcutting biological borders with transporting live animals, fresh greens and seafood across the world, we tend to homogenize the global food supply into a monoculture. Long-term consequences of monocultures may be unpredictable, but if we give foodborne microbes the upper hand, it could be best not to be around when the next scandal hits.

References

Altekruse, S.F., Cohen, M.L., Swerdlow, D.L., 1997. Emerging foodborne diseases. Emerg. Infect. Dis. 3, 285-293.

Andersson, A., Ronner, U., Granum, P.E., 1995. What problems does the food industry have with the spore-forming pathogens *Bacillus cereus* and *Clostridium perfringens*? Int. J. Food. Microbiol. 28, 145-155.

Armstrong, G.L., Hollingsworth, J., Morris, J.G., Jr., 1996. Emerging foodborne pathogens: *Escherichia coli* O157:H7 as a model of entry of a new pathogen into the food supply of the developed world. Epidemiol. Rev. 18, 29-51.

Bar-Yam, Y., 1997. Dynamics of complex systems. Reading, MA, Addison-Wesley, 848 p.

Benenson, A.S., (Ed.), 1995. Control of Communicable Diseases in Man, v. 1: Washington DC, American Public Health Association, 532 p.

Chamberlin, W., Graham, D.Y., Hulten, K., El-Zimaity, H.M., Schwartz, M.R., Naser, S., Shafran, I., et al., 2001. Review article: *Mycobacterium avium* subsp. *paratuberculosis* as one cause of Crohn's disease. Aliment. Pharmacol. Ther. 15, 337-346.

Cruz, I. de la, Davies, I., 2000. Horizontal gene transfer and the origin of species: lessons from bacteria. Trends. Microbiol. 8, 128-133.

Evans, M.R., Lane, W., Ribeiro, C.D., 1998. *Salmonella enteritidis* PT6: another egg-associated salmonellosis? Emerg. Infect. Dis. 4, 667-669.

Foster, E.M., 1997. Historical overview of key issues in food safety. Emerg. Infect. Dis. 3, 481-482.

Gould, G.W., 2000. Preservation: past, present and future. Br. Med. Bull. 56, 84-96.

Grant, I.R., 1998. Does *Mycobacterium paratuberculosis* survive current pasteurization conditions? Appl. Environ. Microbiol. 64, 2760-2761.

Granum, P.E., 1994. *Bacillus cereus* and its toxins. Soc. Appl. Bacteriol. Symp. Ser. 23, 61S-66S.

Henderson, I., Yu, D., Turnbull, P.C., 1995. Differentiation of *Bacillus anthracis* and other '*Bacillus cereus* group' bacteria using IS231-derived sequences. FEMS Microbiol. Lett. 128, 113-118.

Holsinger, V.H., Rajkowski, K.T., Stabel, J.R., 1997. Milk pasteurisation and safety: a brief history and update. Rev. Sci. Tech. 16, 441-451.

Kapperud, G., Gustavsen, S., Hellesnes, I., Hansen, A.H., Lassen, J., Hirn, J., Jahkola, M., et al., 1990. Outbreak of *Salmonella typhimurium* infection traced to contaminated chocolate and caused by a strain lacking the 60-megadalton virulence plasmid. J. Clin. Microbiol. 28, 2597-2601.

Kapperud, G., Stenwig, H., Lassen, J., 1998. Epidemiology of *Salmonella typhimurium* O:4-12 infection in Norway: evidence of transmission from an avian wildlife reservoir. Am. J. Epidemiol. 147, 774-782.

Keswani, J., Frank, J.F., 1998. Thermal inactivation of *Mycobacterium paratuberculosis* in milk. J. Food. Prot. 61, 974-978.

Koenraad, P.M., Jacobs-Reitsma, W.F., Van der Laan, T., Beumer, R.R., Rombouts, F.M., 1995. Antibiotic susceptibility of *Campylobacter* isolates from sewage and poultry abattoir drain water. Epidemiol. Infect. 115, 475-483.

Laconcha, I., Baggesen, D.L., Rementeria, A., Garaizar, J., 2000. Genotypic characterisation by PFGE of *Salmonella enterica* serotype Enteritidis phage types 1, 4, 6, and 8 isolated from animal and human sources in three European countries. Vet. Microbiol. 75, 155-165.

Lederberg, J., 1997. Infectious disease as an evolutionary paradigm. Emerg. Infect. Dis. 3, 417-423.

Meyer-Broseta, S., Bastian, S.N., Arne, P.D., Cerf, O., Sanaa, M., 2001. Review of epidemiological surveys on the prevalence of contamination of healthy cattle with *Escherichia coli* serogroup O157:H7. Int. J. Hyg. Environ. Health. 203, 347-361.

Morse, S.S., 1995. Factors in the emergence of infectious diseases. Emerg. Infect. Dis. 1, 7-15.

National Center for Infectious Diseases (U.S.), 1994. Addressing emerging infectious disease threats : a prevention strategy for the United States, Atlanta Ga., Centers for Disease Control and Prevention.

National Institute of Allergy and Infectious Diseases (U.S.), 1992. Emerging infectious diseases research : meeting the challenge. Bethesda, Md., U.S. Dept. of Health and Human Services Public Health Service National Institutes of Health, 8 p.

Notermans, S., Hoogenboom-Verdegaal, A., 1992. Existing and emerging foodborne diseases. Int. J. Food. Microbiol. 15, 197-205.

Onions, D.E., Witt, C.J., 2000. Xenotransplantation: an overview of microbiological risks and potentials for risk management. Rev. Sci. Tech. 19, 289-301.

Pawelec, D.P., Korsak, D., Wyszynska, A.K., Rozynek, E., Popowski, J., Jagusztyn-Krynicka, E.K., 2000. Genetic diversity of the *Campylobacter* genes coding immunodominant proteins. FEMS Microbiol. Lett. 185, 43-49.

Perna, N.T., Mayhew, G.F., Posfai, G., Elliott, S., Donnenberg, M.S., Kaper, J.B., Blattner, F.R., 1998. Molecular evolution of a pathogenicity island from enterohemorrhagic *Escherichia coli* 0157:H7. Infect. Immun. 66, 3810-3817.

Rabsch, W., Hargis, B.M., Tsolis, R.M., Kingsley, R.A., Hinz, K.H., Tschape, H., Baumler, A.J., 2000. Competitive exclusion of *Salmonella enteritidis* by *Salmonella gallinarum* in poultry. Emerg. Infect. Dis. 6, 443-448.

Ridley, A., Threlfall, E.J., 1998. Molecular epidemiology of antibiotic resistance genes in multiresistant epidemic Salmonella typhimurium DT 104. Microb Drug Resist 4, 113-118.

Rodrigue, D.C., Tauxe, R.V., Rowe, B., 1990. International increase in *Salmonella enteritidis*: a new pandemic? Epidemiol. Infect. 105, 21-27.

Skjerve, E., 2000, Causal relations in epidemiological research. Consequences of evaluating degrees of complexity and non-linearity. International Society of Veterinary Epidemiology and Economics.

Stanley, J., Goldsworthy, M., Threlfall, E.J., 1992. Molecular phylogenetic typing of pandemic isolates of *Salmonella enteritidis*. FEMS Microbiol. Lett. 69, 153-160.

Tauxe, R.V., 2001. Food safety and irradiation: protecting the public from foodborne infections. Emerg. Infect. Dis. 7, 516-521.

Threlfall, E.J., Frost, J.A., Ward, L.R., Rowe, B., 1994. Epidemic in cattle and humans of *Salmonella typhimurium* DT 104 with chromosomally integrated multiple drug resistance. Vet. Rec. 134, 577.

Tu, Z.C., Dewhirst, F.E., Blaser, M.J., 2001. Evidence that the *Campylobacter fetus* sap locus is an ancient genomic constituent with origins before mammals and reptiles diverged. Infect. Immun. 69, 2237-2244.

van den Braak, N., van Belkum, A., van Keulen, M., Vliegenthart, J., Verbrugh, H.A., Endtz, H.P., 1998. Molecular characterization of vancomycin-resistant enterococci from hospitalized patients and poultry products in The Netherlands. J Clin Microbiol 36, 1927-1932.

van Netten, P., van De Moosdijk, A., van Hoensel, P., Mossel, D.A., Perales, I., 1990. Psychrotrophic strains of *Bacillus cereus* producing enterotoxin. J. Appl. Bacteriol. 69, 73-79.

Vold, L., Klungseth Johansen, B., Kruse, H., Skjerve, E., Wasteson, Y., 1998. Occurrence of shigatoxinogenic *Escherichia coli* 0157 in Norwegian cattle herds. Epidemiol. Infect. 120, 21-28.

Whittam, T.S., Reid, S.D., Selander, R.K., 1998. Mutators and long-term molecular evolution of pathogenic *Escherichia coli* 0157:H7. Emerg. Infect. Dis. 4, 615-617.

Whittam, T.S., Wachsmuth, I.K., Wilson, R.A., 1988. Genetic evidence of clonal descent of *Escherichia coli* 0157:H7 associated with hemorrhagic colitis and hemolytic uremic syndrome. J. Infect. Dis. 157, 1124-1133.

Wren, B.W., Linton, D., Dorrell, N., Karlyshev, A.V., 2001. Post genome analysis of *Campylobacter jejuni*. Symp. Ser. Soc. Appl. Microbiol. 30, 36S-44S.

Veterinary drug use in farm animal production and the antibiotic resistance problem

Frank Møller Aarestrup
Danish Veterinary Laboratory, Bülowsvej 27, DK-1790 Copenhagen V, Denmark; E-mail: faa@svs.dk

Summary

In the production of food animals antimicrobial agents are used for therapy and prophylactics of bacterial infections and when added to feed to promote growth. The use of antimicrobial agents for food animals may cause problems in that they select for resistance among bacteria pathogenic for animals or humans (zoonotic bacteria). The emergence of resistant bacteria and -genes promoted by the use of antimicrobial agents is well-documented and, generally, a close association exists usage of- and resistance against a certain drug. Among the various European countries the dosage of antimicrobial agents used to produce the same amount of meat varies greatly. This leaves room for considerable reductions in antimicrobial usage in at least some countries. In Denmark their application for food animals has been significantly reduced, through banning usage for purposes of growth promotion and by restricting the usage for therapy. In general, resistance rates have decreased following the reduction in usage. However, we do not yet know enough about the occurrence of antimicrobial resistance in food animals, the quantitative impact of the use of different antimicrobial agents on selection for resistance, and about the most appropriate treatment regimes for limiting the development of resistance. Programmes monitoring the occurrence and development of resistance and consumption of antimicrobial agents are needed, as is research into the most appropriate application modes in veterinary medicine. Guidelines for prudent use of antimicrobial agents may extend their life span. Such guidelines should be based on knowledge regarding the normal susceptibility patterns of the causative agents and should consider potential problems for human health.

Key words: antimicrobial agents, resistance, food animals, monitoring, veterinary antibiotic policy

1. Introduction

The first who observed the inhibitory effect of *Penicillium* mould on bacteria have reportedly been Sir John Burden-Sanderson in 1871 and Joseph Lister in 1872 (Fraser-Moodie, 1971; MacFarlane, 1984). In 1928, Alexander Fleming made similar observations (Fleming, 1929) and with the identification of a method to purify penicillin (Abraham et al., 1941), the use of penicillin as a therapeutic was established. Since then a large number of other antimicrobial agents have been discovered and introduced to become the key stone in the therapy of bacterial infections in humans and animals.

The discovery of antimicrobial agents in human clinical medicine and animal husbandry has been one of the most significant achievements of the 20th century. The first antimicrobial agents were introduced in the 1930's, and a large number of new compounds was discovered in the following decades. Unfortunately, resistance against the various compounds has invariably emerged, in all reported cases shortly after their introduction (Levy 1982). Over the years bacteria have acquired resistance to all known antimicrobial agents (Jacoby & Archer, 1991; Skolnick, 1991).

Modern food animal production depends on the use of large amounts of antibiotics for disease control. However, over the past few decades awareness has grown that this application creates favourable conditions for selection, spread and persistence of antimicrobial-resistant bacteria capable of causing infections in animals and humans. It has thus become clear that antimicrobial resistance poses a threat to public and animal health and is a reason for serious concern.

The following overview deals with the use of antimicrobial agents for food animals, the subsequent selection of antimicrobial resistance, its spread from food animals to humans and the potential preventive interventions. Special attention is given to data and examples from Denmark.

2. Consumption of Antimicrobial Agents for Food Animals

In modern food animal production antimicrobial agents are normally used in one of four different ways, i.e. i) *therapy*: treatment of infections in clinically affected animals, preferably based on a bacteriological diagnosis, ii) *metaphylactics*: treatment of clinically healthy animals belonging to the same flock or pen as animals showing clinical signs; In this way infections may be treated before they become clinically apparent and the entire treatment period may thereby be shortened. In fact, in view of the modern productions systems, such may often be the only effective approach to treat for instance large broiler flocks through water medication, iii) *prophylactics*: treatment of healthy animals in a period of stress (e.g. early weaning) to prevent disease; in such cases the use of antimicrobials is indicative for general management problems, and hence in most countries is either illegal or considered imprudent; and, finally, iv) *growth promotion*: the continuous inclusion of antimicrobials in animal feed to prevent (subclinical) infections and hence promote growth; such usage is under serious recent debate.

It is difficult to obtain good information about the consumption of antimicrobial agents for medical and growth promoting purposes. Exact figures are scarce and estimates are only available for a few countries. In the United States the consumption of antimicrobial agents increased tremendously from 1950 to 1978. In 1951, a total of 110 tonnes were produced for addition to animal feed and for other applications, whereas 580 tonnes were produced for purely medical use in humans and animals (Black 1984). In 1978, 5,580 tonnes were produced as feed additives, 6,080 tonnes for medical use in humans and animals: a 50- and 10-fold increase, respectively. British estimates on the 1996 situation mention approximately 650 tonnes antimicrobials for animal therapy, and 100 tonnes for growth promotion (Harvey and Mason 1998). For humans approximately 470 tonnes were used in 1997 (Harvey and Mason 1998).

The pharmaceutical industry has estimated the world market for animal health products at 11 billion ECU in 1995. Therapeutic pharmaceuticals accounting for 44% (4.8 bill.) and feed additives for 41% (4.5 bill.) (FEDESA, 1997). Figures were not given for growth promoters and it is not known whether therapeutic pharmaceuticals include agents that are used for both therapy and growth promotion.

In a recent report the European Agency for the Evaluation of Medical Products (EMEA, 1999) estimated the amounts of antimicrobial agents used in 1997 for treatment and growth promotion of food animals in the different EU countries. The estimates for the production of animals from 1996 were also given. In Figure 1 the usage of antimicrobial agents to produce one kilogram of meat in the different EU-countries is given.

Although the validity of some of these data is difficult to assess, and in the realisation that in different countries differences in production systems exist, it is obvious that in the various countries there are major differences in the amounts of antimicrobial agents used to produce the same amount of meat. This provides room for major reductions in those countries that comparatively overdose antimicrobial agents.

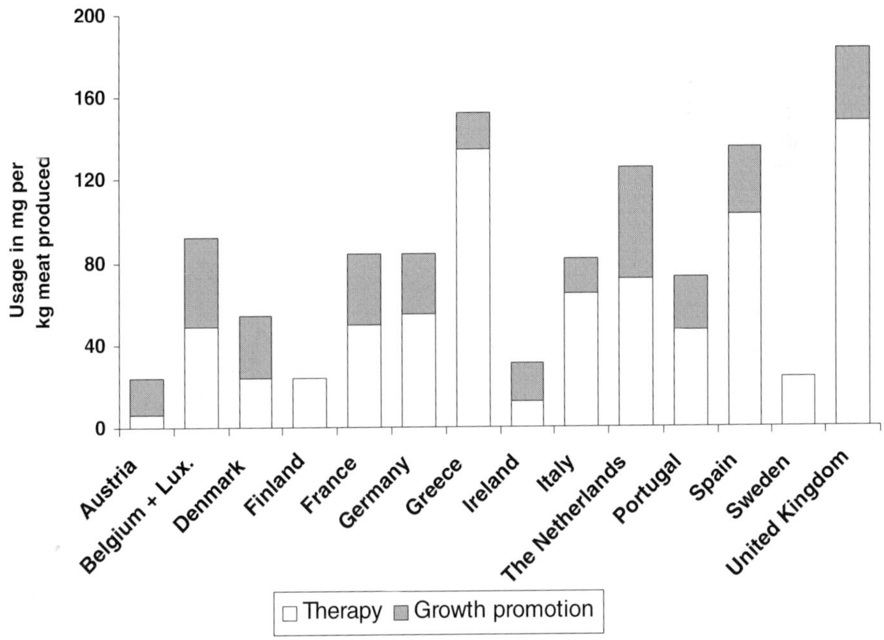

Figure 1. Milligram of antimicrobial agents used in 1997 per kilogram of produced meat in the different countries in the European Union (source EMEA 1999).

Until recently, more antimicrobials were used as growth promoters than as therapeuticals (Aarestrup et al., 1998a; 1998b). Concerns, that such use of antibiotics might destroy their therapeutic efficacy because of increased resistance led in 1969 the Swann committee (Swann, 1969) to recommend antibiotics to be exclusively used for treatment of animal and human infections. This recommendation was only partly adopted by the European Union in their Guidelines for approval of antimicrobial growth promoters.

In 1995, the Danish authorities observed an increase in the use of antibiotics for treatment of animals. Furthermore, their use as growth promoters came under increased scrutiny. Avoparcin was banned in Denmark because they select vancomycin resistant enterococci, and in view of the potential risk of resistance transmission to humans over the food chain. In 1997, avoparcin was banned in all EU-countries.

In January 1998, virginiamycin was banned in Denmark because of cross-resistance to Synercid, a streptogramin used for human treatment. In December 1998, the European Commission decided to ban the use of bacitracin, spiramycin, tylosin and virginiamycin for growth promotion from the 1st of July 1999 onwards.

In Denmark the food animal industries have decided to voluntarily stop all use of antimicrobial growth promoters from the end of 1999, and for the last 2 years they have worked to achieve this goal. Consequently, the use of growth promoters has decreased significantly during 1998 and 1999.

Table 1. Consumption of growth promoters in Denmark during 1990 through 1998.

Growth promoter	1990	1992	1994	1995	1996	1997	1998	1999	2000
Avilamycin	10	853	433	1,665	2,740	670	7	91	-
Avoparcin[1,3]	13,718	17,210	24,117	5,690	-	-	-	-	-
Bacitracin[3]	3,983	5,657	13,689	7,910	8,399	8,544	3,945	63	-
Carbadox[3]	850	7,221	10,012	1,181	1,985	4,153	1,803	293	-
Flavomycin	494	1,299	77	48	18	93	6	665	-
Monensin	2,381	3,700	4,755	5,007	4,741	3,08	935	0	-
Olaquindox[3]	11,391	21,193	22,483	16,213	13,486	17,595	28,445	9,344	-
Salinomycin	12	0	213	850	759	460	113	0	-
Spiramycin[3]	0	0	95	507	15	3	0	0	-
Tylosin[3]	42,632	26,980	37,111	52,275	68,350	62,009	13,148	1,827	-
Virginiamycin[2,3]	3,837	15,537	2,801	2,590	5,055	10,644	892	-	-
Total	79,308	99,650	115,786	93,936	105,548	107,179	49,294	12,283	0

[1] Banned in May 1995 in Denmark

[2] Banned in January 1998 in Denmark

[3] Approval suspended in EU in 1999

Table 2. Estimated consumption of antimicrobial agents for treatment of infections in food animal production in Denmark.

Antimicrobial agent	1986	1988	1990	1992	1994	1996	1997	1998	1999
Aminoglycosides	7,800	7,400	7,700	8,500	8,600	7,100	7,000	7,800	7,500
Macrolides	10,100	9,300	10,900	12,900	11,400	7,600	6,700	7,100	5,300
Penicillins	3,700	3,800	5,000	6,700	9,400	7,200	13,100	14,300	14,700
Semisyntetic penicillins	850	1,000	1,200	2,500	4,400	5,800	6,200	6,700	6,600
Sulfonamides	22,300	24,200	8,700	5,900	5,600	2,100	1,400	1,000	1,000
Sulfa/TMP	2,500	2,200	3,800	7,900	9,500	4,800	6,900	7,700	6,800
Tetracyclines	3,800	3,600	9,300	22,000	36,500	12,900	13,700	12,100	16,200
Others	13,800	6,900	6,700	6,800	4,400	600	650	3,800	3,800
Total	64,800	58,400	53,400	73,200	89,900	48,000	55,700	57,300	61,900

As stipulated by Danish regulations all sales of veterinary medicines must take place by prescription of a veterinarian. The use of antimicrobials increased between 1986 and 1994 (Table 2). This concurred with a simultaneous increase in the production of pigs in Denmark. However, this production increase alone could not explain the increase. The fluctuations in the total quantity of antimicrobials used were not associated with animal health problems. In the mid-1990s large amounts of tetracyclin were used as prophylactics. In 1995, the regulations for delivery of therapeutic antibiotics to farmers and for sale of antimicrobial for therapy were changed, removing the economical incentive for veterinarians to sell antimicrobials to farmers. Consequently, the consumption decreased. During 1996 and 1997, the total consumption of antimicrobial agents has increased. However, this increase correlates rather well with the simultaneous increase that has taken place in the production of food animals. Following the total stop in using growth promoters for pigs at the end of 1999, an increase in the use of antimicrobial agents for therapy seems to have taken place. However, it is still too early to tell whether this holds true in the long run.

Concurrently with the institution of the ban on growth promoters, the Danish Veterinary Laboratory and the agricultural organisations have provided extensive information to farmers and veterinarians on how to react if they experience problems with diarrhoea or other diseases. This is a very important step to avoid problems with infections when one refrains from the use of antibiotic growth promoters.

3. Antimicrobial Resistance

A very high frequency of resistance to therapeutical and/or growth promoting antimicrobials has been observed in several studies. However, most reports are based on research studies, severely hampered by the lack of proper epidemiological selection criteria, and mostly based on pathogenic isolates from clinically ill and perhaps medicated animals. It is very difficult and sometimes impossible, therefore, to compare results of studies performed in different countries or at different times.

Information on the occurrence of resistance is needed at both the local, regional and international level to guide policy and detect changes that require intervention strategies. To fulfil this requirement, systems for the continuous monitoring of the changes in the occurrence of resistance are needed.

3.1 Monitoring of Antimicrobial Resistance

Monitoring of antimicrobial resistance is a prerequisite for assessing the magnitude of the problem. When establishing monitoring programmes several factors have to be taken into consideration (O'Brien, 1997; Livermore et al., 1998). These include deciding on the bacterial species to be included, sampling strategies, susceptibility testing methods, data recording, computing and reporting.

To collect data on resistance in pathogenic bacteria it is easiest to rely on laboratory tests studies on antimicrobial susceptibility. However, results obtained from such isolates are greatly biased because i) requisition varies among different veterinarians and medical doctors, ii) some infections are more likely to generate isolates, and, iii) the isolates from some infections are more likely to be sent in for susceptibility testing. Furthermore, especially among animals, in several cases isolates are not collected until after initial treatment and will often include several isolates from the same herds. Reporting of data from these types of samplings has to be evaluated with great caution and it is essential to consider the value of each individual sample. A better method is the collection of isolates from randomly selected animals and humans, based on epidemiological considerations and aimed at covering the target population.

Data can be collected from different regional laboratories and compiled centrally. However, differences in methods to test susceptibility might include testing biases. Careful standardisation and inter-laboratory calibration can render such programmes useful.

Data recording, computing and reporting are essential for an appropriate monitoring system. Thus, the database has to include information on the sample population, origin down to specific patient or herd, time of isolation, and testing procedure. Furthermore, it is advisable to centrally store the isolates to allow later studies on the emergence of resistance.

Systems for monitoring antimicrobial resistance among pathogenic bacteria isolated from food producing animals (Wray et al., 1993; Martel et al., 1995; Anon., 1996b; Tollefson et al., 1998) and for both pathogenic and bacteria isolated from healthy animals (Aarestrup et al. 1998a, 1998b, Moreno et al. 2000, Wray and Gnanou 2000) have been described. These programmes are not co-ordinated and only limited exchange of information and bacterial strains takes place. A central collection and analysis of data from different monitoring programmes is badly needed.

3.2 Development and Spread of Resistance Against Microbials Following their Introduction; Examples

A zoonosis is an infection or an infectious disease that is generally transmissible from vertebrate animals to man (Acha & Szyfres, 1994). Well-known foodborne zoonotic agents are *Salmonella, Campylobacter, Yersinia, Listeria* and enterohaemorrhagic *E. coli*. Several studies have shown that zoonotic bacteria will acquire resistance among food animals, which they can subsequently transfer to man over the food chain causing infections (Holmberg et al., 1984; Endtz et al., 1991; Wall et al., 1995; Mølbak et al., 1999; Helmuth, 2000).

One of the most striking examples in recent years relates to the resistance to fluoroquinolones. In several countries fluoroquinolones are the drugs of choice for treatment of gastrointestinal infections in man and the emergence of resistance among zoonotic organisms such as *Salmonella* and *Campylobacter* is a matter of increasing concern. The first reported study was from The Netherlands where water medication with the fluoroquinolone enrofloxacin in the poultry production was followed by an emergence of fluoroquinolone- resistant *Campylobacter* species among both poultry and humans (Endtz et al., 1991). Since then several studies worldwide have documented an increase in the occurrence of resistance to fluoroquinolones among *Campylobacter* from food animals and humans following their introduction for the treatment of infections in food animals (Engberg et al., 2001, Smith et al., 2000). In Germany an increase in the occurrence of fluoroquinolone resistance among *Salmonella typhimurium* DT204c was observed after the introduction of enrofloxacin for veterinary use in 1989 (Helmuth, 1997). Most recently, in the UK substantial increases in the resistance to fluoroquinolones in *Salmonella hadar* and *Salmonella virchow*, and in multiresistant *Salmonella typhimurium* DT104 have followed the licensing for veterinary use of enrofloxacin in 1993 and danofloxacin in 1996 (Threlfall et al., 1999).

In several cases genes encoding resistance are transmissible between different bacteria and in those cases the transmission of the resistance gene between the food animal and human reservoir is of importance. A number of observations on the spread of resistance genes from bacteria in food animals to bacteria in humans have been reported. For instance, in 1983, the streptothricin antibiotic nourseothricin was introduced as a growth promoter in the former German Democratic Republic. No similar compound had been used prior to its introduction and resistance was only observed at a very low frequency. After the introduction *E. coli* isolates with transferable resistance plasmids emerged among pigs (Tschäpe et al., 1984). This plasmid was subsequently found in isolates from the pig farmers and their families and was, furthermore, found in *E. coli* isolates of the gut flora or causing urinary tract infections among humans living in the same regions (Hummel et al., 1986). Streptothricin resistance has also been found in *Shigella* isolates (Witte, 1998), among *Campylobacter* (Bottcher, 1992) and spread of resistant *Campylobacter coli* from pig farms to humans has been observed (Bischoff, 1996).

After the introduction of the aminoglycoside antibiotic apramycin for veterinary use in the beginning of the 1980's resistance to apramycin emerged among *E. coli* isolates found in cattle and pigs in France and the United Kingdom (Chaslus-Dancla et al., 1985; Wray et al., 1986). Apramycin has never been used for treatment of infections in humans. The resistance gene encoded also resistance to gentamicin, but had not been observed previously (Hedges et al., 1984). The same apramycin resistance gene and similar resistance plasmids were subsequently found in *S. enterica* from animals and in human clinical isolates of *E. coli, S. enterica* and *Klebsiella pneumoniae* (Chaslus-Dancla et al., 1986; Threlfall et al., 1986; Chaslus-Dancla et al., 1991; Hunter et al., 1992, 1993; Pohl et al., 1993; Johnson et al., 1994, 1995). These observations strongly suggest that this resistance gene primarily emerged because of the selection by the use of apramycin for food animals, was then transmitted to humans and perhaps selected for by the use of gentamicin for treatment of humans.

More recently, several studies have incriminated growth-promoting antimicrobials in the selection of resistant bacterial strains and resistance genes that are transferable to man (Aarestrup, 2000; Witte, 1997).

3.3 Changes in the Occurrence of Resistance Following Withdrawal of Antimicrobial Agents; Examples

Only limited information is available on the effects of terminating the usage of an antimicrobial agent on the occurrence of antimicrobial resistance. Following the ban of tetracycline for growth promotion in The Netherlands in 1974, a decrease in the occurrence of tetracycline resistance in salmonellae isolated from food animals and humans was observed (van Leeuwen et al., 1979).

In Denmark the DANMAP program for surveillance of antimicrobial resistance was established in 1995 (Aarestrup et al., 1998a; 1998b). The program determines the susceptibility of antimicrobial resistance among selected pathogenic bacteria and bacteria isolated from healthy food animals, and are co-ordinated with similar programs determining resistance among bacteria from food and humans. The establishment of this programme has made it possible to follow the changes in occurrence of antimicrobial resistance as a consequence of changes in usage.

The occurrence of resistance among *E. faecium* from pigs and broilers has largely followed the consumption of the various antimicrobials. Conversely, since the ban of avoparcin in 1995 the occurrence of vancomycin resistance has decreased significantly among enterococcal isolates from broilers, whilst no significant change occurred in pigs (Bager et al., 1999). One explanation for the persistence of vancomycin resistant enterococci (VRE) among the pig population in Denmark could be that the production of broilers in Denmark is performed in an all-in all-out system, including thorough cleaning of housing facilities after each flock; in contrast, the production of pigs takes place in facilities in continuous use. However, all VRE isolated from pigs were also found to be resistant to macrolides and tetracycline. These two antimicrobials have in Danish pig production been commonly used as growth promoter and therapeutical, respectively. It was since shown that all VRE isolated from pigs in Denmark belonged to the same clone and that the genes encoding resistance to macrolides (*ermB*) and glycopeptides (*vanA*) were located on the same mobile

DNA-element (Aarestrup, 2000). As the administration of the growth promoter tylosin decreased substantially during 1998, so did during the last half of 1998 and first half of 1999 the occurrence of VRE among *E. faecium* isolates from pigs (Aerestrup et al., submitted). These findings strongly suggest that the persistence of VRE among the pig population was caused by the continued use of macrolides, mainly tylosin, for growth promotion and therapy.

In Germany a decrease in the VRE carrier rate among healthy humans in the community in the period 1994 to 1997 has been observed (Klare et al., 1999). In 1994 the carrier level was found to be 13%. This level has decreased to approx. 4% in 1997. Similarly, a decrease in the occurrence of VRE among poultry products in Italy has been observed during 18 months after the ban of avoparcin (Pantosi et al., 1999). Also in The Netherlands a decrease has been observed among broilers, pigs and humans from 1997 to 1999 (van den Bogaard et al., 2000). A plausible explanation for this is a reduced exposure to VRE from food as a result of the withdrawal of avoparcin from animal feed.

Avilamycin has primarily been used for growth promotion in the broiler production in Denmark. The consumption of avilamycin for growth promotion increased from 10 kg in 1990 to 2,740 kg 1996 and decreased in the following years to only 7 kg in 1998. The occurrence of resistance among isolates from broilers increased from 63.6% at the end of 1995 to a maximum of 80.7% during the last half of 1996. Since then, the occurrence of resistance decreased to 23.3% in the last half of 1998 (Aarestrup et al., 2000). Thus, the occurrence of resistance has closely followed the consumption.

The use of virginiamycin for growth promotion was banned in Denmark in January 1998 and in all of EU from June 1999. In Denmark the first monitoring results indicate that a decrease in resistance has already taken place (Aarestrup et al., submitted).

Taken together these observations show that the occurrence of resistance will decrease after the selective pressure has been removed. However, in the case of resistance to vancomycin among isolates from pigs it has also been shown that if the resistance genes are located on the same genetic element, resistance can persist as a consequence of co-selection.

4. Prudent Use of Antimicrobial Agents

Prudent use of antimicrobial agents is a way to extend their therapeutic lifespan. However, 'prudent use' is defined in various ways, and the term should be used with care.

In Denmark and other Scandinavian countries guidelines for use of antimicrobial agents have been in place in human medicine for more than 30 years. These guidelines recommend that antimicrobial therapy should in general only be initiated after the causative agent has been identified, and preferably only after susceptibility testing. Furthermore, narrow spectrum antimicrobials should be drugs of first priory. These simple guidelines have proven their value over time, and today Denmark has a very low consumption of antimicrobial agents and a low occurrence of antimicrobial resistance among bacterial isolates from humans as compared to most other countries around the world.

4.1 Veterinary Antibiotic Policy

The use of antimicrobial agents for food animals should take into account the potential selection of antimicrobial resistance in bacteria of importance for both animals and humans. As a helpful tool, the Danish Veterinary Laboratory in 1996 issued a set of guidelines for prudent use of antibiotics, for Danish veterinary practitioners (Pedersen et al., 1999). These were based on recommendations from The Netherlands (van Bogaard, 1993). The guidelines include some overall principles for a veterinary antibiotic policy and some specific guidelines for selection of antimicrobial agents in veterinary practice. General guidelines for veterinary antibiotic policy may be of value as a part of an overall strategy for limiting the emergence and spread of antimicrobial resistance. However, to be of everyday value the policy has to give concrete guidelines for the choice of antimicrobial agents for specific infections.

The specific priorities are based on general knowledge regarding the antimicrobial susceptibility of the most common bacterial pathogens causing disease in Danish food animals and they also take into account whether the antimicrobial is of importance in human medicine. For example, the use of fluoroquinolones in food animals is discouraged.

4.2 Main Principles for Prudent Use of Antimicrobial Agents for Food Animals

Prudent use of antimicrobial agents in veterinary medicine should ensure efficient clinical treatment of sick animals, but also limit the negative consequences as much as possible. The instigation of antimicrobial treatment should be based on an aetiological diagnosis. This can in many cases be based on clinical signs, or general knowledge of the specific herd. However, often a laboratory diagnosis will be necessary. The choice of antimicrobial agents for treatment can in many cases be based on general knowledge regarding the susceptibility of the infecting agent. For some bacterial species the susceptibility can be predicted with great certainty. This includes *Erysipelotrix rhusiopathiae* and β-haemolytic streptococci. For other species, such as *Escherichia coli*, *Salmonella* spp. and staphylococci, the susceptibility patterns vary.

During acute infections it is important to instigate antimicrobial treatment immediately. The choice of antimicrobial agents can therefore not be based on laboratory examinations but has to be based on a general knowledge regarding the susceptibility pattern of the different aetiological agents. It is the veterinarian's responsibility to have knowledge regarding the susceptibility patterns in the different food animal herds and farms. The effect of empirically initiated treatment should be followed and evaluated. The choice of antimicrobial treatment should be changed if indicated by the susceptibility test or lack of effect.

4.3 Choice of an Antimicrobial Agent

The following considerations play a major role in the choice of an antimicrobial agent for each individual infection: i) narrow spectrum antimicrobials. Old antimicrobials are prioritised over newer compounds, ii) general occurrence of resistance to the given bacterial species, iii) expected clinical effect. Mode of administration, and iv) only antimicrobial agents that are approved for treatment of the given food animal species are included

For each disease included in the Danish antibiotic policy, antimicrobials are suggested, which are classified as either 1st, 2nd or 3rd priority. In prophylaxis vaccination it is inappropriate to give antimicrobial agents.

1st priority antimicrobials are suggested as drugs of choice for that specific infection. 2nd priority antimicrobials can be used when the mode of administration or resistance problems do not allow 1st priority agents. 3rd priority antimicrobials should only be used after susceptibility testing and when specific circumstances exclude that 1st and 2nd priority antimicrobials can be used. Under 3rd priority only some examples have been included. The veterinarian can also choose other approved antimicrobial agents.

In relation to the general occurrence of resistance the following criteria have been used: All antimicrobial agents where less than 30% of the bacteria are susceptible have been excluded and antimicrobials where less than 60% are susceptible are not placed higher than as 3rd priority.

For information of practising veterinarians about the susceptibility of common animal pathogens among Danish food producing animals, a list of susceptibility is given for each animal species. The recommendations are evaluated and adjusted annually.

5. Exclusion or Downgrading of Certain Antimicrobial Agents

Public Health considerations dictate that certain antimicrobial agents be given low priority. In Denmark the occurrence of antimicrobial resistance is generally low and it is in many cases possible to exclude certain antimicrobial agents from the list without limiting the possibilities to treat animals.

Fluoroquinolones and gentamicin are important antimicrobials in treatment of gastrointestinal infections, septicaemia and meningitis in humans. Methicillin is important in treatment of infections with staphylococci in humans. Thus, quinolones, gentamicin, cloxacillin and nafcillin are totally excluded of the list of suggested antimicrobial agents.

A number of other antimicrobial agents have been downgraded. These are ceftiofur, cefoperazone, tetracycline and lincospectin. Ceftiofur and cefoperazone are downgraded because they are 3rd generation cephalosporins and may select for broadspectrum penicillin resistance. Tetracyclines are downgraded because this group of antimicrobial can select for multiple resistance. Lincospectin is downgraded because it is not suitable to use combination treatment when single drug can be used.

The Danish antibiotic policy cannot simply be adopted by other countries without changes. Differences in the importance of the different bacterial species causing diseases exist, as do differences in the most common resistance patterns and antimicrobial agents available. However, the principles could be duplicated and a veterinary antibiotic policy taking into account the development of resistance and the importance of the different antimicrobial agents for treatment of infections in humans should be implemented in all countries.

6. Discussion

It is generally accepted that the use of antimicrobial agents for treatment of infections is the major driving force in the selection for resistance. Even though there are no major differences in the concentrations used, whether the use of antimicrobial agents for growth promotion selects for resistance remains a subject of controversy.

When the Swann Committee's recommendations were released, their strategies were mainly directed against the use of tetracycline. By and large their recommendations were implemented in the European countries, whereas countries such as USA continued to use tetracycline and penicillin as growth promoters. However, despite the general acceptance of the recommendation from the Swann Committee, macrolides such as spiramycin and tylosin were approved for use in the mid 1970's. It can be questioned how large an impact this use has had on human health, because resistance to macrolides was at that time already widespread among human pathogens. However, in the case of macrolide resistant zoonotic bacteria such as *Campylobacter* the use of tylosin and/or spiramycin may have had consequences for human health.

Before 1986 enterococci were in general considered susceptible to glycopeptides, but soon after the first observations of VRE this resistance spread rapidly world-wide. Thus, like in several other cases antimicrobial resistance can emerge very rapidly in bacteria in response to the selective pressure. The only way to limit the spread of antimicrobial resistance and thereby extend the usefulness of antimicrobials is through a restricted use of antimicrobials (WHO, 1997).

It has been recommended that antibiotics that select for resistance against antibiotics used for human therapy should no longer be allowed in animal husbandry. Especially the use of growth promoters should be limited to agents that are of no interest for therapeutic use (WHO, 1997). For instance, when avilamycin, avoparcin and virginiamycin were approved for use as growth promoters in animals, these groups of antibiotics were not seen as potentially important human antibiotics. Due to resistancy problems with other antibiotics this is no longer the view. However, the resistance that has already been created by the use of avoparcin and virginiamycin as growth promoters will most likely reduce the efficacy and shorten the lifespan of vancomycin and quinupristin/dalfopristin as therapeutics in humans. The only way to avoid or limit the consequences of such situations is to avoid the usage of antimicrobial agents for food animals as much as possible and more in particular the use of those classes that are considered of major importance for treatment of infections in humans.

The Danish programme for monitoring and control of antimicrobial resistance has shown that it is possible at least to some degree to limit the occurrence of resistance by removing the selective pressure (Bager et al. 1999, Aarestrup et al., 2000, Aarestrup et al. submitted). It is, however, not known whether the occurrence of resistance will ever reach the same low level as before the antimicrobials were introduced. Most probably a low frequency of resistant bacteria will persist and make up a reservoir from where resistance problems easily can be selected if antibiotics are introduced again.

Furthermore, it has generally been expected that the presence of resistance mechanisms would reduce the fitness of the bacteria and that resistance, thus, would decline if the selective pressure were removed. However, recent studies have indicated that resistant bacteria might adapt and thereby uphold the same fitness as susceptible bacteria (Schrag et al., 1997; Andersson et al., 1999).

7. Conclusions

To control the development of resistance to antimicrobial agents, the safest way seems to be to limit their use as much as possible. Thus, *all unnecessary use of antimicrobial agents should be strongly discouraged*. This includes usage to increase production or as prophylactics. However, it is and will also in the future be necessary to use antimicrobial agents for treatment of infectious diseases in food animals and we will also in the future observe the selection of antimicrobial resistance among bacteria from food animals.

It is not known which resistance problems we may face in the future. However, to implement timely actions, to limit the emergence of antimicrobial resistance and the consequences for human and animal health, it is necessary to obtain and maintain scientific knowledge regarding factors affecting the occurrence, emergence and spread of resistance. Some suggestions for programmes that could be implemented are given in Table 3.

At present, knowledge of antimicrobial resistance in food animals is incomplete in most countries. An organised *monitoring of antimicrobial resistance* carried out by an international network of laboratories should be implemented in order to be able to identify and report emerging resistance problems at the earliest possible stage. In this way we might be able to implement interventions before emerging resistance causes major problems.

Knowledge regarding the *consumption of antimicrobial agents* for different food animal species is generally unavailable in most countries. Such knowledge is needed for assessing its impact on the occurrence of resistance and for determining where and for which infections most antimicrobials are used. Programmes to collect data on the consumption of antimicrobial agents are therefore also strongly needed.

Table 3. Suggestions for programmes to implement in all countries producing food animals.

Suggestion	Content and benefit
Monitoring of antimicrobial resistance	• Data on the occurrence of antimicrobial resistance • Used to guide the veterinary antibiotic policy and to follow the effects of changes of usage of antimicrobial agents • Used as an early warning system for emergence or changes in the occurrence of resistance that requires further studies or interventions
Monitoring of usage of antimicrobial agents	• Used to follow the association between usage and resistance • Gives information regarding which infections that require most antimicrobial agents
Veterinary antibiotic policy	• Must take into account the need to treat animals, but also potential negative consequences for human health • Should be based on the resistance monitoring and scientific data • Should give concrete guidelines and not overall statements

Research	• Research should be used to form the veterinary antibiotic policy and should use data and bacterial strains from the monitoring systems
	• Research could for example be performed within the following areas:
	    • The effect of treatment strategies on resistance
	    • The genetic background for resistance
	    • The spread of resistance and resistance genes between reservoirs
	    • The evolution and adaptation of resistance

It appears that one of the mechanisms that may contribute to slow down the emergence of antimicrobial resistance is a more limited and prudent use of antimicrobial agents. It should be stressed that antimicrobial agents are extremely valuable drugs in the treatment of infections and they should only be used when needed and in an appropriate way. Some antimicrobial agents are more important than others for treatment of infections in humans. Furthermore, some antimicrobial agents are probably more prone to select for resistance than others.

It is recommended that those responsible for medical treatment of food animals adopt a *policy for prudent use of antimicrobial agents for food animals*. This policy should take into account the need to treat infections in animals, but also the potential negative consequences for human health. The policy should also be based on knowledge regarding the normal susceptibility patterns of the different causative agents. Furthermore, genes encoding resistance can be located on the same mobile DNA-elements and the potential co-selection by use of different antimicrobial agents should be taken into account when implementing an antibiotic policy.

Much knowledge regarding the most appropriate treatment regime in terms of achieving cure of infections are available. However, virtually no knowledge regarding the most appropriate treatment in relation to limiting the development of resistance is at hand. *Research* in this area is highly needed.

The impact on human and animal health of the selection for antimicrobial resistance in the animal reservoir is not sufficiently known. Large-scale epidemiological and experimental studies including all relevant information on the use of antimicrobial agents, modes of administration, occurrence of resistance and spread of resistant clones and resistance genes in all relevant bacterial reservoirs exposed to antimicrobial agents or that might act as reservoirs, are strongly encouraged.

In Denmark a continuous monitoring of the occurrence of antimicrobial resistance among food animals was implemented in 1995 (Aarestrup et al., 1998a). Furthermore, a programme monitoring the consumption of antimicrobial agents at the farm level was implemented during 2000. A veterinary antibiotic policy has recently been described in Denmark (Pedersen et al., 1999). This includes concrete guidelines for use of specific antimicrobial agents for infections with specific infectious bacterial agents and furthermore, guidelines on which antimicrobial agents should be reserved for special cases only.

These programmes and guidelines are at present not fully operational and the long-term effects must await future studies. However, the emergence of antimicrobial resistance has to be limited. Thus, to avoid further deterioration of available and future antimicrobial agents we need to implement timely interventions based on scientific knowledge before it is too late.

References

Aarestrup F.M., 2000. Occurrence, selection and spread of resistance to antimicrobial agents used for growth promotion for food animals in Denmark. APMIS Suppl. 101: 1-48.

Aarestrup, F.M., Bager, F., Madsen, M., Jensen, N.E., Meyling, A. and Wegener, H.C., 1998a. Surveillance of antimicrobial resistance in bacteria isolated from food animals to antimicrobial growth promoters and related therapeutic agents in Denmark. APMIS 106: 606-622.

Aarestrup F.M., Bager, F., Jensen, N.E., Madsen, M., Meyling, A. and Wegener, H.C., 1998b. Resistance to antimicrobial agents used for animal therapy in pathogenic-, zoonotic- and indicator bacteria isolated from different food animals in Denmark: A baseline study for the Danish Integrated Antimicrobial Resistance Monitoring Programme (DANMAP). APMIS 106: 745-770.

Aarestrup, F.M., Bager, F. and Andersen, J.S., 2000. The association between the use of avilamycin for growth promotion and the occurrence of resistance among *Enterococcus faecium*. Microb. Drug Resist. 6: 71-75.

Aarestrup, F.M., 2000. Characterization of glycopeptide resistant *Enterococcus faecium* from broilers and pigs in Denmark. Genetic evidences that the persistence of GRE among the pig population is associated with co-selection by resistance to macrolides. J. Clin. Microbiol. 38: 2774-2777.

Aarestrup, F.M., Seyfarth, A.M., Emborg, H.D., Pedersen, K., Hendriksen, R.S., and Bager, F. The effect of a national intervention program on antimicrobial resistance in food animals. Antimicrob. Agents Chemother. (Submitted).

Acha, P.N., Szyfres, B., 1994. Zoonoses and communicable diseases common to man and animals. 2nd Ed. Pan American Health Organisation, Pan American Sanitary Bureau, regional office of the World Health Organization, Washington D. C., USA.

Abraham, E.P. and Chain, E., 1940. An enzyme from bacteria able to destroy penicillin. Nature 146: 837-839.

Andersson, D.I. and Levin, B.R., 1999. The biological cost of antibiotic resistance. Curr. Opin. Microbiol. 2: 489-493.

Bager, F., Aarestrup, F.M., Madsen, M. and Wegener, H.C., 1999. Glycopeptide resistance in *Enterococcus faecium* in broilers and pigs following discontinued use of avoparcin. Microb. Drug Resist. 5: 53-56.

Bischoff, K. and Jacob, J., 1996. The sat4 streptothricin acetyltransferase gene of *Campylobacter coli*: its distribution in the environment and use as epidemiological marker. Zentralbl. Hyg Umweltmed. 98: 241-257.

Black, W.D., 1984. The use of antimicrobial drugs in agriculture. Can. J. Physiol. Pharmacol. 62: 1044-1048.

Bottcher, I. and Jacob, J., 1992. The occurrence of high-level streptothricin resistance in thermotolerant campylobacters isolated from the slurry of swine and the environment. Int. J. Med. Microbiol. Virol. Parasitol. Infect. Dis. 277: 467-473.

Chaslus-Dancla, E. and Lafont, J.P., 1985. Resistance to gentamicin and apramycin in *Escherichia coli* from calves in France. Vet. Rec. 117: 90-91.

Chaslus-Dancla, E., Martel, J.L., Carlier, C., Lafont, J.P. and Courvalin, P., 1986. Emergence of aminoglycoside 3-N-acetyltransferase IV in *Escherichia coli* and *Salmonella typhimurium* isolated from animals in France. Antimicrob. Agents Chemother. 29: 239-243.

Chaslus-Dancla, E., Pohl, P., Meurisse, M., Marin, M. and Lafont, J.P., 1991. High genetic homology between plasmids of human and animal origins conferring resistance to the aminoglycosides gentamicin and apramycin. Antimicrob. Agents Chemother. 35: 590-593.

Emborg, H.D., Tornøe, N., Heuer, O.E. and Madsen, M., 1999. The effect of discontinued use of antimicrobial growth promoters in broilers. In proceedings of 39th Interscience Conference on Antimicrobial Agents and Chemotherapy, San Francisco, California, USA. Abstract C712, pp. 98.

EMEA., 1999. Antibiotic resistance in the European Union associated with therapeutic use of veterinary medicines. Report and qualitative risk assessment by the Committee for Veterinary Medical Products, The European Agency for the evaluation of Medical products, 14 July 1999.

Endtz, H.P., Ruijs, G.J., van Klingeren, B., Jansen, W.H., van der Reyden, T. and Mouton, R.P., 1991. Quinolone resistance in *Campylobacter* isolated from man and poultry following the introduction of fluoroquinolones in veterinary medicine. J. Antimicrob. Chemother. 27: 199-208.

Engberg, J., Aarestrup, F.M., Smidt, P.G., Nachamkin, I. and Taylor, D.E., 2001. Quinolone and macrolide resistance in *Campylobacter jejuni* and *coli*: A review of mechanisms and trends over time of resistance profiles in human isolates. Emerg. Infect. Dis. 7: 24-34.

FEDESA., 1997. Facts and figures about the European animal health industry. Animal Health Dossier 14. The European Federation of Animal Health (FEDESA), Brussel, 1997.

Fleming, A., 1929. On the antibacterial action of cultures of a penicillum, with special reference to their use in the isolation of *B. influenzæ*. Br. J. Exp. Path. 10: 226-236.

Fraser-Moodie, W., 1971. Struggle against infection. Proc. R. Soc. Med. 64: 87-94.

Harvey, J. and Mason, L., 1998. The use and misuse of antibiotics in UK agriculture. Soil Association, Bristol, UK.

Hedges, R.W. and Shannon, K.P., 1984. Resistance to apramycin in *Escherichia coli* isolated from animals: detection of a novel aminoglycoside-modifying enzyme. J. Gen. Microbiol. 130: 473-482.

Helmuth, R., 2000. Antibiotic resistance in *Salmonella*. In: Wray C, Wray A (eds.). *Salmonella* in domestic animals. CAB International, 89-106.

Helmuth, R. and Protz, D., 1997. How to modify conditions limiting resistance in bacteria in animals and other reservoirs. Clin. Infect. Dis. 24: 136-138.

Holmberg, S.D., Osterholm, M.T., Senger, K.A. and Cohen, M.L. 1984. Drug-resistant *Salmonella* from animals fed antimicrobials. N. Engl. J. Med. 311: 617-622.

Hummel, R., Tschäpe, H. and Witte, W., 1986. Spread of plasmid-mediated nourseothricin resistance due to antibiotic use in animal husbandry. J. Basic Microbiol. 26: 461-466.

Hunter, J.E., Shelley, J.C., Walton, J.R., Hart, C.A. and Bennett, M., 1992. Apramycin resistance plasmids in *Escherichia coli*: possible transfer to *Salmonella typhimurium* in calves. Epidemiol. Infect. 108: 271-278.

Hunter, J.E., Hart, C.A., Shelley, J.C, Walton, J.R. and Bennett, M., 1993. Human isolates of apramycin-resistant *Escherichia coli* which contain the genes for the AAC(3)IV enzyme. Epidemiol. Infect. 110: 253-259.

Jacoby, G.A. and Archer, G.L., 1991. New mechanisms of bacterial resistance to antimicrobial agents. N. Engl. J. Med. 324: 601-612.

Johnson, A.P., Burns, L., Woodford, N., Threlfall, E.J., Naidoo, J., Cooke, E.M., and George, R.C., 1994. Gentamicin resistance in clinical isolates of *Escherichia coli* encoded by genes of veterinary origin. J. Med. Microbiol. 40: 221-226.

Klare, I., Badstubner. D., Konstabel, C., Bohme, G., Claus, H. and Witte, W., 1999. Decreased incidence of VanA-type vancomycin-resistant enterococci isolated from poultry meat and from fecal samples of humans in the community after discontinuation of avoparcin usage in animal husbandry. Microb. Drug Resist. 5: 45-52.

Levy, S.B., 1982. Microbial resistance to antibiotics. An evolving and persistent problem. Lancet, July 10: 83-8.

Livermore, D.M., Macgowan, A.P. and Wale, M.C., 1998. Surveillance of antimicrobial resistance. Centralised surveys to validate routine data offer a practical approach. BMJ 1998; 317: 614-5.

MacFarlane, G., 1984. Alexander Fleming: The man and the myth. Harvard University Press, Cambridge.

Martel, J.L., Chaslus-Dancla, E., Coudert, M., Poumarat, F. and Lafont, J.P. 1995. Survey of antimicrobial resistance in bacterial isolates from diseased cattle in France. Microb. Drug Resist. 1: 273-283.

Moreno, M.A., Domínguez, L., Teshager, T., Herrero, I.A, Porrero, M.C. and the VAV network., 2000. Antibiotic resistance monitoring: the Spanish programme. Int. J. Antimicrob. Agents 14: 285-290.

Mølbak, K., Baggesen, D.L., Aarestrup, F.M., Ebbesen, J.M., Engberg, J., Frydendahl, K., Gerner-Smidt, P., Petersen, A.M., and Wegener, H.C., 1999. From the stable to the table: an outbreak of multiresistant, quinolone-resistant *Salmonella* Typhimurium DT104 in Denmark. N. Engl. J. Med. 341: 1420-1425.

O'Brien, T.F., 1997. The global epidemic nature of antimicrobial resistance and the need to monitor and manage it locally. Clin. Infect. Dis. 24 Suppl 1: 2-8.

Pantosti, A., Del Grosso, M., Tagliabue, S., Macri, A., Caprioli, A., 1999. Decrease of vancomycin-resistant enterococci in poultry meat after avoparcin ban. Lancet 1999; 354: 741-2.

Pedersen, K.B., Aarestrup, F.M., Jensen, N.E., Bager, F., Jensen, L.B., Jorsal, S.E., Nielsen, T.K., Hansen, H.C., Meyling, A. and Wegener, H.C., 1999. The need for a veterinary antibiotic policy. Vet. Rec. 144: 50-53.

Pohl, P., Glupczynski, Y., Marin, M., Van Robaeys, G., Lintermans, P. and Couturier, M., 1993. Replicon typing characterization of plasmids encoding resistance to gentamicin and apramycin in *Escherichia coli* and *Salmonella typhimurium* isolated from human and animal sources in Belgium. Epidemiol. Infect. 111: 229-238.

Schrag, S.J., Perrot, V. and Levin, B.R., 1997. Adaptation to the fitness costs of antibiotic resistance in *Escherichia coli*. Proc. R. Soc. Lond. B. Biol. Sci. 264: 1287-1291.

Skolnick, A., 1991. New insights into how bacteria develop antibiotic resistance. JAMA 265: 14-16.

Smith, K.E., Bender, J.B. and Osterholm, M.T., 2000. Antimicrobial resistance in animals and relevance to human infections. In: Nachamkin, I., Blaser, M. J. (ed.), *Campylobacter* (2nd ed.), ASM Press, Washington D. C., pp 483-495.

Swann, M.M., 1969. Joint committee on the use of antibiotics in animal husbandry and veterinary medicine. H.M.S.O., London.

Threlfall, E.J., Rowe, B., Ferguson, J.L. and Ward, L.R., 1986. Characterization of plasmids conferring resistance to gentamicin and apramycin in strains of *Salmonella typhimurium* phage type 204c isolated in Britain. J. Hyg. Camb. 97: 419-426.

Threlfall, E.J., Ward, L.R., Skinner, J.A. and Rowe, B., 1997. Increase in multiple antibiotic resistance in nontyphoidal salmonellas from humans in England and wales: a comparison of data for 1994 and 1996. Microb. Drug Resist. 3: 263-266.

Tollefson, L., Angulo, F.J. and Fedorka-Cray, P.J., 1998. National surveillance for antibiotic resistance in zoonotic enteric pathogens. Vet. Clin. North Am. Food Anim. Pract. 14: 141-150.

Tschäpe, H., Tietze, E., Prager, R., Voigt, W., Wolter, E. and Seltmann, G., 1984. Plasmid borne streptothricin resistance in gram-negative bacteria. Plasmid 1984; 12: 189-196.

van den Bogaard, A.E., 1993. A veterinary antibiotic policy: a personal view on the perspectives in The Netherlands. Vet. Microbiol. 35: 303-312.

van den Bogaard, A.E., Bruinsma, N. and Stobberingh, E.E., 2000. The effect of banning avoparcin on VRE carriage in The Netherlands. J. Antimicrob. Chemother. 46: 146-147.

van Leeuwen, W.J., van Embden, J., Guinee, P.A., Kampelmacher, E.H., Manten, A., van Schothorst, M. and Voogd, C.E., 1979. Reduction of the number of tetracycline-resistant strains of *Salmonella* in the Netherlands. Tijdschr. Diergeneeskd. 104: 923-927.

Wall, P.G., Morgan, D., Lamden, K., Griffin, M., Threlfall, E.J., Ward, L.R. and Rowe, B., 1995. Transmission of multi-resistant strains of *Salmonella typhimurium* from cattle to man. Vet. Rec. 136: 591-592.

WHO., 1997. The medical impact of the use of antimicrobials in food animals. Report of a WHO meeting, 13-17 October, Berlin, Germany. World Health Organization, Geneva, Switzerland.

Witte, W., 1997. Impact of antibiotic use in animal feeding on resistance of bacterial pathogens in humans. Ciba Found. Symp. 207: 61-71.

Witte, W., 1998. Medical consequences of antibiotic use in agriculture. Science 279: 996-997.

Wray, C. and Gnanou, J.C., 2000. Antibiotic resistance monitoring in bacteria of animal origin: analysis of national monitoring programmes. Int. J. Antimicrob. Agents 14: 291-294.

Wray, C., Hedges, R.W., Shannon, K.P. and Bradley, D.E., 1986. Apramycin and gentamicin resistance in *Escherichia coli* and salmonellas isolated from farm animals. J. Hyg. Camb. 97: 445-456.

Wray, C., McLaren, I.M. and Beedell, Y.E., 1993. Bacterial resistance monitoring of salmonellas isolated from animals, national experience of surveillance schemes in the United Kingdom. Vet. Microbiol. 35: 313-319.

An overview of current research on animal transmissible *Spongiform encephalopathy* (TSE)

Raymond Bradley
Veterinary Laboratories Agency, New Haw, Addlestone, KT15 3NB, United Kingdom; E-mail: raybradley@btinternet.com

Summary

This paper describes the history of BSE in the context of public and animal health, the measures applied to enable effective protection and how, via research and experience, the necessary knowledge to bring this about has been achieved. A full account of the UK BSE epidemic until 20 March 1996, has been published by the BSE Inquiry (Phillips, Bridgeman and Ferguson-Smith, 2000). More recent information is provided in Progress Reports provided by the Ministry of Agriculture, Fisheries and Food (MAFF) for example in MAFF (2000) and its predecessors.

Keywords: transmissible spongiform encephalopathy, BSE, variant Creutzfeldt-Jakob disease, prion disease, research, public health protection, animal health protection

1. Introduction

Until 1985, animal transmissible spongiform encephalopathies (TSE) were regarded as posing no problems for public health. Scrapie, a TSE of sheep (and more rarely goats) known in Europe since the 18th century was the only farmed food animal species known to be affected by TSE. However, rare animal TSE had also been reported in farmed mink (transmissible mink encephalopathy or TME) and various species of North American deer (chronic wasting disease or CWD). Several extensive epidemiological studies such as those by Brown (1980) and Chatelaine et al., (1981) supported the generally held view that scrapie was not a zoonosis. The disease was transmitted from sheep to sheep and generation-to-generation, making it very difficult, in the absence of a test, to eradicate. No cases of an equivalent, rare TSE of humans called Creutzfeldt-Jakob disease (CJD) could be attributed to exposure to infected material from scrapie-affected animals either by consumption or occupational exposure.

Rare familial forms of CJD occur as a result of a variety of inherited mutations of the *PrP* gene. Even these inherited forms are experimentally transmissible but will not be considered further. There have been no recorded familial or inherited TSE in animals. However, it is important to recognise that polymorphisms in the *PrP* gene influence the length of the incubation period in experimental TSE in mice and in sheep (but not in cattle exposed to BSE infection). Furthermore, some allelic combinations of these polymorphisms in sheep appear to protect them from developing disease following natural exposure. In some countries, use is being made of this property by selecting 'scrapie-resistant' rams to increase the proportion of sheep with the protective alleles. In man heterozygosity at codon 129 of the *PrP* gene that occurs in about 50% of the Caucasian population, appears also to confer a degree of resistance to developing CJD. Furthermore all new variant CJD (vCJD – see below) patients so far tested are homozygous for methionine at this position.

Sporadic CJD occurs in Australia and New Zealand at the normal worldwide, annual incidence of one to two cases per million and neither scrapie nor bovine spongiform encephalopathy (BSE) is reported in either country. There is thus an intriguing mystery as to how the commonest form of CJD (sporadic CJD, about 85% of all cases) is maintained in the human population, since person to person spread of the disease has been reported only by rare iatrogenic means. A small number of historical, iatrogenic transmissions have occurred as a result of using incompletely decontaminated neurosurgical instruments or electrodes, transplants of cornea or dura mater or administration of human growth and gonadotrophic hormone derived from pituitary glands extracted from human cadavers. Presumably undetected CJD existed in patients operated upon previously (or used as tissue donors), were the source of these rare infections. New exposures to sporadic CJD agents by these means are now most unlikely as risks are now effectively managed.

The whole situation changed following the discovery of BSE in dairy cattle in the UK in November 1986 (Wells et al., 1987) and most particularly on 20 March 1996 when the Secretary of State for Health in Great Britain announced the discovery of vCJD in ten young patients. In the absence of any credible alternative the most likely explanation given was that the cases were linked to exposure to BSE infection before the introduction of the specified bovine offals (SBO) ban in 1989 (see below). The consequences of this discovery have reverberated round the world and continue to this day. The following is an account of the salient features of the epidemic and how protection of public and animal health has been achieved in the light of research results and experience.

2. History of TSE

The diseases we now know as TSE may be ancient, but, because for the most part they are rare, they did not receive sufficient attention to identify them beyond the clinical level until the late XIXth and early XXth century. Transmissibility was not determined until the late 1930's.

2.1 Scrapie

Scrapie is the oldest known TSE. It dates from perhaps biblical and Roman times but one of the earliest, accurate, clinical accounts, was given by Leopoldt (1759) in Germany, in 1759. Early accounts in the English, French, German, Hungarian and Spanish literature refer only to the disease in sheep, though it is now known to less commonly affect goats and rarely moufflon (*Ovis musimon)*. Sheep can be regarded as the principal natural reservoir of infection for all three species. The clinical signs are mostly confined to adults. They are insidious in onset, distinctive and include behavioural changes, ataxia and loss of condition. Pruritus is a common, but not invariable sign and results in rubbing and loss of wool. Once established in a country or flock scrapie is exceedingly difficult to get rid of, even by draconian culling of family lines. This is attributed to the fact that the disease is maintained endemically by sheep to sheep transmission, maternally and horizontally. The placenta is infected and is thought to be a prime vehicle for direct transmission of the disease, or indirectly via contaminated lambing pens and ground. It is believed most new infections occur in young lambs. Detectable infectivity (by mouse inoculation) appears first in the lymphoreticular system before one year of age and subsequently (in Suffolk sheep) in the central nervous system (CNS) at around two years of age, (Hadlow et al 1982). The CNS has the highest titres during the clinical phase. Neither milk nor muscle show detectable infectivity, even in the clinical phase of disease.

2.2 Creutzfeldt-Jakob Disease (CJD) and Kuru

CJD was reported for the first time in the early 1920s and was recognised then as a spongiform encephalopathy. The earliest transmission (to chimpanzees) was reported by Gibbs et al (1968). The pathological features, as in other TSE, include spongiform generation of grey matter, astrocytosis and neuronal loss. In specific instances amyloid plaques are found in the brain and it is now known that these are formed of aggregates of the disease specific prion protein or PrP^{Sc}. The clinical signs typically are of a rapidly developing dementia with ataxia being a late stage sign, though there is considerable variation related to the particular form of CJD, the *PrP* genotype of the host, the route of exposure and the strain of agent. CJD is distributed throughout the world and occurs at an incidence of between 1 and 2 cases per million *per annum*. Infectivity is much less widely distributed in peripheral and non-neural tissues than is scrapie in sheep.

Kuru is a geographically localised TSE of humans affecting the Fore-speaking tribe in the Eastern highlands of Papua New Guinea. It was at one time responsible for the majority of deaths in the tribe but is now declining to extinction though with rare incubation periods of over 40 years. Also known as the laughing death, because the facial expression of affected patients resembles a smile, kuru was not investigated by westerners until the 1950s. Following much epidemiological and pathological research, kuru was finally identified as a TSE by Gajdusek et al (1966). Transmission studies in a laboratory primate were recommended by Hadlow (1959) on the basis that the neuropathology of kuru resembled that in sheep with scrapie, a disease he had extensively studied. Transmission to Chimpanzees was successful and was reported by Gajdusek et al (1966). Kuru, in contrast to CJD, presented clinically as an ataxia with dementia being a late sign. Kuru appeared to be maintained in the Fore because transmission was effected by consumption of all body parts of deceased relatives during funeral rites. Thus, kuru was orally transmitted, though cutaneous infection through jungle sores was also possible. The disease is dying out following the cessation of endocannibalism in the late 1950s following decree by the Australian Government.

2.3 Other Animal TSE until 1985

Apart from scrapie, only two other animal TSE were known in 1985. These were a feed-induced TME of farmed mink and CWD of mule deer and elk in North America. Neither of these diseases has occurred in countries affected subsequently with BSE so it is most unlikely that TME or CWD are the origin of BSE. TME affects and kills virtually all the adult mink on affected farms but has not been reported in the wild. CWD is more sporadic affecting animals in some wildlife parks but more recently has been found in the wild.

2.4 BSE and New TSE in Animals after 1985

In the summer of 1986 a solitary case of a spongiform encephalopathy was identified in a captive nyala (*Tragelaphus angasi)*, an African antelope, in a zoo in Great Britain. At the time it was regarded as an interesting curiosity but in reality it was the herald of something far more sinister - the first cases of BSE in cattle. These were identified in November 1986 although epidemiological studies later revealed that the first clinical case of BSE had probably occurred as early as April 1985. During and subsequent to 1987 the epidemic of BSE in cattle increased and sporadic cases of spongiform encephalopathy were found in a limited range of captive wild ruminants in zoos and wildlife parks (Kirkwood and Cunningham, 1994). In 1990 the first cases of feline spongiform encephalopathy (FSE) were reported in domestic cats in Great Britain (Wyatt et al. 1990, Pearson et al. 1993). A case each was also subsequently reported in Norway and Liechtenstein but without any connection to the cases in Britain.

FSE has been reported now in several species of captive wild cats including cheetah, lions, tigers, pumas and an ocelots (Bradley, 1997b, MAFF,2000 Personal Communication). All these species showed clinical neurological signs and spongiform encephalopathy, with presence of PrP^{Sc} in the brain when tested.

3. Origin of BSE and New TSE

Epidemiological studies were initiated in 1987 and by the end of the year it had been established that BSE was a new disease, the mean incubation period was 60 months, and the first effective exposures occurred in 1981-1982. The cause was an increased exposure to a scrapie-like agent via concentrate feed. The vehicle in the concentrate feed was meat-and-bone-meal (MBM) (Wilesmith et al 1988). An increased exposure was proposed because MBM had been fed to farmed, food animals, including cattle, for decades previously without apparent mishap. The reason for the timing of the epidemic was related to changes to the processing of animal waste to prepare MBM occurring around the time of putative first effective exposure (Wilesmith, Ryan and Atkinson 1991). The epidemiological studies also explained the occurrence of disease in captive wild BOVIDAE because these were essentially fed the same rations as domestic cattle. MBM was included in some feed for domestic cats too. Captive wild FELIDAE were however most likely exposed following consumption of BSE-infected cattle heads or vertebral columns from fallen stock. These could have contained central nervous tissue and thus, potentially high doses of infectivity. Furthermore, results from experimental transmission of BSE to cattle, sheep, goats, mink and mice, from brain material via the oral route (Bradley, 1996), and experimental rendering studies supported this view (Taylor, Woodgate and Atkinson, 1995; Taylor et al. 1997). Wilesmith et al. (1988) and Wilesmith, Ryan and Atkinson (1991) proposed three hypotheses for the origin of BSE. First from a mutant form of scrapie but this was not supported by the epidemiological findings that were indicative of an extended common source epidemic rather than of a propagative epidemic. Second, it could have been from a scrapie-like agent from sheep. As sheep were known to be a reservoir of TSE infection this seemed scientifically plausible. Third, BSE could have arisen from a cattle-adapted scrapie-like agent even though there was no evidence to suggest that cattle were even a sporadic reservoir. These last two hypotheses were nevertheless compatible with the epidemiological findings.

4. Connections between Animal and Human TSE

Although man and a small number of animal species (sheep, goats, deer and mink) had been identified with TSE before 1985 there was no evidence of any epidemiological connection between the animal and human diseases or *vice versa*. However, in the early days of the BSE epidemic it was considered that the risks from cattle with BSE might be different than those from sheep with scrapie, partly because cattle with BSE might harbour more infectivity than sheep with scrapie, and partly because the host range for the BSE agent might have been different from that of the scrapie agent. This range might have included man, though there was no evidence until 1995-1996 to suggest that the BSE agent from cattle was indeed a human pathogen.

5. Risks to Humans from BSE

In April 1988 the UK Government set up a Working Party to examine the implications of BSE in relation to animal and human health hazards and to advise the Government on any necessary measures. The Working Party reported in February 1989 having made several valuable interim recommendations during the course of investigation (Report, 1989). During the period from June 1988 until February 1989 the Government had made BSE a notifiable disease, had banned the use of ruminant protein in ruminant rations (mainly because of the epidemiological findings mentioned above), and had introduced a compulsory slaughter, destruction and compensation policy (as a result of advice from the Working Party) for clinically suspect cattle. The brains had to be examined for evidence of disease.

The principal outstanding issue was to protect public health from the risk of being exposed to the BSE agent in food via infected tissues of clinically healthy incubating cattle, killed for human consumption. The Working Party did not recommend specific action but nevertheless stimulated thought in this direction that culminated in the specified bovine offals ban in November 1989. The offals chosen to be excluded compulsorily from the human food chain, were selected on the basis of the knowledge of the pathogenesis of scrapie in goats and Suffolk sheep (Hadlow et al. 1980,1982). They were the brain, spinal cord, tonsil, thymus, spleen and intestine, from duodenum to rectum inclusive. The idea was to remove from the human food chain those tissues that were, on scrapie evidence, most likely to harbour significant levels of BSE infectivity. However, it was essential to replace the scrapie data with data from field cases of BSE. So in order to do that, a series of transmission studies using up to 50 tissues from cattle affected with BSE were inoculated into susceptible mice. The Working Party also recommended that a Consultative Committee on Research (The Tyrrell Committee) be set up. This was done, and subsequently this Committee became, in 1990, the Spongiform Encephalopathy Advisory Committee or SEAC. The SEAC was to play a most important role in the subsequent events that led to the announcement of vCJD in 1996 (see below).

6. The TSE Agents and Strains

Just like conventional microbiological agents, those causing TSE exist in a number of strains that can be distinguished biologically following strain typing in five strains of inbred laboratory mice. Typing is done by comparison of the length of the incubation periods and the neuropathological lesion profiles for each isolate with that for known, cloned strains. It transpired that the BSE agent strain was distinct from all known strains of scrapie. Indeed it was unique. Interestingly, strain typing of three isolates from domestic cats with FSE, one each from the nyala and a great kudu with TSE, and from cases of experimental BSE in sheep, goats and pigs all had indistinguishable strain properties from the BSE agent from cattle (Bruce et al., 1994).

This was a serious challenge to the widely held prion theory of agent structure, which argues that the agent is an infectious protein, prion protein or PrP. This is because the amino acid sequence of the PrP from each of these species is distinct so how could proteins with different amino acid structures communicate a single biological phenotype in experimental mice? A hypothesis was established, and supported by research, that prion protein could code for different biological and phenotypic properties by virtue of its shape or conformation, particularly at certain important domains in the protein. In other words proteins with different amino acid sequences could nevertheless exist in similar conformations. These proteins could imprint their properties on to the normal host protein in the infected individual, even of a different species, and thus converting the normal host protein PrP^{C} into the pathological and disease-specific form PrP^{Sc}.

The agent structure hypothesis is not yet resolved (Schreuder, 1994). Some believe it is a virino (a small unidentified nucleic acid protected by prion protein). Others believe it is an unconventional virus. The virino hypothesis provides a simple explanation for the coding of strain properties by conventional molecular biological laws, nucleic acids (that have yet to be detected), communicating the essential information.

6.1 Host Range for the BSE Agent

The natural and experimental host range for the BSE agent is shown in Table 1.

Table 1. BSE agaent: Natural and experimental host range

PRIMATES	RUMINANTIA	FELIDAE	MUSTELIDAE	RODENTIA	OTHER ARTIODACTYLA
Man	Cattle	Domestic cat	Mink	Mice	Pigs
Lemur?	Nyala	Puma			
Rhesus Monkey?	Gemsbok	Cheetah			
Monkeys:	Greater kudu	Ocelot			
Marmoset	Arabian oryx	Tiger			
Macaque	Eland	Lion			
Squirrel *	Scimitar-horned oryx				
Capuchin *	Ankole				
Lemurs?	*Bison bison*				
	Sheep				
	Cattle				
	Goats				

Hamsters & chickens challenged by the i/c route did not succumb

Chickens challenged by the oral route did not succumb

* UNPUBLISHED INFORMATION COURTESY C J GIBBS Jnr .

7. Detection of Infected Animals

There is no practical test to detect infected animals. The only way this can be done is by bioassay in a susceptible species, and this can take years. Tests to detect PrP^{Sc} (the disease specific form of prion protein) are available and these are now used in cattle and sheep. In cattle with natural disease no peripheral tissues show infectivity or PrP, so the tests are essentially confined to post mortem examination of central nervous tissue (Moynagh and Schimmel, 1999).

In sheep, several tests have been reported such as *PrP* detection in the tonsil (Schreuder et al. 1996) and the third eyelid (nictitating membrane) (O'Rourke et al., 1998).

8. BSE in Cattle

BSE occurs as a natural disease and it can be generated experimentally by parenteral or oral challenge with high titre brain material.

8.1 Natural BSE

Natural BSE is a disease mainly of adult dairy cattle since it is mainly dairy calves destined for breeding that received concentrate rations containing MBM and live long enough to express the incubation period. All breeds appear to be susceptible, though in practice by far the majority of cases occur in Holstein Friesians or crosses of these, since this breed numerically dominates the dairy breed structure in affected countries. The UK has had by far the most cases (c.180,000). The total number of cases elsewhere is close to 1,500 (MAFF, Personal Communication).

Geographical Distribution
Originally confined to the UK, BSE was later discovered in some other countries of the EU (including Benelux, Denmark, France, Ireland, Portugal and Liechtenstein) and Switzerland. This has led to a Europe-wide approach to public and animal health protection, notably and separately, through the EC for all Member States of the EU and the Swiss Government for Switzerland. A small number of isolated cases has occurred in far flung lands including Canada, the Falkland Islands and the Sultanate of Oman and in Europe as a result of importing unknowingly infected dairy calves or heifers destined for breeding from the UK.

Tissue Distribution of Infectivity
Research has demonstrated that by contrast with scrapie in sheep, infectivity is restricted to the brain, spinal cord and retina in natural cases (Bradley, 1999). Around 50 other tissues have been bioassayed in mice and no detectable infectivity has been found (Bradley, 1996b, MAFF, 2000).

Semen and Embryos
No detectable infectivity has been detected in any male or female reproductive tissues, placenta or in semen or embryos (Bradley 1996a). Embryos derived from BSE-affected cattle (that are not used in practice) have not transmitted BSE to recipient heifers or to the offspring kept for seven years, though the experiment does not finish until 2001, (Wrathall, 2000).

Edible Products
In natural BSE, outside the CNS tissues mentioned above, no tissue, of about 50, from clinically affected confirmed cases of BSE that have been tested has demonstrated infectivity after bioassay in mice (Bradley, 1996b, 1999; MAFF, 2000). Neither has infectivity been found in tallow derived by rendering animal waste spiked with brains from cattle affected with BSE, either before or after filtration (Taylor, Woodgate and Atkinson, 1995; Taylor *et al* 1997).

8.2 Experimental BSE

Since it is not possible to identify cattle incubating natural BSE, the pathogenesis of the disease can only be studied experimentally. This has been done following oral challenge of 4 months old calves with untreated brain material from confirmed cases of natural BSE. Parallel studies have investigated the minimum infective oral dose and attack rate. The results of these experiments are briefly summarised below.

Pathogenesis
In this study 30 calves were each challenged with 100g of brain (Wells et al., 1996). There were ten un-dosed controls. At six months of age (two months after dosing), and thereafter at approximately four monthly intervals, three case calves were killed and about 44 tissues

were collected from each. One control animal was similarly killed and sampled. Tissues were intracerebrally inoculated into susceptible mice. The mice were observed clinically and brains were examined at death, or termination, for evidence of TSE (Wells et al. 1998).

No tissues collected two months after challenge showed evidence of infectivity. Clinical signs in cattle were first observed at 35 months of age, though no animals were killed at this time point. The distal ileum showed consistent infectivity from 10 to 16 months after challenge and again from 32 months post-challenge. Otherwise only CNS tissues and ganglia showed detectable infectivity and then not before 32 months post-challenge (three months before clinical onset). The results are summarised in Table 2.

Table 2. Clinical signs, brain pathology and tissue infectivity by interval form challenge during the pathogenesis of experimental BSE in cattle following oral exposure to infected brain.

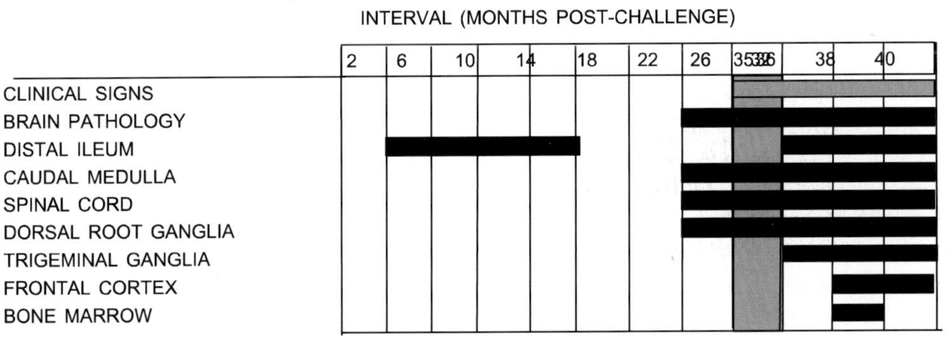

No animals were killed at 35m Post Challenge, the time of onset of clinical signs in survivors

NOTE some studies are incomplete

Data courtesy of Mr G A H Wells and Mr S A C Hawkins September2000

It can be concluded from this study that during the incubation period of experimental BSE, the tissue distribution of BSE infectivity is very much more restricted than the distribution of scrapie agent in Suffolk sheep with natural scrapie, (Hadlow *et al,* 1982). Furthermore, other than in the distal ileum, infectivity is not detectable in other tissues until about 3 months before the onset of clinical signs. Current research is repeating some of the tissue transmissions from the same source material by using cattle, thus eliminating the species barrier and increasing the sensitivity of the assay. To date (September, 2000) no tissue negative in the mouse bioassay has become positive in the cattle bioassay, though studies are as yet incomplete (S.A.C. Hawkins, personal communication).

Attack Rate and Minimum Infective Oral Dose

In this study four groups of ten, 4 month old calves were fed: 100g of BSE-infected bovine brain on three occasions; 100g of BSE-infected bovine brain on one occasion; 10g on one occasion; or 1g on one occasion. Cattle were kept until clinical signs were observed and then were killed and the brains were examined for evidence of TSE.

Cattle in all four groups have developed clinical and pathological evidence of BSE (Bradley, 1997a; Collee and Bradley, 1997). It therefore can be concluded that 1g of brain from clinically-affected cattle can contain an oral infectious dose of BSE for 4 month old calves. This has important messages for control authorities. First, even less than 1g may carry an infective dose (further studies are in progress to test this hypothesis) and second, it supports the view that cross contamination of ruminant diets with small amounts of feed or ingredients containing MBM destined legally for non-ruminant species (Wilesmith, 1996, 1998) is probably responsible for the majority of cases of BSE born after a feed ban was in place.

Comparative Bioassay
This study (Bradley, 1999) was aimed at determining the relative efficiency of cattle and mice for detecting infectivity in brain, spleen and a pool of lymph nodes following intracerebral challenge. Five source cattle with confirmed BSE provided a pool of tissues for inoculation. For brain only, an end-point titration was conducted. Both mice and cattle detected infectivity in brain but neither study has detected infectivity in either spleen or the lymph node pool though this part of the study is not yet quite complete. Cattle were about 500 times more efficient at detecting infectivity in brain than mice, as determined by titration (G.A.H Wells and S.A.C. Hawkins, Personal Communication).

This confirms that lower levels of infectivity can be detected by using cattle for bioassay than by using mice. However, it also supports the view that the pathogenesis of BSE in cattle is different from that of scrapie in sheep. In particular, the spleen and lymph nodes appear to be devoid of detectable infectivity at any stage of incubation or in the clinical phase of the experimental disease. Thus it can be concluded that the risk to public health from these tissues that inadvertently may have been included in human food or might contaminate that food is likely to be very small. By contrast, consumption of central nervous tissue from an affected animal, or from one late in the incubation period, whether by accident or design, may result in exposure to the BSE agent and thus present a significant health risk.

9. Control Measures

9.1 Principles

The principle for protecting animal and public health is to eliminate or reduce exposure to the infectious agent to a level where infection and disease cannot occur. It is convenient to discuss the measures under two headings, namely public health and animal health. It is important to recognise if public health is to be protected from an actual risk the measures must be applied immediately. However, it is also important to recognise that animal health measures, that may take some time to be fully effective, can also be regarded as a long-term and valuable public health protection measure.

9.2 Public Health Control Measures

Two principle measures have been applied, namely compulsory slaughter, compensation and destruction of clinically affected animals, and removal and destruction of infected tissues from dead animals – the specified bovine offals (SBO) ban.

Slaughter, Compensation and Destruction of Clinically Suspect/Affected Animals

This measure was adopted in June1988 in the UK and was followed throughout the EU and Switzerland in 1990. For it to be effective those whose job involves cattle contact must be aware of the clinical signs of BSE and be responsible enough to report their suspicion to the appropriate authority. This demands that a continuous awareness campaign is established and realistic compensation is offered to maximise reporting. At the same time there should be no disincentives such as significant charges for official diagnosis and destruction or unjustified secondary events such as compulsory herd slaughter. Incineration is the preferred route of destruction of diseased animals, but some countries bury them.

The Specified Bovine Offals Ban

This second measure, adopted in the UK in November 1989 was not followed comprehensively throughout the EU until October 2000 (though some BSE-affected countries, including Switzerland, adopted some form of SBO ban soon after the first occurrence of BSE in their country). The initial ban specified six tissues that must be removed for destruction from all cattle over six months of age. These were the brain, spinal cord, thymus, tonsil, spleen and intestines from duodenum to rectum inclusive. The list was based on the available knowledge of sheep and goat scrapie but has been progressively amended in response to new information and risk analyses (see Table 3). The ban originally

Table 3. Direct protection of public health from BSE (GB).

DATE	CLINICALLY HEALTHY CATTLE		BSE SUSPECT CATTLE
PRE BSE	NO SBO (EXCEPT THYMUS) IN UNCOOKED MEAT PRODUCTS		
JUN 1988			NOTIFICATION COMPULSORY SLAUGHTER & DESTRUCTION
AUG1988			
	CATTLE > 6M	CALVES <6M	
NOV 1989	SBO BAN		
NOV 1994		NO INTESTINE OR THYMUS	
AUG 1995	NO SKULLS		
DEC 1995	NO MRM from vertebral column		
MAR 1996	NO CATTLE >30M (OTMS)		
	NO HEADS		
	Export ban on all UK live cattle and products except milk imposed for consumer confidence reasons		
DEC 1997	NO BONES		
AUG 1999	**DBES INTRODUCED - EXPORT BAN ON DE-BONED MEAT LIFTED** *(Except for France)*		
DEC 1999	BONE-IN BAN LIFTED		
JUL 2000		CALF INTESTINE/THYMUS BAN LIFTED	

THE SPECIES BARRIER

MAN

implemented to protect public health was applied to all species of animal and bird from September 1990 following the experimental transmission of BSE to a pig. The SBO ban was incompletely enforced, particularly in regard to animal health protection until after1995 when the Meat Hygiene Service was formed.

SBO = specified bovine offals
OTMS = over 30 months scheme
MRM = mechanically recovered meat
DBES = date based export scheme

9.3 Animal Health Control Measures

A wide range of animal health control measures have been adopted in the UK. These include isolation of parturient cows suspected to have BSE until 72 cows after calving and destruction of the placenta and contaminated materials and buildings (since 1988) and the SBO ban and its modifications, the specified bovine materials ban (SBM) and the specified risk materials (SRM) ban (see above). More widely adopted controls have been applied to the parameters necessary to reduce the TSE risk in rendered products. However, the principle measure to protect animal health has been the ruminant feed ban and its modifications.

The Ruminant Feed Ban
A ban on the feeding of ruminant protein to ruminants was applied in the UK from July 1988 and though some countries adopted a similar measure from 1990 onwards some countries that later reported BSE in their cattle, did not. However, a mammalian MBM ban for ruminants was introduced throughout the EU from 1994. Following the announcement of a new variant form of CJD in March 1996 in the UK, this country adopted a ban on the feeding of mammalian MBM to all food animal species including horses and fish. Had the initial 1988 ban been completely effective from the date of application, BSE in the UK would have already been eliminated. Weaknesses in this ban and the SBO ban will be discussed in the next section.

10. Leakiness of Bans and Further Action

From 1994 onwards, audits on the removal of SBO by the State Veterinary Service in the Great Britain revealed significant deficits of enforcement. This created a particular risk for animal health (though there was some risk to human health, particularly as a result of the incomplete removal of spinal cord) because SBO that should have been destroyed were in fact sent for conventional rendering and thus illegally entered the animal feed chain in the form of MBM. At this time new rendering parameters had not been introduced so that BSE infectivity, if present, would not be completely inactivated in all plants. Additionally it transpired that some enterprises were removing the brain from skulls as SBO, as determined by the law, but residual brain material remained in the cranium and went for conventional rendering thus also creating a BSE risk. This practice was stopped by specifying the skull instead of brain as SBM (Table 3) and additionally removed the eye, in the retina of which from a case of natural BSE infectivity had been found. Later still, in March 1996 after new variant CJD was announced, the head instead of skull was specified as SBM, to further improve the disease security (Table 3).

The feed ban was also leaky. Initially it was expected that some feed would have been in the production pipeline but would soon be exhausted and so the number of cases of BSE that developed in cattle born after the ban (BAB cases) would diminish and be totally small. However, over 40,000 of the c.180,000 total cases of BSE in Great Britain have been BAB cases. A more sinister continuing problem, identified by diligent epidemiological detective work, was also discovered. This was the accidental cross contamination of ruminant diets with MBM destined for non-ruminant species (Wilesmith, 1998).

10.1 Cross-Contamination of Ruminant Diets

Two independent studies revealed the problem in all its detail. Epidemiological studies (Wilesmith, 1998) revealed that in regions of Great Britain with high pig and poultry populations the regional incidence of BSE was increasing whilst in all other areas it was decreasing. Feed mills that prepared feed for ruminant and non-ruminant species in the former regions ran a higher risk of accidentally cross-contaminating ruminant diets with small quantities of MBM legally incorporated into pig and poultry diets than mills in other regions. This occurred *via* vehicles, auger pits, augers, other mill equipment, or sometimes, subsequently on farm. This explanation was strongly supported by the attack rate study results (Bradley, 1997a, Collee and Bradley, 1997, see above) that showed that 1g of wet brain tissue is sufficient to cause BSE by the oral route. An even smaller amount of dried brain material might therefore carry an infectious dose, especially if rendering was ineffective at inactivating any BSE infectivity present. It was a practical impossibility to prevent this cross contamination completely so long as common equipment was being used. This was the main reason that the SEAC advised, in March 1996, that no mammalian MBM should be fed to any food animal species.

11. Mechanically Recovered Meat (MRM)

Following de-boning of the carcase, part of the meat trade recovered the remnants of meat on cattle bones in the form of MRM. MRM is a slurry of soft tissues separated from the bone by use of pressure. Skulls that might have been a high BSE-risk source were not used as the teeth damaged the expensive equipment. However, the vertebral column bones were used and so the slurry would likely have contained dorsal root ganglia (that showed infectivity in the pathogenesis study (Table 2) and any remnants of spinal cord that had been incompletely removed after carcase splitting. If the source bones contained infectivity in the attached nervous tissue this could end in the MRM and thus create a BSE risk for consumers. Although the experimental pathogenesis study showed that sternal bone marrow contained detectable infectivity at one stage of the clinical period, this tissue is not regarded as presenting a significant risk to the consumer (see Wells *et al*, 1999 for a full interpretation of the findings) MRM is generally used for products like pies and burgers. Once this potential risk was identified the SEAC advised that bovine vertebral column should not be used to prepare MRM (Table 3).

11.1 CJD

No increase in the incidence of CJD occurred that was attributable to exposure to BSE during the BSE epidemic up until 1995. However things were to change during the period 1995 - March 1996.

Variant CJD (vCJD)
This is a new acquired form of CJD, mainly, and unusually, affecting young people (mean age 26 years versus 65 years for sporadic CJD), with distinctive neurological signs, long duration of illness (mean 13 months, versus 5 months for sporadic CJD) and pathognomic neuropathology (Ironside, 1999). Presenting signs are usually psychiatric or sensory and are followed by ataxia, myoclonus and other movement disorders. Dementia occurs late in the clinical course. The neuropathology is characterised by large and numerous PrP positive amyloid plaques surrounded by spongiform change, arranged like petals of a flower and occurring throughout the cerebral cortex and cerebellum, (Ironside, 1999). All patients so far reported and examined are homozygous for methionine at polymorphic codon 129 of the *PrP* gene.

Initial Announcement
On 20 March 1996 the Secretary of State for Health of the British Parliament announced the occurrence of ten cases of vCJD (Will et al., 1996). He said, quoting SEAC "Although there is no direct evidence of a link, on current data and in the absence of any credible alternative, the most likely explanation at present is that these cases are linked to exposure to BSE before the introduction of the SBO ban in 1989".

The Consequences
The consequences for the UK were immediate and severe. There was a loss of consumer confidence in beef and beef products stimulated by the concern over human health. This concern spilled over into other countries in Europe. The European Commission, at the behest of the Member States of the EU, imposed an immediate ban on the export of all UK live cattle, beef and beef products except milk that was regarded as presenting a negligible BSE risk. Conditions for lifting the ban were established. These demanded various culls that ultimately resulted in several million cattle being killed and removed from the food and feed chains. More funding for research was provided.

Relationship to BSE
It was subsequently shown by Bruce et al., (1997), that the biological strain type of the agent isolated from three brains of patients with vCJD were indistinguishable from the BSE agent isolated from cattle. Further support for the close similarity between the agents was determined by molecular typing of the agent (Collinge et al., 1996). Despite the clear close association between the agents, the precise source of infection for humans is still unknown.

Geographical Distribution
vCJD has occurred to date only in countries with BSE in native-born cattle. The majority of cases >80 have been in the UK with 2 or perhaps 3 in France and one in the Republic of Ireland.

Case Control Study
Case control studies conducted by the UK CJD Surveillance Unit have revealed no evidence of a direct association between the occurrence of any form of CJD and occupations, including those more likely to associated with higher risk of exposure to a TSE agent. There is also no consistent pattern between the occurrence of any form of CJD and dietary history. However the researchers point out the fragility of the data, the problem of recall bias and some transient statistical associations. Specifically there is no evidence to suggest an increased risk of vCJD through surgery, blood transfusion, iatrogenic means, occupation or a range of dietary factors (CJD, 1999).

12. The Role of the World Health Organisation (WHO)

Along with the OIE (see below), WHO has had a significant role in informing the remote world about TSE; vCJD and BSE in particular. It has held workshops in all its regions collaboratively with the OIE, an important association, as veterinarians and medical doctors need to interact and understand each other's problems. Regional Centres of expertise have been established where advice and assistance with diagnosis can be obtained. Importantly there has been encouragement to undertake surveillance and monitoring for CJD in all its forms in all countries of the world.

By means of a series of Consultations with appropriate world experts, documents giving guidance on the important issues have been published. These include the public health issues related to human and animal TSE including epidemiology, risk and research requirements; the criteria for the diagnosis of CJD, including the differentiation of vCJD; global surveillance, diagnosis and therapy of human TSE; measures to minimise risks to humans from medicinal products and medical devices derived from bovine material and infection control guidelines.

Two simple, but profound statements recorded in the publication immediately following the announcement of vCJD in 1996 should be learned by heart by all concerned with public and animal health protection.

"No part or product of any animal which has shown signs of a TSE should enter any food chain (human or animal), and

Countries should not permit tissues that are likely to contain the BSE agent to enter any food chain (human or animal) (WHO, 1996).

13. The Role of the Office International des Épizooties (OIE)

This world association for animal health, or OIE, has contributed enormously to the international understanding of BSE, partly in collaboration with the WHO (see above) but also on its own account at its own Regional meetings, *ad hoc* meetings and at the annual General Assembly of the International Committee held in Paris each May. Of many contributions two of the most important are the OIE *(International Animal Health Code* (OIE, 2000a) chapter on BSE (and its Supporting Document) and the chapter on BSE in the OIE *Manual of Standards for Diagnostic Tests and Vaccines* (OIE, 2000b). The former provides the recommendations for the safe trading between nations of live cattle and bovine products of all kinds. To achieve such a comprehensive document dealing with a very complex disease and to agree it with Member countries has been a meritorious task of immense value. The latter provides up to date diagnostic techniques acceptable internationally for the investigation and confirmation of BSE.

14. European Commission Decisions

The EC, until shortly after the 1996 crisis, through its Scientific Veterinary Committees, sought and obtained advice on BSE. Sub-groups were set up, comprising experts in the subject under discussion. From these discussions a formal report was prepared. This would be presented for adoption, sometimes after amendment, by the main Committee. Subsequently, veterinary legislation was drafted, often in the form of Commission Decisions. Directives and Regulations were also prepared. All, once agreed with Member States, were assimilated into national legislation, thus harmonising the approach on important animal and public health issues. One important report (EC, 1994) determined the protocols for the laboratory diagnosis and confirmation of BSE and scrapie that formed the original basis for the BSE chapter in the OIE *Manual* (OIE, 2000b).

Following the 1996 crisis the Scientific Committees were disbanded and then regenerated. In the new organisation matters pertaining to TSE were dealt with by a new, all embracing committee, the Scientific Steering Committee (SSC). Reports on TSE topics were generally prepared, as before, by Expert Groups convened for the purpose and then submitted to the *Ad hoc* BSE and TSE Committee prior to final submission to the SSC. The SSC usually provided a published Scientific Opinion on the issue under discussion. Legislation, if appropriate was then drafted by a separate division, for agreement and implementation by Member States, sometimes after significant amendment and delay. In this way, for example, Community law was secured to establish consistent feed and offals bans within the Union and enabled Member States to protect their consumers from BSE risks that theoretically might have existed in third countries. Such legislation was thus harmonised. One piece of legislation that was not was the ban on the export of live cattle and most cattle products from the UK (and later Portugal) immediately following the 1996 crisis. Subsequently rules were agreed with Member States at the Florence Summit whereby the stringent restrictions imposed could be methodically relaxed and removed. Restrictions were soon lifted on semen and medicines and subsequently on de-boned beef that was derived from cattle in a certified herd scheme or a date-based export scheme.

15. Other Actions on BSE

A number of other actions based on the science of BSE have been taken by the authorities in the UK stimulated by the advice given by SEAC or by the need to satisfy the terms of the Florence Agreement. Importantly they include the Over Thirty Months Scheme (OTMS), various culls of at-risk UK cattle and a complete ban on the feeding of mammalian MBM to any food animal species including horses and fish.

15.1 Over Thirty Months Scheme

On 20th March 1996, amongst other things, the SEAC advised that beef derived from over 30 months old cattle should be de-boned, with all visible and nervous tissue removed, in licensed premises, supervised by the Meat Hygiene Service. The waste derived should be classed as SBM.

However, due to difficulties in meeting these requirements and in agreement with the EC an alternative scheme, the OTMS was substituted. All UK cattle over thirty months old (except for some under 40 months old from very low BSE-risk herds in a Beef Assurance Scheme) that were fit for sale for human consumption would be destroyed and compensation paid. Because incineration capacity was insufficient to meet the demand of over 1 million cattle *per annum*, it was permitted to render all parts (except the hide that was authorised for use in the leather industry) and securely store the resulting MBM and tallow pending availability of incineration capacity. Around 0.5 million tons of such material has been in store. Some new facilities for the generation of electric power from this material have been introduced in recent years so the end products of defunct, but otherwise healthy cattle that originally used to produce meat and meat products for human consumption are now used to produce electric light and power (Figure 1). Some proscribed tallow is diverted for use as a fuel, specifically to generate heat in rendering plants. This process is closely audited.

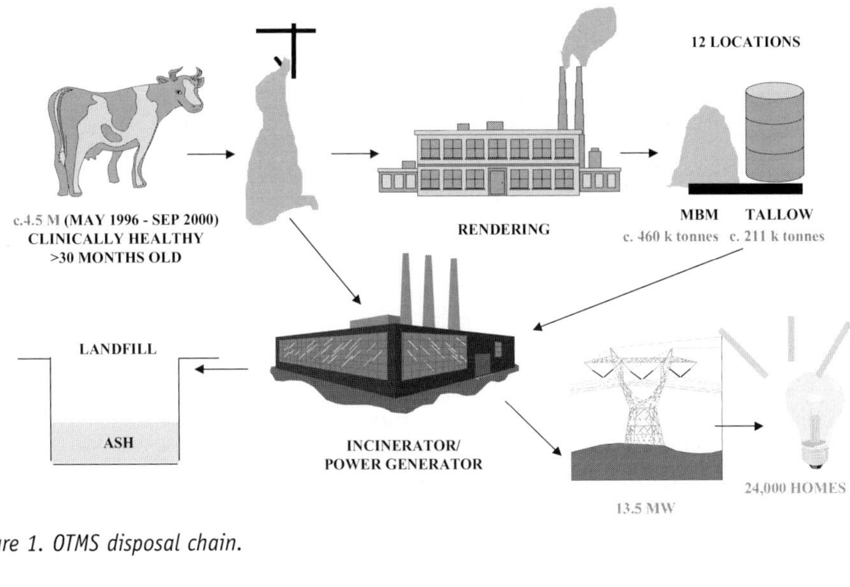

Figure 1. OTMS disposal chain.

Thus a total of around 6 million UK cattle that before BSE appeared, would have been eaten, have been destroyed in the UK as a result of BSE in addition to all the suspect cases.

15.2 Culling

A once and for all selective cohort cull (= cattle in the same birth cohort as a case and therefore likely to have been exposed to the same batch of infected feed) was a requirement of the Florence agreement. Nearly 80,000 cattle have been compulsorily killed and destroyed in this cull. There has been a voluntary and subsequently a compulsory cull of the offspring of cattle that develop BSE and that were born after 1 August 1996, the date when the feed ban was determined to be completely enforced. The total number of cattle slaughtered has been about 10,000 animals. Finally, before 20th March 1996 the UK was legally exporting to EU Member States, some 0.5 million head of calves under 6 months old for rearing to produce veal before they reached 6 months of age. This trade was prohibited by the EC export ban so a Calf Processing Aid Scheme was introduced to remove the surplus unwanted calves from the market. Almost 2 million calves were destroyed until 31 July 1999 when the scheme was abandoned.

16. Network for Control

Figure 2 summarises the control measure network as applied in the UK that reduces the BSE risk to negligible precautions if effectively managed. Account must be taken of the effects on the environment that result from the cultural change of destroying massive numbers of cattle and banned offals that just twelve short years ago were largely consumed by man or animals. However, discussion of this issue is beyond the scope of this paper.

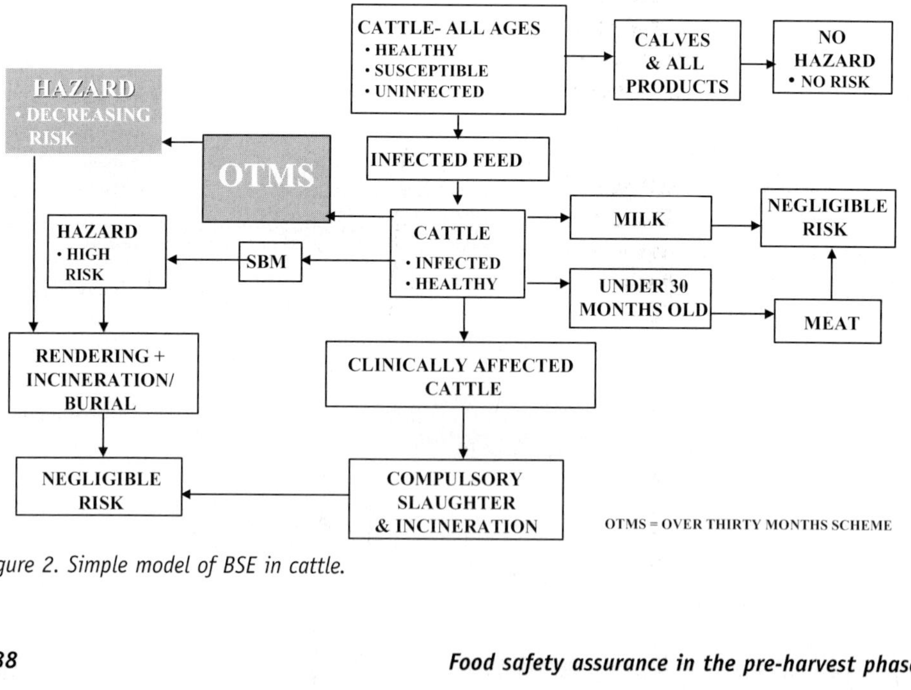

Figure 2. Simple model of BSE in cattle.

17. Lessons from the BSE Epidemic

Of paramount importance is to learn the lessons from the BSE epidemic. They can be summarised thus:

1. The safety of animal feed
2. The safety of food for humans
3. The safe disposal of animal waste to avoid environmental contamination and damage.

Furthermore, the safety of all animal products in regard to TSE risks must be guaranteed by safe souring of staring materials, introduction of effective processing methods and with due regard to the final use of the product.

18. Conclusions

Currently, based on reports to the OIE, BSE in native-born cattle, is confined to Europe. However, this does not mean that a BSE risk is absent everywhere else. Because infected cattle and MBM have been exported from countries with confirmed BSE it is necessary to conduct risk analyses and surveillance to determine if BSE risks exist everywhere cattle are kept.

If a BSE risk is identified appropriate measures should be applied. If these are complied with and enforced the BSE risk will be effectively managed and the disease will be eliminated.

The most important risk management strategies include, for animal health protection, the introduction of a mammalian MBM ban and for public health protection, a slaughter and destruction policy for suspect clinical cases and a specified risk materials ban.

In regard to humans the risk of exposure to the BSE agent mainly depends on occupation and the number and age of BSE-infected animals in the food chain and the consumption habits of individuals.

The BSE risk for an individual, following consumption depends first on the dose consumed (mass of infected tissue (g) x titre/g). Central nervous tissue from an infected animal towards the end of the incubation period or in the clinical phase of disease presents the greatest risk. Second, it depends on the susceptibility of the recipient and the strain of agent involved. Susceptibility appears to be related to the *PrP* genotype of the host.

Worldwide surveillance and monitoring for BSE and CJD is essential.

The guidelines presented by the WHO and the recommendations in the OIE *Code* chapter on BSE are essential to follow if BSE and CJD are to be effectively managed.

18.1 Future

The evidence that derives from the UK, in which country the epidemic started, and Switzerland shows that BSE is controllable and that it is likely to be eliminated unless there is an undiscovered route of infection. If all other BSE-affected countries adopt the same kind of risk management procedures and ensure they are enforced they too will eliminate BSE from there herds. It is vital that all countries in the world follow the OIE *Code* and particularly ensure that ruminant feed cannot be cross-contaminated with MBM under any circumstances. The outstanding issue is that if BSE the disease is eliminated, will the infection also be eliminated?

In regard to vCJD, the only kind of CJD that might be attributable to exposure to infected materials from animals, it is currently not possible to accurately predict the size and extent of the current epidemic. This is because the precise source of infection, the extent and level of exposure, the variation in patient susceptibility and the length of the vCJD incubation period are all uncertain. Surveillance and monitoring for CJD is an essential part of the protection of public health because once cases have occurred it indicates exposure several years or even decades previously. It is possible that human to human transmission of vCJD could occur as a result of blood transfusion, organ transplantation or *via* ineffectively decontaminated surgical instruments used on unidentified infected individuals. This is because unlike sporadic CJD, infectivity is more widely distributed in lymphoreticular tissues. To date no such cases have been identified but several countries have nevertheless taken steps to minimise any risk that there may be.

18.2 What has been achieved

Our knowledge of TSE how they may be transmitted, controlled and eliminated has been enormously extended during the BSE era. Both BSE and vCJD were discovered as new diseases in a short space of time by effective surveillance. Skilled epidemiological investigation led to the discovery of the vehicle of BSE namely MBM. Methods that do and do not inactivate TSE agents in general and BSE in particular have been identified. Earlier weaknesses in the compliance with, and enforcement of, regulations have been identified and corrected, but at the expense of an increase in the size and duration of the BSE epidemic and prolonged risks to public and animal health.

18.3 What has been neglected

Insufficient diligence has been applied to enforce measures that have been correctly applied in countries with BSE. Some countries have now overcome this setback but others still have to learn from their own experience and the experience of others. Insufficient efforts have been applied to communicate, in a user-friendly way, BSE risks to consumers and others.

18.4 What needs to be done

As XIXth century scientists noted, what is needed (at that time for scrapie and now for BSE as well) is a practical, sensitive, reliable and cheap test to identify infection at an early stage of incubation in an infected animal. This would enable a culling programme to be initiated ahead of the disease, instead of behind it, as at present. Present evidence indicates that BSE in cattle can be eliminated and if this is so, the disease could probably be eradicated from the world. This should be the aim in the next decade. The possible occurrence of BSE in other food animal species, notably small ruminants, should be investigated with due speed and diligence in flocks and herds where scrapie exists. In order to do this rapid molecular testing, that should be developed to give equivalent results to biological strain typing, should be employed. Alternative uses for mammalian MBM and tallow from risk sources should be developed that are environmentally friendly and cost effective. Plant-derived substitutes for animal protein that can be consistently grown in Europe should be developed. A close watching brief should be maintained on the human population to detect early occurrences of vCJD in populations without the disease and to detect significant changes in incidence in counties with the disease.

References

Bradley, R., 1996a. Bovine Spongiform encephalopathy distribution and update on some transmission and decontamination studies. In: Bovine spongiform encephalopathy. The BSE dilemma. 1995. Serona Symposia U.S.A. Inc. Ed. Gibbs,C.J.(Jr.). Springer-Verlag, New York. Pp 11-27.

Bradley, R., 1996b. Experimental transmission of bovine spongiform encephalopathy. In: Transmissible subacute spongiform encephalopathies: prion diseases. (Eds. Court, L. and Dodet B. Elsevier, Paris. Pp51- 56.

Bradley, R., 1997a. An overview of the BSE epidemic in the UK. Dev. Biol. Stand. 93, 45-52.

Bradley, R., 1997b. Animal prion diseases. In: Prion diseases. Eds. Collinge J. and Palmer M.S., Oxford University Press, Oxford. Pp 89-129.

Bradley, R., 1999. BSE transmission studies with particular reference to blood. Dev. Biol. Stand. 99, 35-40.

Brown, P., 1980. An epidemiologic critique of Creutzfeldt-Jakob disease. Epidemiologic Rev., 2, 113- 135.

Bruce, M., Chree, A., McConnell,I., Foster, J., Pearson, G. and Fraser, H., 1994. Transmission of bovine spongiform encephalopathy and scrapie to mice: strain variation and the species barrier. Phil. Trans. R. Soc. Lond.B 343, 405-411.

Bruce, M., Will, R.G., Ironside, J.W., McConnell,I., Drummond, D., Suttle, A., McCardle, L., Chree, A., Hope, J., Birkett, C., Cousens, S., Fraser, H. and Bostock, C.J., 1997. Transmissions to mice indicate that 'new variant' CJD is caused by the BSE agent. Nature, 389, 498-501.

Chatelain, J., Cathala, F., Brown, P., Raharison, S., Court, L. and Gajdusek, D.C., 1981. Epidemiologic comparisons between Creutzfeldt-Jakob disease and scrapie in France during the 12-year period 1968-1979. J. Neuro. Sci. 51, 329-337.

CJD,1999. Creutzfeldt-Jakob disease surveillance in the UK. Eighth annual report 1999. National CJD Surveillance Unit, Edinburgh. Pp 53.

Collee, J.G. Bradley, R., 1997. BSE:a decade on – part 2. Lancet, 340, 715-721.

Collinge, J., M.S., Sidle, K.C.L., Meads, J., Ironside, J. and Hill, A.F., 1996. Molecular analysis of prion strain variation and the aetiology of 'new variant' CJD. Nature, 383, 685-690.

EC, 1994. Transmissible Spongiform Encephalopathies. Protocols for the laboratory diagnosis and confirmation of bovine spongiform encephalopathy and scrapie. A report from the Scientific Veterinary Committee. September, 1994. European Commission, Brussels.

Gajdusek, D.C., Gibbs, C.J. and Alpers, M., 1966. Experimental 'kuru' in chimpanzees: a pathological report. Lancet ii, 1056-1059.

Gibbs, C.J., Gajdusek, D.C., Asher, D.M., Alpers, M.P., Beck, E., Daniel, P.M. and Matthews, W.B., 1968. Creutzfeldt-Jakob disease (spongiform encephalopathy): transmission to the chimpanzee. Science, 161, 388-389.

Hadlow, W.J., 1959. Scrapie and kuru. Lancet 2, 289-290.

Hadlow, W.J., Kennedy, R.C., Race, R.E. and Eklund, C., 1980. Virologic and neurohistologic findings in dairy goats affected with natural scrapie. Vet. Pathol. 17, 187-199.

Hadlow, W.J., Kennedy, R.C. and Race, R.E., 1982. Natural infection of Suffolk sheep with scrapie virus. J.Inf.Dis. 146, 657-664.

Ironside, J.W., 1999. nvCJD: exploring the limits of our understanding. Biologist, 46, 172-176.

Kirkwood, J.K. and Cunningham, A.A., 1994. Epidemiological observations on spongiform encephalopathies in captive wild animals in the British Isles. Vet. Rec. 135, 296-303.

Leopoldt, J., 1759. Nützliche und auf die Erfahrung gegründete Einleitung zu der Landwirtschaft, Part 5, chapter 12, Glogau, Berlin. Pp 344-360.

MAFF, 2000. Bovine spongiform encephalopathy: A progress report, June 2000. MAFF, London. Pp239.

Moynagh, J. and Schimmel, H., 1999. Tests for BSE evaluated. Nature, 400, 105.

OIE, 2000a. International Animal Health Code, chapter 3.2.13., Bovine spongiform Encephalopathy. 2000 Edition. OIE, Paris.

OIE, 2000b. Manual of Standards for diagnostic tests and vaccines. Chapter 3.2.13. Bovine spongiform encephalopathy. OIE, Paris (In press).

O'Rourke, K.I., Baszler, T.V., Parish, S.M. and Knowles, D.P., 1998. Preclinical detection of PrP^{Sc} in nictitating membrane lymphoid tissue of sheep. Vet. Rec. 142, 489-491.

Pearson, G.R., Wyatt, J.M., Henderson, J.P. and Gruffydd-Jones, T.J., 1993. Feline spongiform encephalopathy: a review. Vet. Ann. 33, 1-10.

Phillips, Lord of Worth Matravers, Bridgeman, J. and Ferguson-Smith, M., 2000. Report, evidence and supporting papers of the Inquiry into the emergence and identification pf bovine spongiform encephalopathy (BSE) and variant Creutzfeldt-Jakob disease (vCJD) and the action taken in response to it up to March 1996.The Stationery Office, Norwich. Volumes 16.

Report, 1989. Report of the Working Party on Bovine Spongiform Encephalopathy, 1989. Department of Health, Ministry of Agriculture, Fisheries and Food, London. Pp35.

Schreuder, B.E.C., 1994. BSE agent hypotheses. Livest. Prod. Sci. 38,23-33.

Schreuder, B.E.C., van Keulen, L.J.M., Vromans, M.E.W., Langeveld, J.P.M. and Smits, M.A., 1996. Preclinical test for prion diseases. Nature 381, 563.

Taylor, D.M., Woodgate, S.L. and Atkinson, M.J., 1995. Inactivation of the bovine spongiform encephalopathy agent by rendering procedures. Vet. Rec. 137, 605-610.

Taylor, D.M., Woodgate, S.L., Fleetwood, A.J. and Cawthorne, R.J.G., 1997. Effect of rendering procedures on the scrapie agent. Vet. Rec. 141, 643-649.

Wells, G.A.H., Scott, A.C., Johnson, C.T., Gunning, R.F., Hancock, R.D., Jeffrey, M., Dawson, M. and Bradley, R., 1987. A novel progressive spongiform encephalopathy in cattle. Vet. Rec., 121, 419-420.

Wells, G.A.H., Dawson, M., Hawkins, S.A.C., Austin, A.R., Green, R.B., Dexter, I., Horigan, M.W. and Simmons, M.M.,1996. Preliminary observations on the pathogenesis of experimental bovine spongiform encephalopathy. In: Bovine spongiform encephalopathy. The BSE dilemma. 1995. Serona Symposia U.S.A. Inc. Ed. Gibbs,C.J.(Jr.). Springer-Verlag, New York. Pp 28-44.

Wells, G.A.H., Hawkins, S.A.C., Green, R.B., Austin, A.R., Dexter, I., Spencer, Y.I., Chaplin, M.J., Stack, M.J. and Dawson, M., 1998. Preliminary observations on the pathogenesis of experimental bovine spongiform encephalopathy (BSE): an update. Vet. Rec. 142, 103-106.

Wells, G.A.H., Hawkins, S.A.C., Green, R.B., Spencer, Y.I., Dexter, I. And Dawson, M., 1999. Limited detection of sternal bone marrow infectivity in the clinical phase of experimental bovine spongiform encephalopathy (BSE). Vet. Rec. 144, 292-294.

WHO, 1996. Report of a WHO consultation on public health issues related to human and animal transmissible spongiform encephalopathies with the participation of FAO and OIE. Geneva, Switzerland, 2-3 April 1996. WHO, Geneva. Pp 9.

Wilesmith J.W., 1996. Recent observations on the epidemiology of bovine spongiform encephalopathy. In: Bovine spongiform encephalopathy. The BSE dilemma. 1995. Serona Symposia U.S.A. Inc. Ed. Gibbs,C.J.(Jr.). Springer-Verlag, New York. Pp 45-55.

Wilesmith, J.W., Wells, G.A.H., Cranwell, M.P. and Ryan, J.B.M., 1988. Bovine spongiform encephalopathy: epidemiological studies. Vet. Rec. 123, 638-644.

Wilesmith, J.W., Ryan, J.B.M. and Atkinson, M.J., 1991. Bovine spongiform encephalopathy: epidemiological studies on the origin. Vet. Rec. 128, 199-203.

Wilesmith, J.W. 1998. Manual on Bovine Spongiform Encephalopathy. FAO Animal Health Manual No. 2. FAO of the UN, Rome. Pp 51.

Will, R.G., Ironside, J.W., Zeidler, M., Cousens, S.N., Estibeiro, K., Alperovitch, A., Poser, S., Pocchiari, M., Hofman, A. and Smith, P.G., 1996. A new variant of Creutzfeldt-Jakob disease in the UK. Lancet, 347, 921-925.

Wrathall, A.E., 2000. Risks of transmission of spongiform encephalopathies by reproductive technologies in domesticated ruminants. Livest. Prod. Sci. 62, 287-316.

Wyatt, J.M., Pearson, G.R., Smerdon, T.N., Gruffydd-Jones, T.J. and Wells G.A.H., 1990. Spongiform encephalopathy in a cat. Vet. Rec. 126, 513.

Research-based TSE surveillance strategies in europe

Larry G. Paisley[1] and Preben Willeberg[2]
[1]Danish Veterinary Laboratory, Bülowsvej 27, 1790 Copenhagen V, [2]Danish Veterinary and Food Administration, Chief Veterinary Officer, Mørkhøj Bygade 19, 2860 Søborg

1. Introduction

In November 1986, Bovine Spongiform Encephalopathy (BSE) was identified by the Central Veterinary Laboratory, Weybridge, UK. Initial epidemiological studies began in 1987 and on late 1987 it was concluded that ruminant derived bonemeal was the only viable hypothesis for the cause of BSE. In June, 1988 The Bovine Spongiform encephalopathy Order 1988 was made which prohibited the sale, supply and use of certain feedstuffs to ruminants. It also made BSE a notifiable disease. In addition, in 1988, a slaughter and compensation policy for affected and suspect animals was instated. That same year, the Zoonosis Order 1988 (SI 1998/2264) designated BSE as a zoonosis allowing measures to be taken to reduce the risk to human health from BSE. Early measures included the prohibition of the use of milk from affected or suspect cattle for any purpose other than feeding the cow's own calf and the ban on the use of certain specified bovine offal (SBO) for human consumption in 1988 and 1989, respectively (MAFF, 2000). In March 1996, the CJD Surveillance Unit in the UK announced that a previously unrecognized and consistent neurological disease pattern in humans had been identified and concluded that the most likely explanation was that these cases were linked to exposure to BSE before the SBO ban in 1989 (MAFF, 2000).

Since the advent of BSE, the rest of Europe has been challenged with the task to mount effective surveillance programs to demonstrate freedom from or to document likely levels of infection of BSE and other TSEs. The aim of this paper is to describe the development of TSE surveillance in Europe and to discuss the strengths and weaknesses of the various phases.

2. Surveillance vs. Monitoring

Animal disease monitoring describes ongoing efforts directed at assessing the health and disease status of a given population. The term "disease surveillance" describes a more active system and implies some form of directed action will be taken if the data indicate s disease level above a certain threshold (Martin et al., 1987). According to the definitions in the International Health Code of the OIE, monitoring constitutes on-going programmes directed at the detection of changes in the prevalence of disease in a given population and its environment. Surveillance means the continuous investigation of a given population to detect the occurrence of disease for control purposes and may involve testing part of a population (OIE, 2000b).

3. Phases in the Development of Surveillance and Monitoring

Surveillance for TSEs in Europe can be divided into several phases, i.e. i) Passive surveillance consisting of mandatory reporting of BSE suspects and cases, ii) Targeted passive surveillance consisting of examination of a minimum number of brains of livestock exhibiting neurological signs, iii) Targeted active surveillance consisting of testing of higher risk animals for TSE and, finally, iv) Mass testing of all cattle that are slaughtered for human consumption or belonging to a higher risk group

3.1 Mandatory Reporting

Prior to 1998 surveillance for BSE depended primarily on the mandatory reporting of clinical cases. BSE was designated a notifiable disease in the UK in 1988 (MAFF, 2000) and in the EU in 1990 (EC, 1990). This passive method of surveillance had several weaknesses. First, it depended on the ability of farmers and veterinarians to recognize the signs of BSE. In countries assumed to be BSE-free this awareness cannot be considered to have been adequate, whether or not the BSE-free assumption was valid. In addition, typical signs of BSE are not consistently expressed and can be confused with other diseases (Wilesmith et al., 1992; Braun et al., 1999). If the signs were recognized the brain of the animal had to be examined by a laboratory able to diagnose BSE. The negative consequences of having a BSE case on the affected farm which would likely undergo whole herd culling were, at least in some countries, considered as potentially powerful disincentives for reporting suspect cases.

3.2 OIE Rules in 1992

In 1992, the Office International des Epizootiés (OIE) formulated trading conditions for cattle and bovine products from countries affected by BSE. Countries could be considered free of BSE if no cases had been reported or if BSE cases were confined to cattle imported from countries with BSE (MAFF, 2000).

3.3 The World Trade Organization and the SPS Agreement

The establishment of the World Trade Organization (WTO) in 1995 and the implementation of the Sanitary and Phytosanitary Agreement led to important changes in international trade. The members of the WTO agreed to remove barriers to trade in agricultural products except in situations where a particular import can be demonstrated to constitute a risk to the animal, plant or human health of the importing country. Thus, import restrictions applied in the name of animal health must be based on risk assessment methods and should not be disguised barriers to trade. One of the guidelines requires any WTO member state that wishes to apply import restrictions based on freedom from a disease must provide evidence that the disease does not exist in that country. This evidence is usually the result of surveys that establish that the disease is absent or - if present - at or below a certain prevalence with a given level of confidence. Other groups such as the European Free Trade Association (EFTA) have similar requirements.

3.4 The 1996 OIE International Animal Health Code

The 1996 OIE International Animal Health Code chapter on BSE stated that the BSE status of a country can only be determined by continuous surveillance and monitoring. The minimal requirements for surveillance were compulsory notification and clinical investigation of suspect cases, laboratory examination of brain material from suspect cases and registration of the number of investigations and confirmed cases, each of which should be registered and reported as an outbreak (Bradley,1997).

3.5 OIE Classification of BSE Status

Countries were in 1996 classified as having high incidence (UK only), low incidence (France, Portugal, Republic of Ireland, Switzerland, The Netherlands) or being free of BSE. A country could be considered free of BSE if there had been no confirmed BSE case and the requirements for surveillance and monitoring had been met or all confirmed cases were in cattle imported from countries with BSE and the conditions for surveillance and monitoring were satisfied (Bradley, 1997).

3.6 The 1997 OIE Draft Guideline for BSE Surveillance

The 1997 draft OIE guidelines for surveillance for BSE provided tables whereby the number of brain samples from cattle older than 20 months displaying "neurologic" disorders each country was required to examine yearly to maintain a negative BSE status. The sample sizes, adapted from Cannon and Roe (1987), were designed to detect a prevalence of 0.1% with 90% confidence are shown in Table 1. (OIE, 1997).

Table 1. Table showing the minimum number of examinations to be carried out annually to have a 90% chance of detecting BSE occurring in 1% of the cases of neurologic disease in cattle. (Adapted from Cannon and Roe, 1982).

Cattle population 20 months of age or older	Expected neurologic disease 100 cases per 1,000,000 cattle	Minimum number of brains to examine
500,000	50	50
700,000	70	68
1,000,000	100	91
2,500,000	250	151
5,000,000	500	184
7,000,000	700	196
10,000,000	1,000	205
20,000,000	2,000	217
30,000,000	3,000	221
40,000,000	4,000	224

3.7 Risk of BSE from Imported Cattle from the UK

Schreuder et al., 1997 assessed the risk of BSE from the import of cattle from the UK to countries in the European Union. The authors concluded that the number of BSE cases reported in these cattle by the various importing countries were far lower than would have been expected had the animals not been exported and suggested that this could be due to under-reporting and the lack of a uniform system for BSE surveillance (Schreuder et al., 1997).

3.8 Defining the Geographical BSE-Risk (GBR)

On 22 January 1998, the Scientific Steering Committee of the EU Commission issued an opinion on defining the BSE risk for specified geographical areas (SSC, 1998; SSC, 1999). The SSC identified eight factors about which it would need information to assess the geographical risk of countries or regions. The factors included: a) Population structure and dynamics of farmed ruminants, b) Animal trade, c) Animal feed, d) Meat and bone meal (MBM) ban, e) Specified bovine offal (SBO) ban, f) Specified risk material (SRM) ban, g) Surveillance for TSEs, h) Rendering and feed preparation and, finally, i) Culling policy.

For each of the factors, a group of experts was asked to estimate for a given year its impact on the two aspects of the GBR: the propagation risk and the processing (incident) risk. The GBR for a country or region was defined as the combined probability that a) the BSE-agent is currently and in the foreseeable future present in the native cattle herd and b) currently and in the near future one or more BSE-infected native animals per year enter processing in that geographical area.

3.9 OIE Requirements for a BSE Risk Assessment

In May 1998, the OIE General Assembly adopted a revised chapter on BSE in the International Animal Health Code (OIE, 1998). The OIE proposed to determine the BSE status of a country based on a risk assessment in view of measures taken to manage the risk and the BSE incidence. OIE guidelines for surveillance for BSE provided tables whereby the number of brain samples from cattle older than 20 months displaying "neurologic" disorders each country was required to examine yearly to maintain a negative BSE status. The sample sizes were designed to detect a prevalence of 0.1% with 90% confidence. Different requirements for allowing the export or use of bovine-based material were defined based on the BSE status of the country. The OIE guidelines are aimed at facilitating trade so the thrust of surveillance was to document freedom from or a given level of incidence of BSE.

3.10 The First EU Directive Regarding Epidemiosurveillance for Transmissible *Spongiform Encephalopathy*

EU Directive 98/272, the first EU directive regarding epidemio-surveillance for transmissible spongiform encephalopathy, laid down rules for measures to be taken by Member states where a TSE is suspected in an animal, the minimum requirements for monitoring BSE and scrapie and the rules for sampling and laboratory testing for the presence of a TSE (EC, 1998). The directive incorporated the 1997 OIE guidelines for surveillance. In later amendments to Directive 98/272 the minimum sample sizes were adjusted to provide a confidence level of 95% and the list of conditions which might be considered BSE suspects was expanded to include: i) Animals displaying behavioral or neurological signs lasting for 15 days and resistant to treatment, ii) Moribund animals without signs of infectious or traumatic illness and, finally iii) Animals displaying other progressive disease conditions if over 4 years of age.

In addition, other high-risk animals were recommended for testing: i) Animals originating in countries with indigenous TSE, ii) Animals that have consumed potentially contaminated feedstuffs and, finally, iii) Animals born or derived from TSE infected dams and/or sires.

Bovine animals over 24 months of age were to be tested for BSE and sheep and goats over 12 months were to be tested for scrapie. Minimum numbers of animals to be tested were provided in tables based on the total numbers of animals at risk in the countries (Table 2).

This surveillance system was based on the recognition of suspect cases of bovine disorders that might resemble BSE. It was assumed that in adult bovine animals the risk of these conditions appeared with an incidence of 1% per year at risk. Examination of the prescribed number would allow one to conclude that less than one BSE case occurred per 10,000 suspects if none were positive for BSE. This would then appear to warrant an overall incidence of less than 1 case per million adult cattle which is the level of incidence for a country to qualify for the OIE classification of "provisional freedom from BSE".

3.11 The Swiss BSE Epidemic Model

Doherr at al., 1999a developed a model that estimated the expected number of pre-clinical and clinical BSE cases for each year of the Swiss BSE epidemic. Based on unpublished

Table 2. Table showing the minimum number of examinations to be carried out annually under the provisions of 98/272/EC.

Cattle population 20 months of age or older	Expected behavioral neurologic or non-responsive disease cases per 1,000,000 cattle	Minimum number of brains to examine
100,000	10	10
300,000	30	30
500,000	50	50
700,000	70	69
1,000,000	100	99
2,500,000	250	195
5,000,000	500	300
7,000,000	700	336
10,000,000	1,000	367

observations suggesting under-reporting rates up to 75% in Switzerland the authors included under-reporting in the model. The model predicted that the Swiss epidemic would end between 2003 and 2005, assuming there were no new infections after 1997. The model also showed, that in the later stages of a BSE epidemic any surveillance activities targeted at finding clinical cases or animals in the late incubation stages should focus on the older cohorts.

3.12 The Swiss Targeted Surveillance Program

In 1999, Switzerland began a pilot targeted surveillance program, Cattle believed to be at higher risk for having BSE were tested with a new test system for the post-mortem detection of protease resistant prion protein (PrP^{Sc}) (Prionics Check; Prionics AG). The groups targeted included cows older than 24 months that were disposed of as fallen stock, cows older than 24 months presented for emergency slaughter and a sample of cows older than 24 months that were slaughtered. The results of the first nine months of testing showed that the targeted surveillance was able to identify a considerable number of detectable cases particularly in the fallen stock (Doherr et al. 1999b). The prevalence estimates for detectable BSE cases fallen stock older than 24 months was 159 times higher than the prevalence estimate in the general population based on mandatory reporting. The results for the entire year showed an equal number of cases detected by mandatory reporting and the targeted surveillance program (http://www.admin.ch/bvet). Thus, confirming the inefficiency of surveillance based on mandatory reporting of BSE suspects and cases.

3.13 Evaluation of Rapid post-mortem Tests for BSE

During 1999 the European Commission evaluated four tests for the rapid diagnosis of TSE in cattle (EC, 1999). Three were found to have sensitivities > 99.0% and specificities >99.7% at a 95% confidence limit. The three tests approved were the Prionics Check (Prionics AG, Switzerland), Enfer (Enfer Technologies Ltd. Ireland) and CEA (Commissariat à l'Energie Atomique, France).

3.14 The OIE Updated the BSE Chapter of the International Animal Health Code

In May 2000, the OIE updated the BSE chapter of the International Animal Health Code (OIE, 2000). The Code provides rules for importation of animals and animal products based the disease status of a country. It stated that the BSE status of a country or zone could only be determined based on a risk analysis of the following criteria: i) the outcome of a risk analysis identifying all potential factors for BSE occurrence and their historic perspective, ii) on going educational program for all stakeholders to encourage reporting of all neurological disease in adult cattle, iii) compulsory notification and investigation of all cattle showing neurological signs compatible with BSE, iv) a BSE surveillance and monitoring system and, finally v) examination in an approved laboratory of brain or other tissues collected within the framework of the surveillance system.

Countries or zones would be classified as:
- BSE free;
- Provisionally free where no indigenous case has been reported;
- Provisionally free where at least one indigenous case has been reported;
- With a low incidence;
- With a high incidence.

3.15 Final Opinion on the GBR

On 6 July 2000, the SSC adopted a Final Opinion on the Geographical Risk of Bovine Spongiform Encephalopathy (GBR) (SSC, 2000). It contained the overall assessments for 23 countries that had submitted dossiers for assessment. The countries were categorized according to four different GBR levels:
- GBR-level I -highly unlikely that cattle in these countries are (clinically or pre-clinically infected with the BSE agent. This group included Argentina, Australia, Chile, Norway, New Zealand and Paraguay.
- GBR-level II - unlikely that cattle in these countries are (clinically or pre-clinically infected with the BSE agent, but it cannot be excluded.
  The group included Austria, Finland, Sweden, Canada and the USA.
- GBR-level III -likely that cattle in these countries are (clinically or pre-clinically infected with the BSE agent but this is not confirmed (Germany, Italy and Spain[1]) or BSE confirmed at a lower level (incidence <100) (Belgium, Denmark, France, Ireland, Luxembourg, the Netherlands and Switzerland).
- GBR-level IV - BSE confirmed at a higher level (incidence >100). The UK and Portugal were in this group.

3.16 The GBR and the OIE Risk Assessments

The GBR was not strictly in compliance with the recommendations of the OIE regarding BSE risk assessment. In general, OIE identified the elements of the risk analysis process as.
- hazard identification
- risk assessment
- risk management
- risk communication.

The OIE risk assessment process consists of:
- release assessment
- exposure assessment
- consequence assessment
- risk estimation

The GBR assumed that the hazard was the BSE-agent. The release assessment was addressed by assessment of the "external and internal challenge". The SCC method was quite

[1] Germany, Italy and Spain have since detected native-born BSE cases.

compatible with exposure assessment recommendations. The consequence assessment was not addressed by the GBR. The OIE risk estimation consists of integrating the results from the release assessment, exposure assessment, and consequence assessment to produce overall measures of risks associated with the hazards identified at the outset. Thus, risk estimation takes into account the whole of the risk pathway from hazard identified to unwanted outcome. The geographical BSE risk (GBR) for a country or region was defined as the combined probability, that the BSE-agent is currently and in the foreseeable future present in the native cattle herd, and currently and in the near future one or more BSE-infected native animals per year enter processing in that geographical area (OIE, 1998; SSC, 2000).

3.17 The Second EC Decision on Epidemiosurveillance for TSE

Decision 2000/374/ amending Decision 98/272/EC on epidemio-surveillance for transmissible encephalopathies was adopted on 5 June 2000. It provided for a targeted surveillance program similar to that of Switzerland (EC, 2000a; Doherr et al., 1999b). It required that all emergency-slaughtered bovines over 24 months of age be tested for BSE with one of the approved rapid tests for BSE. In addition, a sample of dead animals not slaughtered for human consumption (fallen stock) were to be tested. At least 10% of the samples were to be from fallen stock. Minimum samples sizes based on the total cattle population over 24 months of age were provided in tables. The sample sizes were calculated to detect a BSE prevalence of 0.1% with a 95% confidence in the sub-populations on the assumption that the proportion of animals in the sub-populations in the total population was 1%. The minimum sample size was increased by 500 samples per 500,000 in populations greater than 1.5 million to account for a larger variation in risk of BSE among sub-populations within these populations (Table 3).

Regarding scrapie, the targeted animals should be over 12 months of age. The list of conditions that might be considered scrapie suspects included a) Animals displaying behavioral or neurological signs lasting for 15 days and resistant to treatment, b) Moribund animals without signs of infectious or traumatic illness, c) Animals displaying other progressive disease conditions if over 4 years of age and, finally d) The minimum number of samples to be examined was provided in a table.

In addition, other high-risk animals were recommended for testing: i) Animals originating in countries with indigenous TSE, ii) Animals that have consumed potentially contaminated feedstuffs and, finally iii) Animals borne or derived from TSE infected dams and/or sires.

The Decision was to apply to BSE from 1 January 2001 and January 2002 for Scrapie.

3.18 Requirement to Test all Bovines over 30 Months of Age Slaughtered for Human Consumption

Commission Decision 2000/764/EC of 29 November 2000 required that all bovines over 30 months of age slaughtered for human consumption be tested for BSE with one of the three approved rapid tests as of 1 July 2001, at the latest (EC, 2000b). This was changed through

Table 3. Table showing the minimum number of examinations to be carried out annually under the provisions of 2000/374/EC. The sample size has been calculated to detect a BSE prevalence of 0.1% with 95% confidence in the subpopulations based on the assumption that the proportion of these subpopulations in the total cattle population >24 months is 1% (EC, 2000a).

Total population over 24 months	Minimum sample size*
100,000	950
200,000	1,550
300,000	1,890
400,000	2,110
500,000	2,250
600,000	2,360
700,000	2,440
800,000	2,500
900,000	2,550
1,000,000	2,590
1,500,000	3,000
2,000,000	3,500
2,500,000	4,000
3,000,000	4,500
3,500,000	5,000
4,000,000	5,500
4.500,000	6,000
5,000,000	6,500
5,500,000	7,000
6,000,000	7,500
6,500,000	8,000
7,000,000	8,500
7,500,000	9,000
8,000,000	9,500
8,500,000	10,000
9,000,000	10,500
9,500,000	11,000
10,000,000	11,500
10,500,000	12,000
11,000,000	12,500
11,500,000	13,000
12,000,000	13,500

* Where the size of the total population over 24 months of age is 1,500,000 or more the sample size has been increased by 500 samples per 500,000 animals as a proportionality adjustment to take account of the larger likelihood of variation in risk of BSE within the population.

Commission Decision 2001/8/EC to apply from 1 January 2001 for cattle not being subject to intervention and subsequent destruction (EC, 2001a).

3.19 Determination of BSE Status by the EC

Regulation (EC) 999/2001 of the European Parliament and of the Council lays down rules for the prevention, control and eradication of certain transmissible spongiform encephalopathies (EC, 2001). It was applicable as of July 1 2001. It requires that Member states and third countries that wish to remain approved for export to the EC submit an application for their BSE status to be determined. The BSE status is to be determined based on certain criteria and the outcome of a risk analysis. The requirements are very similar to those of the OIE. In regard to surveillance and monitoring, 999/2001 includes minimum requirements for a programme for monitoring BSE in bovine animals, minimum requirements for a programme for monitoring Scrapie in ovine and caprine animals and monitoring in higher risk animals. The regulation required all bovine animals over the age of 30 months that were routinely or emergency slaughtered be tested for BSE. In addition, a specified sample of fallen stock over the age of 24 months was to be tested. Article 22 of Regulation 999/2001 article states that a conclusive statistical survey shall be used to confirm or overturn the risk analysis conclusions carried out as the first step in determining the BSE status of a country or a region.

Regulation (EC) 999/2001 was amended by Commission Regulation (EC) No. 1248/2001 on 22 June 2001. In part it requires all animals in the sub-population (fallen stock) over the age of 24 months to be tested.

4. The TSE Situation in Denmark

The TSE situation in Denmark can be described in two distinct phases: Prior to and after 27 February 2000 when the first indigenous BSE case was diagnosed. One cow imported from the UK in 1988 developed BSE in 1992 but based on the OIE classification this did not affect Denmark's BSE free status. That status was maintained until the first indigenous BSE case was diagnosed.

Several measures for the prevention of BSE were put into effect prior to the discovery of the first domestic BSE case. The main pillar in epidemiological surveillance in Denmark is the system of notifability of animal diseases (Act no. 351 of 2 June 1999 on Diseases and Infections in Animals). There was a general ban on the import of MBM during the period 1933-1998, with some individual exemptions for imports from Norway, Sweden and Germany.

All farms in Denmark are registered in the computerized national farm and livestock register under the Ministry of Food, Agriculture and Fisheries. The register contains a section on veterinary matters including the name and address of the veterinarian(s) that attend the farm. The animals in Danish herds must be identified since December 1994. Farmers and cattle and pig dealers are required to keep records of all movements of cloven-footed animals.

Feeding ruminant MBM to ruminants was banned in 1990. BSE was made notifiable in 1990. Imports of cattle from the UK were banned in 1990.

Individual monitoring of cattle imported from the UK was begun after 15 May 1996 (Order no. 366 of 17 May 1996). This law laid down rules for marking, moving and destruction of cattle imported from the UK. It banned these imports from entering the food chain, required the cattle to be identified with blue eartags and at death be examined for BSE. Other steps included: i) Ban on feeding mammalian MBM to ruminants- January 1997, ii) Tightening the rule on mixing of cattle feed from January 1997, iv) Adoption of the rules for rendering MBM (133/3/20) in April 1997 and, finally v) Extending surveillance and notifiability in 1998

4.1 TSE Surveillance in Denmark

In Denmark and several other countries in Europe, surveillance for TSE was passive until 1998 when EU Directive 98/272 on epidemio-surveillance came into force. During the period 1990-96, surveillance consisted of examination of brains form clinically suspect animals and after April 1996 clinically suspect animals and cattle imported form the UK. However, during the period 1990-96 only 11 brains were examined for BSE, one being the 1996 BSE case imported from the UK.

In 1996, the OIE International Animal Health Code required continuing monitoring and surveillance for BSE to determine a country's BSE status. Under the OIE guidelines, Denmark was required to examine approximately 100 brains of cattle with neurologic signs, yearly. Assumptions underlying this scheme were, that suspicions should occur in >100 cases per million cow-years. If no positive were found in the sample, the BSE-prevalence would be <1% of suspects with 99% confidence.

The same number of examinations was required by EC Decision 98/272 on epidemio-surveillance for TSE. The numbers of examinations actually done were 6, 24, 39 and 96 in 1997,1998, 1999 and 2000, respectively.

Prior to 27 February 2000, Denmark was presumed free from BSE. This was based on preventive measures that had been taken, the results of the passive surveillance for BSE and an informal risk assessment. The informal risk assessment based on the following elements led to the (spurious) conclusion that Denmark was free of BSE: a) Clinical suspects had been notifiable since 1990, b) MBM import had been banned (with a few exceptions) since 1933, c) Imports of cattle from the UK were banned since 1990, e) No slaughtering or rendering of UK cattle since April 1996 and, finally f) Small numbers of cattle were imported from other countries with BSE after 1990.

A Danish cow exported to Ireland developed BSE in 1996, within 25 months of export. However, it could not be determined with any certainty when and where the cow acquired the infection.

4.2 Assessment of the Geographic BSE-Risk of Denmark

The Final Report on the Assessment of the Geographic BSE-risk of Denmark was published in July 2000 (SSC, 2000b). The conclusion was that current GBR-risk level was Level III, i.e. BSE confirmed at a lower level. It noted that the BSE case diagnosed in February 2000 by a passive surveillance system that was unable to detect all cases. It was assumed that a likely source of the BSE infection had been imported UK cattle during the 1980'es, which before 1996 were slaughtered and rendered as other cattle in Denmark. The number of brains examined for BSE was considerably lower than the OIE recommendations of 100 per year. The report went on to state, that if current risk management measures were continued and well implemented the GBR would be expected to decrease further. Improved surveillance would verify the current risk level and its expected trend.

4.3 Targeted Active Surveillance for BSE in Denmark

Commission Decision 2000/374 of 5 June 2000 mandated targeted surveillance for TSE in Member states beginning 1 January 2001(EC, 2000a). It required that all emergency-slaughtered bovines over 24 months of age be tested for BSE with one of the approved rapid tests for BSE. In addition, a sample of dead animals not slaughtered for human consumption (fallen stock) were to be tested. At least 10% of the samples were to be from fallen stock. Minimum samples sizes based on the total cattle population over 24 months of age were provided in tables. The sample sizes were calculated to detect a BSE prevalence of 0.1% with 95% confidence in the sub-populations on the assumption that the proportion of animals in the sub-populations out of the total population was 1%. Under this scheme, Denmark would be required to test approximately 2500 cattle for BSE yearly. However, the problem for Denmark was not to document "freedom from BSE" but to estimate the prevalence with an acceptable level of confidence. If one assumed that the prevalence of BSE in the emergency slaughtered cattle and fallen stock were 0.1% a sample size of 3000 could not establish that the prevalence was 0.1% but only that it was between 0.02% and 0.28%. That is, the upper confidence bound 0.28% was nearly 3 times higher than the upper limit to document "freedom" from BSE.

4.4 DanishTargeted BSE Surveillance

With this limitation in mind, a new surveillance scheme for Denmark was devised. A total of 15,000 cattle were to be tested. Assuming a BSE prevalence of 0.1% finding no cases in a sample of this size would provide a 95% confidence interval of prevalence of 0.07% to 0.14%, much closer to the desired <0.1%.

The following sub-populations were to be tested: a) All emergency slaughtered cattle over 20 months of age (ca. 2500 per year), b) A sample of fallen stock over 24 months of age (ca. 8000 of an estimated 30,000 fallen stock per year), c) A "feed cohort" (ca. 4500 cattle per year), d) Cattle believed to have consumed the same feed as the February 2000 BSE case (ca. 3100 per year), e) Controls of the same age cohort (ca. 1400 per year), f) Clinical BSE suspects (ca. 100-200 per year) and, finally g) Cattle imported from the UK (ca. 20 per year).

Testing under this scheme began 1 October 2000. Out of ca.21,000 animals tested there have been 3 BSE cases diagnosed since this program began. One case was in a 7-year-old cow that died on the farm while under restrictions for BSE suspicion. One was a 4-year old cow that was a clinical suspect and one was a 5year old cow that died on the farm (fallen stock).

4.5 Goal of the Danish Targeted BSE Surveillance Program

The goal of the BSE surveillance program was to document the prevalence of BSE in the adult cattle population with sufficient precision and validity to enable: i) proper evaluation of the BSE situation for use by international organizations, trade partners and others, ii) evaluation over time the prevalence of BSE with a view to an-going optimization of consumer protection and animal disease prevention and control, iii) identification of other relevant variations such as high risk populations and, finally iv) Collection of evidence of a progressingly low prevalence eventually leading to a renewed status as BSE free.

4.6 Mass Testing for BSE

Commission Decision 2001/8/EC required that all cattle over 30 months of age that are slaughtered for human consumption be tested for BSE with one of the three approved rapid tests as of 1 January 2001 at the latest (EC, 20001a). As of July 2001, the Danish scheme has been revised such that testing is done on: i) All emergency slaughtered cattle over 24 months of age (ca. 2500 per year), ii) All fallen stock over 24 months of age (ca. 35,000 per year), iii) Cattle found to display signs of zoonosis or disease on ante-mortem inspection (ca. 1600 per year) and finally iv) The "feed cohort" (ca. 3100 per year).

5. Discussion

5.1 Mandatory Reporting of BSE Suspects

During the period 1988 when BSE became notifiable in the UK until 1998 when the first EC Decision on epidemio-surveillance came into force, surveillance for TSE in Europe was passive and depended on the mandatory reporting of suspects and cases. For many years, veterinary authorities have used this system for monitoring infectious diseases. The regulations usually require farmers and veterinarians to inform the authorities when an out break is suspected. The officials then proceed with various clinical, pathological and diagnostic investigations to confirm the suspicion. The system works satisfactorily with some diseases and some disease eradication programs have relied primarily on this principle. The system is an important tool for the recognition of introduction of new or exotic diseases. However, under-reporting can be a problem especially if the disease is not considered serious or the consequences of having a case are considered more severe than the disease itself. In addition, it relies on the ability of the farmer to recognize the disease.

5.2 Under-reporting

With TSEs, under-reporting has been considered a serious problem. The efficiency of mandatory reporting is unknown but several reports suggest it is quite low resulting in underestimates of incidence rates Doherr et al., 1999a cited unpublished data suggesting under-reporting rates for BSE of up to 75% in Switzerland. The results of the targeted surveillance in Switzerland suggested that ca. 50% of the detectable cases went unreported (Doherr et al., 1999b, Doherr et al. 2001). Several authors have included under-reporting in BSE epidemic models (Cohen, 1999; Doherr et al.1999a; Donnelly and Ferguson, 2000). Schreuder et al., 1996 presented a model suggesting that many more cases of BSE in cattle imported from the UK were expected than were reported. Prior to 1996 when the putative association with the human disease, variant Cruetzfeldt-Jacob Disease (vCJD) was announced the disease was not considered a human health risk. The whole herd culling policies in most countries could make farmers reluctant to report suspects when slaughtering suspicious animals was a viable option. Because of the adverse effects on trade, countries could be reluctant to admit to the presence of BSE in the national herd.

The situation for scrapie might be worse. Under estimation levels between 30 and 87% have been reported (Schreuder et al., 1993; Hoinville et al., 1999; Baumgarten et al., 2000).

5.3 Active Surveillance for BSE

The OIE requirement for active surveillance that was approved in 1998 was an improvement over the passive system of mandatory notification. Each country would also be required to examine a minimum number of brains of cattle older than 20 months that displayed " neurologic signs". The sample sizes were based on assumptions that neurologic diseases occur at an annual incidence of 0.01% in the population and that BSE prevalence among "neurologic cases" was 1%. Unfortunately, there was little or no scientific evidence to support either of these assumptions. In addition, there was no scientific justification for the level of homogeneity of the incidence of suspicions or the sensitivity of "diagnosing" potential BSE cases as suspects. Countries with large cattle populations were required to sample relatively fewer suspect cases. The result being that smaller countries were at a disadvantage to fulfil the required quota.

While the requirement for examination of BSE suspects and other higher risk animals was expected to increase the probability of detecting BSE cases over mandatory reporting of clinical cases there were problems with this system. These included the lack of scientific data to substantiate the assumptions on the 1% level of suspect conditions across all strata of geography, breeds, production systems etc. The lack of possibilities to police low levels of surveillance when based on reporting suspect conditions. The predictive value of suspect signs depend on the expected incidence, which in countries with no reported cases is considered to be extremely low by those making the original diagnoses, thus tending towards under-reporting of BSE suspects. Statistical assumptions of homogeneous distribution of BSE cases across all strata are unlikely to be valid, especially for large populations, which in turn lead to underestimation of the proper sample size for those countries.

Even with the inherent weakness of mandatory reporting one can gain some information regarding the prevalence in the population. The following model can be used to estimate the BSE prevalence based on reporting of BSE suspects.

- Let N = number of cattle in the population at risk
- Let n = number of suspects reported
- Let x = number of positive BSE cases among the suspects
- p(case) = p(suspect) * p(case|suspect) / p(suspect|case)
- p(suspect|case) = probability a case is reported = sensitivity of reporting = Se
- Prevalence = n/N * x/n/Se = x/(N*Se)

Figure 1 shows the results when 1 BSE case is found among 150 suspects from a population at risk of 1 million cattle. The estimated sensitivity of reporting cases varies between 0.01 to 1.

Decision 2000/374/EC overcame some but not all of these problems. This decision provided for targeted surveillance in "higher risk" sub-populations. The higher risk sub-populations were emergency slaughtered animals and fallen stock. Based on the experience of the Swiss targeted surveillance program in 1999, it was assumed that 1% of the cattle populations over 24 months would end up in one of these groups. At least 10% of the samples must be from fallen stock. The number of samples to be tested was calculated to detect a BSE prevalence of 0.1% with 95% confidence in the sub-populations.

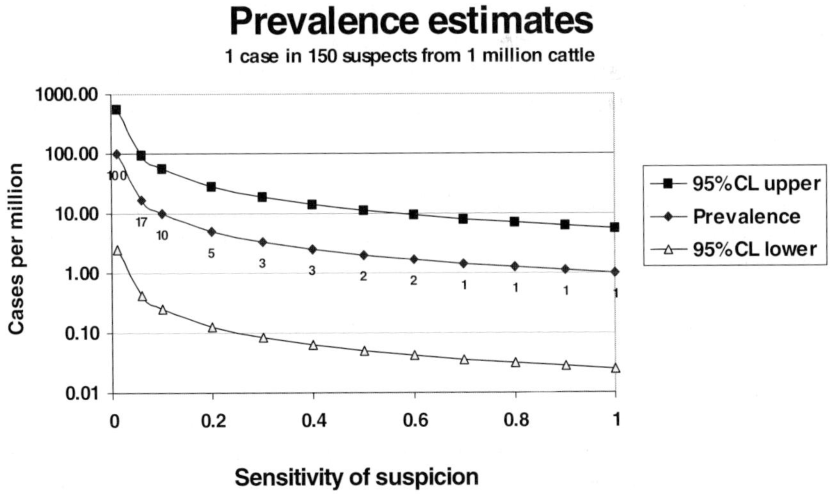

Figure 1. Chart showing the estimated prevalence of BSE cases in a population of 1 million cattle at risk based on finding 1 BSE case among 150 "suspects".

The Decision raised the minimum number samples to be tested in Denmark from 99 to approximately 2500 thus increasing the probability of detecting BSE. However, the assumption that 1% of the population over 24 months would either become fallen stock or emergency slaughtered in all countries has not been validated. In Denmark, 2-4% of the cattle population over 24 months of age become fallen stock (Paisley, unpublished). It was also noted that finding no cases in this sample size would not establish a 95% confidence bound for prevalence of BSE of less than 0.1% in that sample. Furthermore, proper statistical models also incorporating the variation due to sampling of animals in the risk category should be applied in establishing the table.

5.5 Danish Targeted Surveillance for BSE

For this reason, Denmark devised a targeted surveillance program that required 15,000 samples from higher risk sub-populations, which if no cases were found would better justify a classification of "preliminary freedom from BSE". The program began in October 2000 but was soon made moot by the Decisions requiring testing of all emergency slaughtered animals and fallen stock (EC, 2000b; EC, 2001b; EC 2001c).

5.6 EU TSE Research Project

As part of the EC research project PL 98-7021, Surveillance and Diagnosis of Ruminant TSE several groups developed models for surveillance for BSE and Scrapie in the EU. The BSE models suggest that, in terms of cost-effectiveness, targeting specific age groups among the "high risk" sub-populations could greatly increase the efficiency of the surveillance (de Koeijer et al. 2000, Paisley, in preparation). Both models show that with the low BSE prevalence expected in most countries the probability of finding a BSE case in an animal less than 4 years old is extremely small. In Denmark, fallen stock less than 4 years of age comprise approximately 40% of that sub-population. Thus, approximately 40% of the cost of testing would be used with virtually no probability of finding a case. The model of de Koeijer et al., 2000 also demonstrates that the age group with the highest prevalence will progressively become older as the time after effective controls (MBM ban) became effective increases.

The advent of mass testing in routinely slaughtered cattle and high-risk groups made the question of sampling procedures irrelevant. However, the question of the cost-effectiveness of testing cattle less than 4 years of age is still relevant and will become more relevant as the time interval from the MBM ban increases.

5.7 Monitoring the Mass Testing for BSE

When testing of all routinely slaughtered animals became a reality in Denmark in January 2001 the need to analyze and interpret the results became apparent. A classical statistical model and a Monte Carlo simulation model using Bayesian inference were developed to analyze the data and to provide estimates of the BSE prevalence and likely number of cases to be found in 2001. The estimates are based on the current test results (Paisley, submitted). Both approaches have strengths and weaknesses so it was of interest to compare the results. The estimates are revised monthly or when an additional case is diagnosed. The estimates are based on assumptions that 250,000 slaughtered cattle and 25,000 high-risk cattle will be tested in Denmark in 2001. As of 23 August 2001, 147,256 routinely slaughtered cattle had been tested with on2 positive result. The observed prevalence was 0.00068% and the 95% upper bounds for BSE prevalence in the routinely slaughtered population was 0.0024% and for the additional number of cases to be found it was 5. Since testing of the high-risk sub-populations began in October 2000, 12,764 tests have been done with 2 positive results. The 95% upper bounds for BSE prevalence in the high-risk population was 0.044% and for the number of additional cases to be found it was 9.

The results of BSE testing in Denmark are shown in Figures 2-5. Figure 2 and 3 show the prevalence estimates for the routinely slaughtered cattle and risk group cattle, respectively. Figure 4 and 5 show the BSE case estimates for the routinely slaughtered cattle and risk group cattle, respectively.

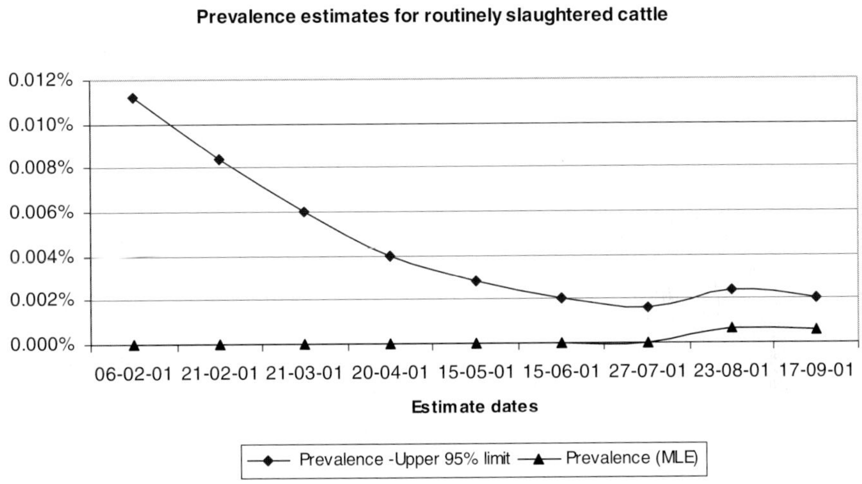

Figure 2. Chart showing the BSE prevalence estimates for routinely slaughtered cattle during the period February-September 2001. The estimates were based on the number of samples tested in a total population of 250,000.

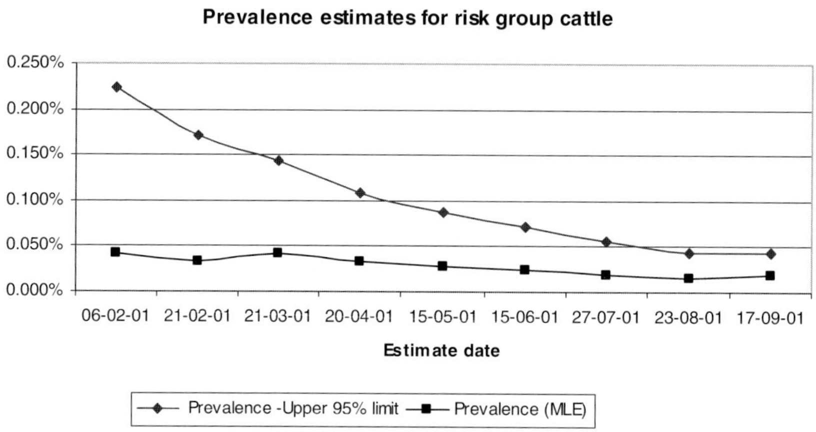

Figure 3. Chart showing the BSE prevalence estimates for risk group cattle during the period October 20 - September 2001. The estimates were based on the number of samples tested in a total population of 25,000.

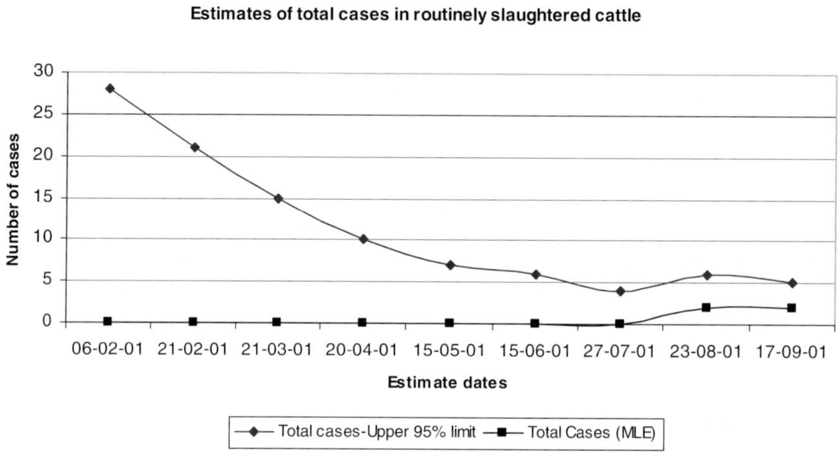

Figure 4. Chart showing the BSE case estimates for routinely slaughtered cattle during the period February-September 2001. The estimates were based on the number of samples tested in a total population of 250,000.

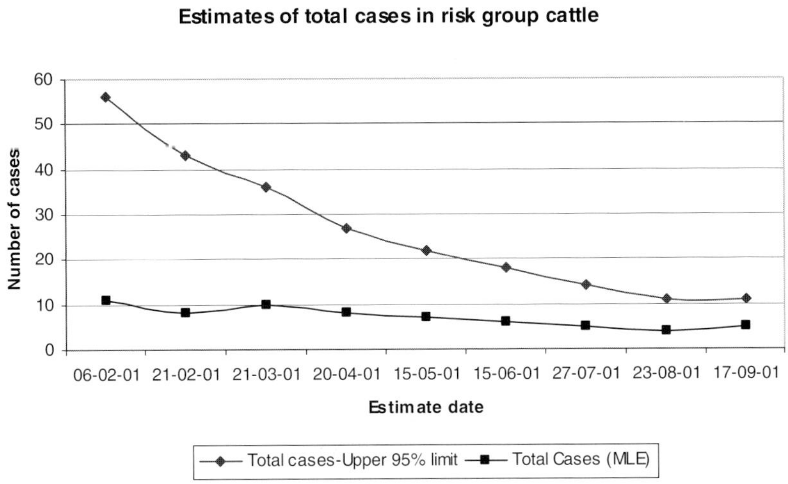

Figure 5. Chart showing the BSE case estimates for risk group cattle during the period October 2000 - September 2001. The estimates were based on the number of samples tested in a total population of 25,000.

6. Conclusions

6.1 What has been achieved?

Surveillance for TSE in Europe has evolved from the phase of mandatory reporting of cases, through active surveillance, active targeted surveillance and finally to the present phase characterized by mass testing of all cattle over 24 months of age, that are slaughtered for human consumption or are among designated high risk groups. Surveillance and testing for BSE is well developed but surveillance for scrapie in sheep and goats has been largely ignored despite regulations to the contrary.

6.2 What has been neglected?

The sampling procedures used until mass testing began were not necessarily scientifically justified. Still, proper statistical models also incorporating the variation due to multi-stage sampling of risk animals should be applied in establishing the table of minimum numbers to be tested.

In addition, the goals of the surveillance and testing programs are not clearly stated and may vary between the OIE and the EC. The OIE requirements for surveillance for TSE have the primary goal of facilitating trade and justification of trade barriers based on the TSE status of the countries. The EC goals are not so clear-cut, especially today. However, consumer protection seems to be a major thrust. It appears that in some countries verifying freedom from TSE is the goal. In others, it is to establish prevalence estimates that justify a classification of "preliminary freedom from BSE/TSE. In others with a higher

TSE prevalence, the goal might to be monitoring the progress of eradication efforts and to identify herds with possible TSE infections.

The EU surveillance program requires testing a minimum sample of brains from scrapie suspect sheep and goats by histopathology. Rapid diagnostic procedures for scrapie need to be developed. To date, none of the rapid tests used to detect BSE in cattle is approved for detection of scrapie in sheep and goats. However, they are approved for monitoring purposes. Positive or inconclusive test results must be confirmed by immunocytochemistry or immuno-blotting (EC, 2001c).

6.3 What needs to be done?

If the BSE epidemic ends within a period of a few years as predicted by several models the surveillance methods will need revision (Doherr, 1999a; Donnelly and Ferguson, 2000). Questions such as "how much is enough surveillance"? "How can we document freedom from TSE?" and "Can TSEs be eradicated?" need to be answered.

Acknowledgement

This project was funded in part by FAIRJ-CT98-7021: Surveillance and diagnosis of ruminant TSE.

References

Bradley, R., 1997. Aetiology, pathogenesis and surveillance for BSE. 27 Aug 2001. <http://senasa.mecon.gov.ar/animal/bsetrab1.htm>.

Braun, U., Amrein, E., Estermann, U., Pusterla, N., Schönmann, M., Schweizer, T., Ehrensperger, F., Vandevelde, M. and Kihm, U., 1999. Reliability of a diagnosis of BSE made on the basis of clinical signs. Veterinary Record 145:198-200.

Cohen, C., Cesbron, J-Y. and Valleron, A-J., 1999. Cost - effectiveness of Bovine Spongiform Encephalopathy screening. Veterinary Record 144:703-706.

Cannon, R.M. and Roe, R.T.,1982. Livestock Disease Surveys, A Field Manual for Veterinarians. Bureau of Rural Science, Department of Primary Industry, Australian Government Publishing Service, Canberra; Cited in: Office International des Epizootiés (OIE)(1997). Bovine spongiform encephalopathy . In: International Animal Health code, draft: 28 Aug 2001. <http://www.maf.govt.nz/tse-conference/970506-02/htm.>

European Commission (EC) 1990. Commission Decision 90/134/EEC of 6 March 1990 on compulsory notification of bovine spongiform encephalopathy. Cited in: Schreuder, B E C 1994. Animal spongiform encephalopathies: an update. Veterinary Quarterly 16:182-192.

European Commission (EC), 1998. 98/272/EC: Commission Decision of 23 April, 1998 on epidemio-surveillance for transmissible spongiform encephalopathies. Official Journal L 122, 24/04/1998.

European Commission (EC), 1999. The evaluation of tests for the diagnosis of transmissible spongiform encephalopathy in bovines.1-36. 28 Aug 2001.
< http://europa.eu.int/comm/food/fs/bse/bse12_en.html>.

European Commission (EC), 2000a. 2000/374/EC: Commission Decision of 5 June, 2000 amending Decision 98/272/EC on epidemio-surveillance for transmissible spongiform encephalopathies. Official Journal L 135, 08/06/2000.

European Commission (EC), 2000b. Commission Decision 2000/764/EC of 29 November 2000 on the testing of bovine animals for the presence of bovine spongiform encephalopathy and amending Decision 98/272 EC on epidemio-surveillance for transmissible bovine spongiform encephalopathies. Official Journal L 305, 06/12/2000.

European Commission (EC), 2001a. Commission Decision of 29 December 2000 amending Decision 2000/764/EC on the testing of bovine animals for the presence of bovine spongiform encephalopathy and updating Annex IV of Decision 98/272 EC on epidemio-surveillance for transmissible bovine spongiform encephalopathies. Official Journal L 002, 05/01/2001.

European Parliament and Council of the European Union (2001b) Regulation(EC) No.999/2001 of the European Parliament and of the Council of 22 May 2001 laying down rules for the prevention, control and eradication of certain transmissible spongiform encephalopathies. Official Journal L147, 31/05/2001.

European Commission (EC)(2001c). Commission Regulation (EC) No. 1248/2001 on 22 June 2001 amending Annexes III, X and XI to Regulation (EC) No 999/2001 of the European Parliament as regards epidemio-surveillance and testing of transmissible spongiform encephalopathies. Official Journal L173 27/06/2001.

de Koeijer, A., Schreuder, B. and, de Jong, M.C.M., 2000. Active surveillance for BSE, targeting risk groups. Poster presentation. EU conference, "Food Safety Assurance in the Pre-harvest phase", Vienna, 30 September-2 October, 2000.

Doherr, M. G., Oesch, B., Moser, M., Vandevelde, M and Heim, D., 1999a. Modelling the expected number of preclinical and clinical cases of bovine spongiform encephalopathy in Switzerland. Vet Rec 155: 160.

Doherr, M. G., Oesch, B., Moser, M., Vandevelde, M and Heim, D., 1999b. Targeted surveillance for bovine spongiform encephalopathy. Vet Rec 145: 672.

Doherr, M.G., Heim, D., Fatzer. R., Cohen, C.H., Vandevelde, M. and Zurbriggen, A., 2001. Targeted screening of high-risk cattle populations for BSE to augment mandatory reporting of clinical suspects. Prev Vet Med (in press).

Donnelly, C.A. and Ferguson, N.M. Statistical Aspects of BSE and vCJD, 2000. Monographs on Statistics and Applied Probability 84.Chapman & Hall/CRC. 2000:81-82.

Griffen, J., Collins, J. D., Nolan, J. P.and Weavers, E. D., 1997. Bovine spongiform encephalopathy in the Republic of Ireland: epidemiological observations 1989-1996. Irish Vet J 1997;50:593-600.

Ministry of Agriculture, Fisheries and Food (MAFF) 2000. Chronology of events. In: MAFF BSE information. 31 July 2001. http://www.defra.gov.uk/animalh/bse/chronol.pdf>.

Martin, S. W., Meek, A.H. and Willeberg, P., 1987. Veterinary Epidemiology. Ames: Iowa State University Press

Office International des Epizootiés (OIE), 1992. Bovine spongiform encephalopathy . In: International Animal Health code; Cited in: Ministry of Agriculture, Fisheries and Food (MAFF) 2000. Chronology of events. In: MAFF BSE information. 27 Aug 2001. <http://www.defra.gov.uk/animalh/bse/chronol.pdf>.

Office International des Epizootiés (OIE), 1996. Bovine spongiform encephalopathy . In: International Animal Health code; Cited in: Ministry of Agriculture, Fisheries and Food (MAFF) 2000. Chronology of events. In: MAFF BSE information.. 27 Aug 2001. <http://www.defra.gov.uk/animalh/bse/chronol.pdf>.

Office International des Epizootiés (OIE), 1997. Bovine spongiform encephalopathy. In: International Animal Health code, draft: August 21, 2001. <http://www.maf.govt.nz/tse-conference/970506-02/htm>.

Office International des Epizootiés (OIE), 1998. Bovine spongiform encephalopathy . In: International Animal Health code; Cited in: Ministry of Agriculture, Fisheries and Food (MAFF) 2000. Chronology of events. In: MAFF BSE information. August 21, 2001. < http://www.defra.gov.uk/animalh/bse/chronol.pdf>.

Office International des Epizootés (OIE), 2000a. Bovine spongiform encephalopathy . In: International Animal Health code, 2000 edition. 21 Sept 2000. <http://www.oie.int/Norms/MCode/A_00066.htm>.

Office International des Epizootés (OIE), 2000b. Surveillance and monitoring systems for bovine spongiform encephalopathy. In: International Animal Health code, 2000 edition. . 21 Sept 2000. <http://www.oie.int/Norms/MCode/A_000154.htm>.

Paisley, L.G. Monitoring and evaluation of BSE testing in Denmark using statistical models (Submitted).

Scientific Steering Committee(SSC), 1998. Opinion of the SSC on defining the BSE-Risk for specified geographical regions:1-6. 21 August 2001. <http://europa.eu.int/comm/food/fs/sc/ssc/out06_en.html>

Scientific Steering Committee(SSC), 1999. Opinion of the SSC on a method to assess the Geographic BSE-Risk (GBR) of Countries or Regions. 21 August 2001. <http://europa.eu.int/comm/food/fs/sc/ssc/out51_en.html>.

Scientific Steering Committee(SSC), 1999. Opinion on: Actions to be taken an the basis of (1) the September 1998 SSC Opinion on the Risk of infection of sheep and goats with the BSE agent and (2) the April 1999 SEAC Subgroup report on Research and Surveillance for TSEs in sheep. 30 Aug 2001. <http://europa.eu.int/comm/food/fs/sc/ssc/out48_en.html>.

Spongiform Encephalopathy Advisory Committee (SEAC), 1999. Report of 14 April 1999 of the SEAC Subgroup on Research and Surveillance for TSEs in Sheep. 30 Aug 2001. <http://www.defra.gov.uk/animalh/bse/bse-publications/seac/sub-rep.pdf>.

Scientific Steering Committee(SSC), 2000a. Final opinion of the Scientific Steering Committee on the Geographic Risk of Bovine spongiform encephalopathy(GBR). 1-60. 15 Aug 2001. <http://europa.eu.int/comm/food /fs/sc/ssc/out113_en.pdf>

Scientific Steering Committee(SSC), 2000b. Report on the Geographical BSE-risk (GBR) of Denmark. 30 Aug 2001. <http://europa.eu.int/comm/food/fs/sc/ssc/out117_en.pdf>.

Simmons, M.M., Ryder, S.J., Chaplin, M.C., Spencer, Y.I., Webb, C.R., Hoinville, L.J., Ryan, J., Stack, M.J., Wells, G.A.H. and Wilesmith, J.W., 2000. Scrapie surveillance in Great Britain: results of an abattoir survey, 1997/98. Vet. Rec, 146:391-395.

Schreuder, B.E.C., Wilesmith, J.W. Ryan, J.B.M. and Straub, 1997. Risk of BSE from the import of cattle from the United Kingdom into countries of the European Union. Vet Rec, 141: 187-190.

Wilesmith, J. W., Wells, G. A. H., Cranwell, M. P. and Ryan, J. M., 1988. Bovine spongiform encephalopathy: epidemiological studies, Vet Rec,123:638-644.

Wilesmith, J.W., Hoinville, L.J., Ryan, J.B. and Sayers, A.R., 1992. Bovine spongiform encephalopathy: aspects of the clinical picture and analysis of possible changes 1986-1990. Vet Rec. 130: 197-201.

Safety aspects associated with preharvest conditions of aquatic food products

Hans Henrik Huss
Danish Institute for Fisheries Research, Dept. of Seafood Research, Søltofts Plads, DTU, Building 221, DK-2800 Kgs. Lyngby, Denmark

Summary

Aquatic food products are high on the list of foods transmitting diseases. In countries having a well established surveillance system, up to 10% of all foodborne disease outbreaks are caused by aquatic food. With exception of histamine that is released through post mortem bacterial activity, most other hazards (biotoxins, pathogenic bacteria, viruses, parasites and chemicals) are completely or partly associated with pre-harvest conditions of the animals. The presence of biotoxins in fish and shellfish is a serious problem. Diseases include paralytical shellfish poisoning (PSP), diarrhoetic shellfish poisoning (DSP), amnesic shellfish poisoning (ASP), neurotoxic shellfish poisoning (NSP), ciguatera and puffer fish poisoning. Most of these cases are caused by the fish or shellfish ingesting toxic dinoflagellates and accumulating toxin in the body or flesh. The presence of poisonous fish and shellfish is often related to specific geographical areas. The closure of shellfishing areas during algae-blooms has reduced exposure to the toxins, but fish and shellfish may become toxic long before algae-blooms are visible. The toxins are heat-stable and hence not destroyed by cooking. A range of pathogenic bacteria is found naturally in the aquatic environment *(Clostridium botulinum*, pathogenic *Vibrio* spp., *Aeromonas hydrophila, Plesiomonas, Listeria monocytogenes)*. Consequently, these bacteria may also be found in low numbers on live fish from certain areas. Low numbers of these organisms on the raw fish is only a potential hazard, since growth is required to produce a real risk and cause disease. The proper application of well-known technologies allows the control of this hazard, the only exception being the possible accumulation of pathogenic vibrio's in shellfish for raw consumption. In contrast, the presence in water, fish and shellfish of low numbers of pathogenic bacteria from the animal/human environment (e.g. pathogenic *E. coli, Salmonella, Shigella*) is a real hazard, when these foods are eaten raw. Monitoring of exposed harvesting areas for possible faecal pollution is essential therefore, and consumption of raw fish and shellfish from polluted areas should be avoided. The presence of enteric viruses in shellfish is one of the major problems facing the industry today. For instance, studies in the USA, Germany and the UK have estimated that up to 25% of all hepatitis cases may be due to consumption of contaminated shellfish. The problem is that many consumers prefer to eat shellfish raw or undercooked (steamed) whilst preharvest control options are not reliable. There are still great difficulties in detecting some viruses, and the use of *E. coli* as an indicator organisms of faecal pollution is doubtful, as viruses are generally more resistant to environmental stress than is *E. coli*. It is well known that fish may harbour *parasites* pathogenic to man. In northern marine waters mainly the nematodes (*Anisakis* spp.) and a cestode *(Diphyllobotrium latum)* are often found in freshwater fish. In Asia a large number of trematodes are found in freshwater fish, and they cause severe problems. It is estimated that more than 50 million Asians are infected with trematodes. One reason for this high rate of infection is the Asian habit of eating raw fish and uncooked fish products. Application of freezing is effective in controlling this hazard, but consumers of for instance raw fish (sushi, ceviche) prefer to eat the fresh raw product rather than frozen and thawed items. In contrast to the naturally occurring biotoxins, the presence of chemical residues in the environment and the fish is a man-made problem. Risks resulting from chemical contaminants (heavy metals, pesticides, drug residues) in commercially harvested marine fish and shellfish are low. Potential risks are associated with freshwater fish, aquaculture products, fishing from certain coastal areas, and subsistence and recreational fishing in polluted rivers. In conclusion, the pre-harvest condition of aquatic food products is of great importance, although difficult to control in many cases. Problem areas include the presence of heat-stable biotoxins in fish and shellfish, the traditional common consumers' preference for consuming raw molluscan shellfish, where pathogenic bacteria and viruses may have accumulated and, finally, consumption of raw fish causing transmission of parasites and pathogenic bacteria.

Keywords: seafoods, biotoxins, pathogenic bacteria, viruses, parasites, chemical residues

1. Introduction

In many parts of the world, fish constitutes a major part of ingested animal protein. Globally, about 120 million tonnes of fish are marketed annually, of which only 90 million are utilised for human consumption. Most fish and shellfish are extracted from a 'wild' population, but aquaculture is currently the fastest growing food production system in the world, totalling nearly 28 million tonnes in 1997 (FAO, 1999) (Fig. 1). This situation may lead to increased problems both of an environmental and public health nature, because the risk of contamination of products by chemical and biological agents is greater in freshwater and coastal ecosystems than in the open seas.

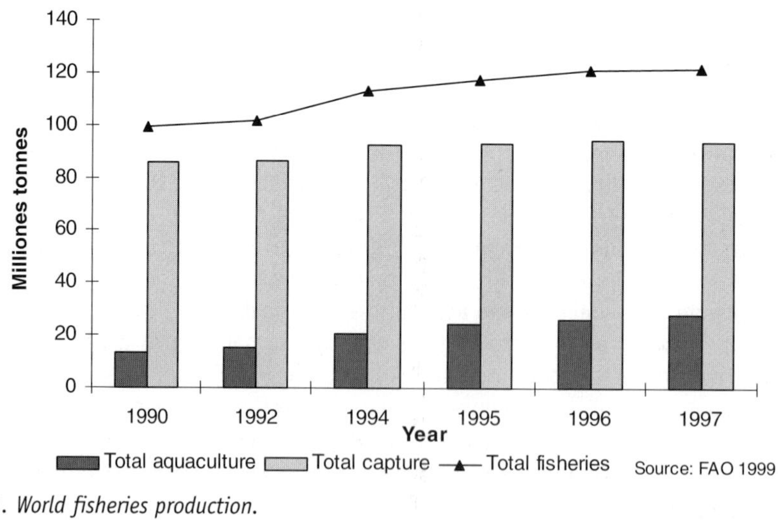

Figure 1. World fisheries production.

On August 7, 2000, CNN released the following news from Reuters : " Seafood is the biggest cause of food poisoning in the United States having caused 237 outbreaks since 1990". It is well established that seafood ranks high on the list of disease-transmitting foods being responsible for up to 10% of all outbreaks (Huss et al., 2000). In a recent publication summarising the food borne disease outbreaks in the United States in the period 1993-1997 nearly 7% of all outbreaks were listed as caused by seafood, but only 3% of cases (people affected) could be related to seafood (Table 1).

Based on published statistics and available literature, the following safety hazards can be identified for aquatic food products (in decreasing order of significance): i) scombrotoxin or histamine, ii) biotoxins, iii) pathogenic bacteria, iv) viruses, v) parasites, and, finally, vi) chemicals.

Formation of histamine is caused by the growth of histidine-decarboxylating bacteria in fish or fish products and thus the only hazard entirely related to post mortem handling and storage conditions. The present contribution gives an overview of the hazards completely or partly related to pre-harvest conditions of fish and shellfish.

2. Biotoxins

Marine biotoxins are responsible for a large number of seafood-borne diseases. With the exception of tetrodotoxins, the primary origin of these biotoxins are toxic marine algae, principally dinoflagellates. (see Table 2).

Table 1. Foodborne Disease Outbreaks, USA 1993-1997.

Vehicle of transmission	Outbreaks		Cases		Deaths	
	No.	%	No.	%	No.	%
Meat (beef, pork, poultry)	168	6.1	6,709	7.7	7	24.1
Shellfish	47	1.7	1,868	2.2	0	0.0
Fish	140	5.1	696	0.8	0	0.0
Eggs, dairy, ice-cream	52	1.9	1,874	2.2	4	13.7
Bakery products	35	1.3	853	1.0	0	0.0
Fruits and vegetables	70	2.5	12,369	14.4	2	6.9
Salads	127	4.6	6,483	7.5	2	6.9
Others	66	2.4	2,428	2.8	0	0.0
Multiple vehicles	262	9.5	25,628	29.8	1	3.4
Total known	967	35.2	58,908	68.5	16	55.2
Unknown	1,784	64.8	27,150	31.5	13	44.8
TOTAL	2,751	100.0	86,058	100.0	29	100.0

Source: MMWR-report, 2000.

Table 2 Aquatic Biotoxins.

Toxin	Where/when produced	Animal(s)/organ involved
Tetrodotoxin	in fishante mortem	puffer fish *(Tetraodontidae)* mostly ovaries, liver, intestines
Ciguatera	Marine algae	> 400 tropical/subtropical fish sp.
PSP-paralytic shellfish poison	Marine algae	filter feeding shellfish, mostly digestive glands
DSP-diarrhoeic shellfish poison	Marine algae	filter feeding shellfish
NSP-neurotoxic shellfish poison	Marine algae	filter feeding shellfish
ASP-amnesic shellfish poison	Marine algae	filter feeding shellfish (blue mussels)

There are over 4000 species of marine algae (phytoplankton) and about 2% (up to 80 species) are known to produce toxins, most of which are *Dinophycae* spp. (dinoflagellates) (Scogin, 1998). Some of these algae have a red-brown or green pigmentation and at blooms of algae they may grow in such abundance that they change the colour of the seawater to red, brown or green (all termed 'red tide'). However, not all toxic algae are coloured – and not all coloured algae are toxic. Visible red tides may contain from 20 000 to > 50 000 algae cells ml^{-1} of seawater, but as low as 200 cells ml^{-1} may produce toxic shellfish (Scogin, 1998). Bivalves can accumulate enough toxin to effect human illness after only 24-28 h. of filter feeding during a bloom.

2.1 Ciguatera poisoning

Ciguatera is caused by ingestion of tropical or subtropical fish species that contain ciguatoxin originating from the dinoflagellate *Gambierdiscus toxicus*. There are at least 400 species of reef-feeding fish that are potentially toxic. Globally, ciguatera poisoning is the most common fishborne poisoning syndrome with an estimated 50 000 cases a year (Taylor, 1988). The toxin accumulates in fish, feeding on the toxic algae or in larger carnivores that prey on these herbivores. Toxin can be detected in the gut, liver and muscle tissue by means of mouse assay and chromatography. Some fish may be able to clear the toxins from their systems (Taylor, 1988).

The clinical picture varies, but the onset time ranges from a few minutes up to 30 h. after consumption. Gastrointestinal and neurological systems are affected (vomiting, diarrhoea, tingling sensations, ataxia, weakness, hot/cold temperature reversal, dull aches and sharp shooting pains). The duration of illness may be 2-3 days, but some neurological effects may persist for weeks, even years. Death results from circulatory collapse. Halstead (1978) has reported a case fatality rate of about 12%.

2.2 Shellfish toxic syndromes

There are four major shellfish toxic syndromes, i.e. i) Paralytic Shellfish Poisoning (PSP), ii) Diarrhoeic Shellfish Poisoning (DSP), iii) Neurotoxic Shellfish Poisoning (NSP), and, iv)

Amnesic Shellfish Poisoning (ASP). They are all caused by toxic pelagic dinoflaggelates having accumulated in bivalve shellfish. Details of each syndrome are summarised in Table 3.

Table 3. Shellfish toxic syndroms.

Disease	PSP	DSP	NSP	ASP
Geographical distribution	Global	Europe Japan US Australia New Zealand	US New Zealand Japan Spain	Canada Western US New Zealand
Major toxin	Saxitoxin	Okadaic acid	Brevetoxin	Domoic acid
Symptoms	Neurologic paresthisia Headache Floating sensation Paralysis Death by asphyxiation	Diarrhoea Nausea Vomiting Non-lethal	Neurological hot/cold reversal Paresthesia Paralysis Vomiting Diarrhoea	Neurological memory loss Gastro-enteritis Death by brain- stem damage
Fatality rate	1-14%	0%	0%	3%

2.3 Puffer fish (tetrodotoxic poisoning)

Unlike all other biotoxins accumulating in live fish or shellfish, tetrodotoxin is not produced by algae. The precise mechanism of production of this very potent toxin is not clear, but apparently quite commonly occurring symbiotic bacteria are involved (Kodama et al., 1995). Tetrodotoxin is mainly found in the liver, ovaries and intestines of various species of puffer fish, while the muscle tissue normally remains free of toxin. Puffer fish poisoning results in neurological symptoms becoming manifest 10-45 min. after ingestion. Symptoms include a tingling sensation in the face and extremities, paralysis, respiratory symptoms and cardiovascular collapse. Fatalities occur within 6 h.

2,4 Legislation and control

The control of marine biotoxins is difficult and disease cannot be entirely prevented. The toxins are all of non-protein nature and extremely stable (Gill et al., 1985). Thus cooking, smoking, drying or salting do not destroy them, and one cannot tell from the appearance of fish- or shellfish flesh whether it is toxic or not.

The major preventive measure is inspection and sampling from fishing areas and shellfish beds and subsequent analysis for toxins. The mouse bioassay is often used for this purpose and confirmatory HPLC is done if death of mice occurs after 15 min. If high levels of toxin are found, commercial harvesting is halted. It seems unlikely that it will ever be possible to control phytoplankton composition in endemic areas or to eliminate toxigenic species, and there is no reliable way to forecast, when a particular phytoplankton will grow and thus no way to predict blooming of toxigenic species (Hall, 1991).

Removal of toxin by depuration techniques may have some potential, but the process is very slow and costly. There is also a risk that a small number of organisms will not open and in the process pump clean water through the system. Hence their original level of toxicity is retained (Hall, 1991).

To be effective, monitoring procedures must include reliable sampling plans and efficient means of toxin detection. At present, reliable chemical methods for detection of all toxins are not available and must therefore be developed. The sampling plan must take into consideration that toxicity of shellfish can increase from negligible to lethal levels in less than one week or, for blue mussels, even in less than 24 h. Also, within a growing location of shellfish the toxicity can vary, depending on geography, water currents and tidal activity.

The present situation regarding tolerances and methods of analysis to be used in a monitoring programme is shown in Table 4.

Table 4. Monitoring of Biotoxins (WHO, 1989).

Toxin	Tolerance	Methods of analysis
Ciguatera	control not possible	Mouse bioassay, HPLC many problems remains [1]
PSP	80 µg/100 g	Mouse bioassay, HPLC
DSP	20-60 µg/100g	Mouse bioassay, HPLC
NSP	any detectable level/100 g is unsafe	Mouse bioassay. No chemical method
ASP	20 µg/g domoic acid	HPLC

[1] see review by Lehane and Lewis, 2000.

3. Pathogenic Bacteria

The pathogenic bacteria associated with seafood are listed in Table 5.

The pathogenic bacteria originating in the aquatic environment are also found on the live fish or fish raw material. Obviously, the water temperature has a selective effect. Consequently, the more psychrotolerant organisms (e.g. *C. botulinum* type E, *Aeromonas hydrophila*) are common in Arctic or colder temperate climates, whereas the more

Table 5. Pathogenic Bacteria Associated with Seafood.

Natural habitat of pathogen	Foodborne disease		
	Minimum Infectious Dosis		Intoxication (growth required)
	High	Low	
Aquatic environment	*Vibrio* spp.		*C. botulinum type E* (non-proteolytic)
	Aeromonas h.		*C. botulinum type A, B* (proteolytic)
	Plesiomonas spp.		
General environment	*Listeria monocytogenes*		
Animal-human reservoir		*Salmonella* *Shigella* *E. coli*	
Preventive measure	Prevention of growth G.M.P./GHP	Hygiene	Prevention of growth

mesophilic species (e.g. pathogenic vibrios) are representing part of the natural flora on fish from coastal and estuarine environments of tropical climate zones.

The pre-harvest contamination of the fish with these organisms is not a safety concern, since they are present in numbers too low to cause disease. An exception is the possible accumulation of high numbers of e.g. *Vibrio* spp. in filter feeding shellfish, particularly since these animals are often eaten raw. It should be noted that the presence of these pathogens in the raw material may be a potential hazard at later stages, if conditions for growth are favourable in the final product – particularly in ready-to-eat products such as smoked fish. Similarly, pre-harvest contamination with pathogens from the general environment(e.g. *Listeria monocytogenes)* is of little or no concern.

Pre-harvest contamination with pathogens from the animal/human reservoir may pose a risk, since in some cases a very low infective dose is required to produce disease (e.g. 1-10 cells for some *Shigella* and *Salmonella* serotypes). Normal cooking procedures will eliminate the risk. Safety concerns are therefore primarily related to the consumption of raw shellfish or raw fish dishes such as ceviche or sushi.

4. Viruses

A number of human enteric viruses transmitted by the faecal-oral route have been associated with shellfish *(Caliciviruses, Astroviruses, Rotaviruses, Adenoviruses, Enteroviruses,* Hepatitis A virus).

Viruses are obligatory intracellular parasites depending entirely on a living host cell for replication. Their presence in seafood is purely the result of contamination, either via infected food handlers or via polluted water. Shellfish are filter-feeders and may therefore concentrate viruses from the water in which they are growing. Large water volumes (up to 1.500 liter/day) are passing through an active shellfish, causing the virus concentration in the shellfish to be much higher than that of the surrounding water.

Virus control during production and processing of bivalve shellfish has concentrated on three areas, viz. i) control and classification of harvesting areas, ii) purification and relaying, and iii) heat treatment. The most effective and reliable approach to controlling the contamination of shellfish is to exclusively harvest from areas with good water quality (Lees, 2000). According to the EC Directive (Anon, 1991), harvesting areas will be assigned to one of three categories as shown in Table 6.

Notably, in the EU faecal indicators are measured in the shellfish flesh. In the USA they also use faecal indicators for classification of harvesting areas, but here these are measured in the water from shellfish growing areas. Thus, if the geometric mean of faecal coli is > 14 per 100 ml water, harvesting is restricted and the shellfish must go through depuration or relaying. If the geometric mean is > 88 faecal coli per 100 ml, the shellfish must be relayed for more than two months and if faecal coli is > 260 per 100 ml water, harvesting is prohibited (Lees, 2000).

It has long been known that monitoring of harvesting areas based on bacterial indicators for faecal pollution represents an unreliable predictor of viral contamination (Richards, 1985; Cliver, 1988). One problem is that viruses can survive for weeks to months in the marine environment, which is much longer than the bacterial indicators (Lees, 2000). For this reason new viral test methods or development of alternative, more reliable, faecal pollution indicators are needed.

Table 6. Classification of Harvesting Areas for Bivalve Shellfish (EC Directive 91/492/EEC).

Classification	Criteria/100g shellfish	Treatment required
A	< 230 *E. coli* or < 300 faecal coli	none, acceptable for consumption
B	230 - 4.600 *E. coli* or 300 - 6.000 faecal coli	depurated, heat-treated or relayed until they meet category A standard
C	6.000 - 60.000 faecal coli > 60.000 faecal coli	relayed for > 2 months or heat-treated harvesting prohibited

Purification and relaying are equally unreliable and ineffective in eliminating viral contamination (Lees, 2000). Even when depuration plants have been functioning satisfactorily by the faecal coliform criteria cited above, they might still fail to fully remove the human enteric viruses.

Heat treatment is a reliable way to eliminate viruses. A commercial heat treatment approved by the EU suggests that the internal temperature of molluscs must be maintained at 90°C for 1.5 min before consumption with a view to eliminate all pathogenic microorganisms (Anon, 1993).

5. Parasites

The presence of parasites in fish is very common, but most of them are of little concern with regard to economics or public health. The most important, which pose a potential health risk for humans, are listed in Table 7.

All parasitic helminths have complex life cycles. They do not spread directly from fish to fish, but must pass through a number of intermediate hosts in their development. Very often sea-snails or crustaceans are involved as first intermediate host and marine fish as second intermediate host, while the sexually mature parasite is found in mammals as the final host. In between these hosts, one or more free living stages may occur. Infection of

Table 7. Pathogenic Parasites Transmitted by Fish and Shellfish.

Parasite	Known geographical distribution	Fish and shellfish
Nematodes or round worms		
Anisakis simplex	North Atlantic	herring
Pseudoterranova dicipiens	North Atlantic	cod
Gnathostoma spp.	Asia	freshwater fish, frogs
Capillaria spp.	Asia	freshwater fish
Angiostrongylus spp.	Asia, South America, Africa	freshwater prawns, snails, fish
Cestodes or tape worms		
Diphyllobothrium latum	Northern hemisphere	freshwater fish
D. pacificum	Peru, Chile, Japan	seawater fish
Trematodes or flukes		
Clonorchis spp.	Asia	freshwater fish, snails
Opisthorchis spp.	Asia	freshwater fish
Metagonimus yokagawai	Far East	
Heterophyes spp.	Middle East, Far East	snails, freshwater fish, brackish water fish
Paragonimus spp.	Asia, America, Africa	snails, crustaceans, fish
Echinostoma spp.	Asia	clams, freshwater fish, snails

humans may be part of this life cycle or it may be a side-track causing disruption of the life cycle, as illustrated in Figure 2.

For this reason, fish raised in net cages suspended in the water and feeding on pelleted feed only are unlikely to harbour parasites. Thus a number of surveys have shown that aquacultured salmon are free from *Anisakis* spp. (Table 8).

The parasites are transmitted to man through the consumption of raw or undercooked fish products. Measures to reduce the risk of parasites prevailing in seafood can be applied prior to harvesting by avoiding capture of infected fish by selecting specific fishing grounds,

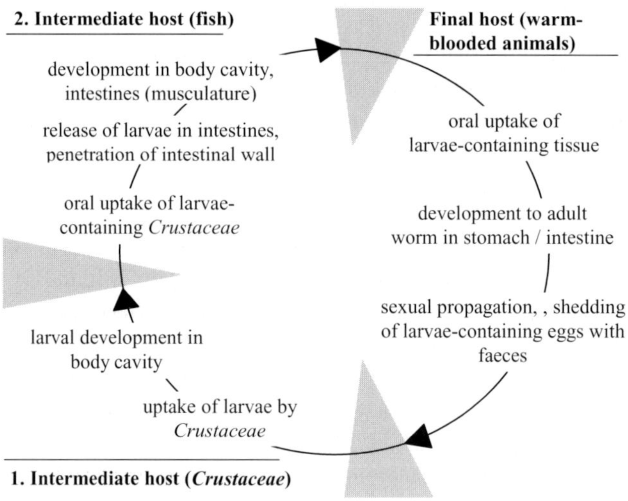

Figure 2. Lifecycle of Anisakis simplex.

Table 8. Prevalence of Anisakis spp. in Farmed Salmon.

Origin	Salmon species	Samples	+ve-samples	Reference
Washington, USA	*S. salar,* *Oncorhynchus kisutch* *O. tshawytscha*	237	0	Deardorff and Kent (1989)
Norway	S. salar	2832	0	Angot and Brasseur (1993)
		2000	0	Bristow and Berland (1991)
Scotland	*S. salar*	867	0	Angot and Brasseur (1993)
Estimated prevalence[a]	<0.017%			

[a] <1/5936

Prevalence in wild salmon 26-100% (Lasse Vigel Jørgensen, 2000, personal communication).

specific fish species or specific age groups of fish. However, this strategy is not likely to be effective (Adams et al., 1997). Complete elimination of parasites from seafood products is not possible. Therefore, post-harvest controls such as proper fish processing and preparation of food are the most effective measures to prevent transmission of parasites.

6. Chemicals

In contrast to the naturally occurring biotoxins, the presence of chemical residues in the environment and in fish is a man-made problem. The ocean-dumping of waste material from industrial processing, sludge from sewage treatment plants, draining into the sea of chemicals used in agriculture and raw unheated sewage from large urban populations and industries all contribute to contaminating the freshwater environment and the coastal marine environment. Special problems are related to aquaculture where chemical fertilisers, water treatment compounds, pesticides, desinfectants and chemotherapeutics are widely used (WHO, 1999). Antibiotics are used in culture systems to control bacterial pathogens, resulting in the development of antibiotic-resistant bacteria and antibiotics residues in the fish flesh.

Fortunately, the high seas are not yet polluted with chemicals to any comparable degree. Thus risk of chemical contaminants being present in commercially harvested marine fish and shellfish is low. Potential risks are associated with freshwater fish, aquaculture products, subsistence and recreational fishing in certain coastal areas or from polluted rivers (Price, 1992). Most countries have a monitoring programme for chemical contaminants of the aquatic environment. Tolerances, action levels and guidance levels are established for some of the most toxic and persistent contaminants, to assist in deciding whether to issue consumption advice or to close waters for commercial harvesting of all or certain species of fish. In aquaculture the use of chemicals can and should be strictly controlled and antimicrobial agents should be confined to emergency use in the event of breakdown of vaccine protection. Basic recommendations for strategies to decrease the emergence and dissemination of resistant organisms in the environment including control of sale and use of antimicrobial agents have been proposed by WHO (1995).

7. Concluding remarks

Most hazards related to consumption of aquatic food products can be controlled by applying GMP, GHP and a well designed HACCP programme. However, there are problems, particularly related to the pre-harvest condition of the aquatic foods. These include: i) presence of heat-stable biotoxins in fish and shellfish, ii) consumption of raw molluscan shellfish, where pathogenic *vibrio* spp. and viruses may have accumulated in the animals, and iii) consumption of raw fish causing transmission of parasites and/or pathogenic bacteria.
Much more emphasis should be directed at control of fish and shellfish at the points of capture. Methods for monitoring of harvest areas for possible faecal pollution, presence of viruses and biotoxins should be improved. Good aquaculture practices should be formulated with much less use of chemicals and antibiotics in the growing or pre-harvest phase.

References

Adams, A.M., Murrell, K.D. and Cross J.H., 1997. Parasites of fish and risks to public health. Rev. Sci. Tech. Off. Int. Epic. 16 (2), 652-660.

Angot, V. and Brasseur, P., 1993. European farmed Atlantic salmon *(Salmon salar)* are safe from anisakid larvae. Aquaculture 118, 339-344.

Anon., 1991. Council Directive of 15 July 1991 laying down the health conditions for the production and placing on the market of live bivalve molluscs (91/492/EEC). Off. J. Eur. Communities, No. 268/1.L

Anon., 1993. Council Decision of 11 December 1992 approving a certain heat treatment to inhibit the development of pathogenic microorganisms in bivalve molluscs and marine gastropods. (93/25/EEC). Off. J. Eur. Communities 16, 22-23.

Bristow, G.A. and Berland, B., 1991. A report on some metazoan parasites of wild marine salmon *(salmon salar l)* from the west-coast of Norway with comments on their interactions with farmed salmon. Aquaculture 98, 311-318.

Cliver, D.O., 1988. Viral transmission via foods. Food Technol. 42, 241-248.

Deardorff, T.L. and Kent, M.L., 1989. Prevalence of larval *Anisakis simplex* in pen-reared and wild-caught salmon (salnidae) from Pugets Sound, Washington. Journal of Wildlife Diseases 25, 416-419.

FAO, 1999. The state of world fisheries and aquaculture. FAO Fisheries Department, Rome.

Gill, T.A., Thompson J.W. and Gould, S., 1985. Thermal resistance of paralytic shellfish poison in soft shell clams. J. Food Protect. 48, 659-662.

Hall, S., 1991. Natural toxins. In Microbiology of marine food products. Eds: Ward, D.R. and Hackney, C. Van Nostrand Reinhold, 301-330.

Halstead, B.W., 1978. Poisonous and venomous marine animals of the World. Princeton Darwin Press.

Huss, H.H., Reilly, A. and Embarek, P.K.B., 2000. Prevention and control of hazards in seafood. Food Control 11, 149-156.

Jørgensen, L.V., 2000. Spoilage and safety of cold-smoked salmon. Personal communication.

Kodama, M., Shimizu, H., Sato, S., Ogatu, T. and Terao, K., 1995. The infection of bacteria in the liver cells of toxic puffer; a possible cause for organisms to be made toxic by tetrodotoxin in association with bacteria. In 'Harmful Marine Algae Blooms', ed.

Lassus P., Arzul G., Erard E., Gention P. and Marcaillou C., 457-462. Paris Lavousier Intercept Ltd.

Lees, D, 2000. Viruses and bivalve shellfish. Int. J. Food Microbiol. 59 (2000), 81-116.

Lehane, L. and Lewis R.J., 2000. Ciguatua: Recent advances, but the risk remains. Int. J. Food Microbiol. 61 (2000), 91-l25.

MMWR report (Morbidity and Mortality Weekly Report), March 17, 2000. 49, No. 55-1. UP Dept. of Human Health and Human Services.

Price, R.J., 1992. Recidue concerns in seafoods. Dairy, Food and Environmental Sanitation 12 (3), 139-143.

Richards, G.P., 1985. Outbreaks of shellfish associated enteris virus illness in the United States: Requisite for development of viral guidelines. J. Food Prot. 48, 815-823.

Scogin, A.C., 1998. Marine biotoxins. J. Appl. Microbiol. Supplement 84, 415-505.

Taylor, S., 1988. Marine toxins of microbial origin. Food Technol. 2, 94-98.

WHO, 1989. Report of WHO consultation of public health aspects of seafood-borne diseases. WHO/CDS/VPH/90.86.

WHO, 1995. WHO Scientific Working Group on monitoring and management of bacterial restistance to antimicrobial agents.

WHO/CDS/BV1/95.7. Genova, WHO.

WHO, 1999. Food safety issues associated with products from aquaculture. WHO Technical Report, series 883.

Food safety in seafood; epidemiological concerns related to the geography of fishing grounds

Sebastian Hielm, Miia Lindström and Hannu Korkeala
Department of Food and Environmental Hygiene, Veterinary Faculty, P.O. Box 57, 00014 Helsinki University, Finland; Corresponding author: sebastian.hielm@helsinki.fi

Summary

Seafood is responsible for a disproportionately large portion of foodborne illnesses occurring in the world today. The reasons for this are manifold, but essentially the preference in many cultures for eating raw or not reheated seafood dishes is the major contributor. The biological hazards associated with seafood as a disease vector can be roughly grouped into bacteria, viruses, parasites, and marine toxins. While many pathogenic bacteria enter seafood at the pre-harvest phase, transmission of infection generally requires post-harvest multiplication as well. Viruses, parasites and marine toxins pathogenic to humans are mostly present in the seafood prior to harvest, and as a rule none of these require any post-harvest product abuse to cause disease. Proper cooking of seafood before consumption, however, can control all biological hazards except the marine toxins. The various marine toxins are not destroyed by heat, and control of these relies solely on regional surveillance of fishing waters for the presence of toxic algal blooms, indicating danger. Likewise, surveying for pathogens or suitable indicator organisms for these in harvesting waters and sometimes the seafood itself is required to ensure the safety of seafood intended for consumption as uncooked.

Keywords: seafood, foodborne illness, food hygiene, fishing areas, aquaculture

1. Introduction

Seafood is known to act as a vehicle in the transmission of a variety of human diseases. According to selected statistics, some 14-24 % of foodborne illness outbreaks with known etiology are estimated to stem from contaminated seafood (Bryan, 1987; Hirn et al., 1992; Lindqvist, 1999; Hatakka and Halonen, 2000). The Finnish outbreak records in Table 1 shows a breakdown of agents rather typical for Northern Europe, and indeed, most developed countries.

Most indigenous fish pathogens are present, but are far outnumbered by post-process hazards such as histamin, *Staphylococcus aureus, Clostridium perfringens* and *Bacillus cereus*. Typically, *Listeria monocytogenes* is low on the list, as the involvement of fish products in 'sporadic' listeriosis is rarely confirmed, although more often suspected. The relatively low frequency (4 %) of viral agents is also noteworthy. This reflects the diagnostic problems traditionally encountered with viruses, which have been alleviated only recently by molecular techniques such as PCR. Most foodborne outbreaks of unknown etiology are today suspected of being caused by caliciviruses (Mead et al., 1999). Recently,

Table 1. Foodborne outbreaks caused by seafood and fish products between 1983-1999 in Finland.*

Causative agent	Number of outbreaks	%
Histamine poisoning	25	24
Staphylococcus aureus	23	22
Clostridium perfringens	14	13
Bacillus cereus	4	4
Calicivirus	4	4
Paralytic shellfish poisoning	4	4
Clostridium botulinum type E	1	1
Listeria monocytogenes	1	1
Vibrio cholerae	1	1
Unknown	28	27
Total	105	101

*Compiled from annual statistics on foodborne and waterborne outbreaks in Finland published by the National Food Administration, Helsinki.

sentinel population or questionnaire-based surveys into the true prevalence of foodborne illness in developed countries have been conducted, and these estimate the percentage of citizens annually succumbing to foodborne disease to lie somewhere between 6 % (in Sweden; Norling, 1994), 14 % (in the United Kingdom; Food Standards Agency, 2001) and 28 % (in the USA; Mead et al., 1999). According to the Food and Agricultural Organisation (FAO, 2001), the percentages are likely to be much higher in most developing countries, due to a combination of poverty, malnutrition, illiteracy and inadequate public facilities. Extrapolating from these data, a cautious assumption of the worldwide incidence of seafood-derived foodborne illness would lie aroud 300-600 million cases 24 *per annum*.

As, in most cultures, seafood (expressed as meals *per capita*) is consumed less often and in substantially smaller quantities (kg *per capita*) than other animal-derived foods such as red meat and poultry products, it seems that seafood is responsible for a relatively high number of foodborne illnesses. In our view, many factors contribute to this, perhaps the most important (1) seafood being an exceptionally good growth surface for many microbes, (2) seafood being eaten raw in several cultures, (3) heat-processed seafood products commonly being consumed without further heating, (4) pathogens present in seafood not necessarily being destroyed in the mild heating processes common in industry, and finally (5) the seafood industry consisting of a proportionately large number of small and medium size- or even one-person enterprises, which are less self-regulating in terms of food safety than the larger units encountered in other food industries. While acknowledging that this list is not exhaustive, these five factors, either alone or in combination, are the likely causes of most problems related to seafood safety. Whereas the physiology, composition and nutrient content of specific seafoods must be taken as a biological fact not easily subject to changes, the other four risk factors mentioned could be at least partially manipulated. Experiences with enterohaemorrhagic *Escherichia coli* (EHEC) in red meat and *Salmonella enteritidis* in eggs show that through education it is entirely feasible to minimize raw food consumption in general, particularly when a clear risk from such behaviour has been established. In the case of seafood processed at exclusively below sterilisation temperatures, it would probably be more effective to rely on self-regulation practices such as in-plant control and HACCP programmes. Also, the recommendations on the shelf-life of heat-processed seafood should be scrutinized and shorter shelf-lives indicated where this seems appropriate. Nevertheless, to achieve improved safety records for all fish products in the future, continued emphasis on the importance of basic hygiene education for fish handlers and consumers is called for.

2. Basic Aspects of Seafood

Seafood can be roughly divided into two groups, finfish and shellfish. The former, largest, group comprises the bony, ray-finned teleosts, representing approximately 30,000 species (12,000 of these living in fresh water); other groups of economic importance include the sharks and rays (elasmobranchs) with some 800 species, and the few surviving species of sturgeons (chondrosteans). Shellfish are divided into molluscan bivalves, with a calcareous shell (e.g., oysters, clams and mussels); and crustaceans, with a chitinous exoskeleton (e.g., shrimp, crayfish and crabs). Out of this multitude of species, about 3000 are captured or cultured for commercial purposes (Martin et al., 1978). To predict, let alone control the microbiological events in raw seafoods originating from so many species appears an almost insurmountable task, if one considers the problems already encountered in similar exercises targeting the most commonly produced mammalian species. Although the microbiota of finfish and shellfish caught in the same waters will to a certain degree differ from each other, the microbial flora of fish in general seems to be more directly related to the fishing grounds' environmental conditions than to other factors.

The muscle composition of different fish species fluctuates considerably and is also dependent of fish size, season, and fishing grounds (Shewan, 1961). In general, seafood is more perishable than meat from terrestrial animals (ICMSF, 1998), primarily because it has a moist surface, a relatively high pH (6.2 to 6.8) and a predominantly psychrotrophic microflora (Hackney and Garrett, 1985). In addition, fresh finfish muscle contains high levels (1.5%, in elasmobranchs 3%) of soluble nutrients, such as sugars, minerals, vitamins, and non-protein nitrogen compounds such as free amino acids, ammonia, trimethylamine oxide (TMAO), creatine, taurine, anserine, uric acid, betaine, carnosine, and histamine (Shewan, 1961; Jackson et al., 1997). Shellfish has a comparable muscle composition, but generally contains more free amino acids whilst TMAO is only present in cephalopods, scallops and cockles. Mollusc flesh contains substantially more carbohydrates (3 to 6%) than most other seafood, including crustaceans (0.5%) and finfish (1%). Thus, generally, shellfish will be subject to a similar type of spoilage as finfish, albeit spoilage occurs more rapid due to excess nutrients. In addition, the carbohydrate-rich molluscan flesh tends to spoil through fermentative processes, whereas other seafoods usually spoil as a result of proteolysis (Jackson et al., 1997). Like all muscle foods, the microflora in fresh ice-stored fish consists largely of pseudomonads, *Shewanella* and *Moraxella* species possibly contributing to the spoilage process (Shewan, 1971).

3. The Initial Microflora of Seafood

Generally, the flesh of freshly caught fish is free of bacteria. However, bacteria may be found in high numbers on the skin (10^2-10^7 cm^{-2}), on the gills (10^3-10^9 g^{-1}) and in intestines (10^3-10^9 g^{-1}). Whereas the intestinal flora is usually (facultatively) anaerobic, skin and gill flora tends to be aerobic. The harvesting environment determines the isolates, fish from clean cold or temperate waters generally carrying fewer organisms than fish from polluted or tropical waters. Paradoxically, fish from colder waters have often been reported to spoil faster: the rapid chilling facilities on modern fishing vessels promote a better survival of psychrotrophic- than of mesophilic species predominantly caught in warmer waters (Disney, 1976; Sumner et al., 1984). In contrast, Lima dos Santos (1981), reports that spoilage patterns during iced storage are usually quite similar, and are universally caused by psychrotrophs. Conversely, if the chilling upon harvest is delayed, tropical fish will spoil at an excessively higher rate than fish from temperate waters (ICMSF, 1998). Gram-negative, rod-shaped bacteria predominate in the psychrotrophic population of cold or temperate waters (<10°C): *Pseudomonas, Moraxella, Acinetobacter, Shewanella, Flavobacterium, Cytophaga* and *Vibrio*. In warm waters up to 50-60% of the flora consists of Gram-positive mesophiles: *Bacillus, Micrococcus*, and *Corynebacterium,* and 40 to 50% are the Gram-negatives also found in colder waters. In freshwater fish the vibrios are replaced by *Aeromonas*. On crustaceans the pseudomonads decline in favour of *Psychrobacter, Acinetobacter, Micrococcus*, and *Corynebacterium* (Hackney and Garrett, 1985; ICMSF, 1998).

The method of catch and its subsequent on-board handling are reported to greatly affect the number and types of bacteria on the raw material. For instance, trawled fish has been shown to carry bacterial loads 10-100 times higher than line-caught fish (Shewan, 1949). During trawling, fish are often dragged along the sea floor, which stirs up sediments contaminating the fish; also, fish are compacted in the trawl, causing rupture and dissemination of gut contents. Obviously, little can be done to avoid these detrimental effects of large-scale commercial fishing, but through development of alternative fishing gear some improvements can be made. However, most research efforts have been focussed on the effects of onboard seafood handling. The Food and Agricultural Organisation's *Code of practice for fresh fish* (FAO, 1973) has stated the basic guidelines for proper handling of fish at sea: as a perishable food fish should be carefully cleaned and cooled down around 0°C in melting ice as quickly as possible. Whereas cleaning in recreational fishery translates into bleeding, gutting and gilling, it would seem that this is not always advisable in the commercial setting. Gutting will invariably expose flesh, and as there is little proof of bacterial passage from the intestine to the flesh during storage, excessive cleaning may actually speed up the spoilage process rather than halt it. Fishing vessels making trips shorter than three days will benefit from delivering the catch to processing plants uneviscerated but properly chilled (Ward and Baj, 1988). However, during long fishing trips, commonly practised in the Northern Atlantic, the fish will undergo autolysis and experience subsequent quality loss unless gutted; consequently, evisceration is recommended under these circumstances (ICMSF, 1998).

Although the detrimental effects of ambient temperatures on freshly caught fish are obvious, the rationale of rapid chilling was substantiated in the work of Barile et al.

(1985). These authors showed that the shelf-life of mackerel was reduced by one day for every one-hour delay in chilling. Most boats today have on-board cooling equipment, either producing ice for the holding tanks or allowing refrigeration by seawater (RSW). However, as the chilling procedures may actually physically damage the fish during prolonged trips, the installation of large freezers aboard fishing vessels has become increasingly popular today. The possibility to freeze all or part of the catch at sea does seem like an ideal solution, effectively halting microbial spoilage (ICMSF, 1998).

4. Fish Safety Concerns

The hazards related to the consumption of seafood can be divided into two distinct categories. Biological hazards, such as parasites, bacteria, viruses, and biotoxins (algal and intrinsic), cause more or less acute clinical symptoms, and the causal link to ingested seafood is fairly easy to appreciate. Chemical hazards include agricultural or other organic pollutants (agrochemicals, chemotherapeutics, metals, feed additives, DDT, dioxins, etc.), that are primarily concentrated in seafood reared in intense aquaculture settings or caught in polluted littoral waters (WHO, 1999). The exact role of fish consumption in contributing to some chronic intoxications is difficult to establish and will be discarded in the present review.

Table 1 includes the foodborne illnesses most commonly associated with seafood. A distinction has to be made between pathogens contaminating the raw materials and those more likely to contaminate the final products, either through soiled equipment or unhygienic practices. Although for instance *L. monocytogenes* and foodborne enteric viruses achieve both, most organisms do not. The mechanisms of contamination in processing plants of sea- or other foods are similar: the variety and severity of contamination largely depends on plant hygiene. Common process-related pathogens include *L. monocytogenes, Salmonella* spp., *E. coli* and *S. aureus*, but essentially any zoonotic food pathogen could contaminate the final products. The contamination of raw material is largely dependent on the geographical spread of various agents (Table 2).

Non-proteolytic *Clostridium botulinum, Vibrio* spp., caliciviruses and hepatitis A viruses, marine toxins and certain helminthic parasites are pathogens indigenous to seafood and are widely distributed in fishing grounds around the world (Huss et al., 1995; ICMSF, 1998). In areas where such hazards prevail, the level of raw material contamination also depends on factors such as the eating habits and habitats of different finfish, on the methods of capture (Chen, 1995), and - in the case of aquaculture - on the type of production system in use (Hielm et al., 1998a). Ultimately, the microbiological condition of raw seafood is improved when harvesting and slaughter practices specifically aim at pathogen risk reduction, preferably following an HACCP approach (FAO, 2001).

Table 2. Seafood pathogens and hazards significantly related to pre-harvest conditions

Causative agent	Special considerations
BACTERIA	*Multiplication in food in most cases prerequisite for disease*
• *Clostridium botulinum* type E	Spores ubiquitous in northern temperate littoral waters, outgrowth in fish causes botulism
• *Vibrio cholerae* 01/0139	Seafoods act as vectors only in endemic or epidemic cholera settings
• V. parahaemolyticus	*Gastro-intestinal infection from finfish especially common in Japan and Southeast Asia*
• *V. vulnificus*	Septicaemia from contaminated oysters leading cause of foodborne death in America
• *Listeria monocytogenes*	Persistent fish process pathogen introduced through contaminated raw materials
VIRUSES	*Feco-oral transmission of disease; viruses do not multiply in foods but infectious dose is low*
• Caliciviruses	Leading cause of GI illness worldwide, sewage-polluted foods (e.g., bivalves) are major vectors
• Hepatitis A	Infections in adults touristing in developing countries can be severe, vectors same as above
HELMINTHIC PARASITES	*Raw seafood contains infectious larvae, which are readily destroyed by cooking or freezing*
• *Anisakis simplex*	Common in many fish species from temperate waters, anisakiasis is most common in Japan
• *Diphyllobothrium latum*	Tapeworm infection is still prevalent in some temperate regions, freshwater fish are vectors
• *Opisthorchis* spp.	Life-cycle of trematode maintained in the intensive agri- and aquaculture of Southeast Asia
MARINE TOXINS	*Algal toxins accumulating in seafood; largely tasteless and cannot be destroyed by cooking*
• PSP, DSP, NSP, ASP, AZP	Present in bivalve molluscs that have been feeding on toxic algae present in harvesting waters
• Ciguatera	Ciguatoxin is accumulated in the flesh of carnivorous fish high in the food chain
• Histamin	Toxin formation caused by bacterial growth in dead fish upon harvesting, before consumption
• Tetrodotoxin	Toxin either endogenous to or formed through bacterial growth in puffer fish (*Sphoeroides*)

5. Risks Related to the Geography of Selected Fish Pathogens

One might argue that the actual geographical spread and prevalence of pathogens indigenous to fish is merely a prerequisite for contamination. Although some safety concerns are indeed solely related to pre-harvest practices, it must be acknowledged that risks can both become apparent and be effectively reduced in the processing. Yet, knowledge about the microbiological status of different fishing grounds is of utmost importance when deciding on risk management strategies related to harvesting or capturing fish (Table 3).

Successful prevention of disease outbreaks - particularly in the case of raw seafoods - depends for a large part on appreciating the fishery's pathogen status. Only on this basis can regulatory agencies impose restrictions on production or the consumer be recommended to refrain from seafood consumption should this be considered prudent. Similarly, post-harvest risk management decisions cannot be taken without sufficient background information on the pathogenicity of organisms possibly prevailing in the raw

Table 3. Methods of risk mitigation for pre-harvest hazards of seafood.

Choice of fishing season and grounds	
	Avoid bivalve harvesting during times of phytoplankton blooms (spring and early summer)
	Avoid reef fishing in tropical waters, which acts as ciguatera 'hot-spots'
	Avoid bivalve harvesting in sewage-polluted waters (as determined by fecal indicator organisms)
Capture and cleaning alternatives	
	Trawling will damage fish tissues more than hook fishing, with ruptures speeding up fish spoilage
	Evisceration onboard fishing vessels recommended, especially if trips exceed 2 days
	Freezing of gutted and washed fish onboard vessels safeguards against pathogenic bacterial growth
Good hygienic practices	
	Rapid chilling of fish and maintaining (2°C until catch delivered for processing or consumption
	Fish gutting procedures aimed at minimizing gut content contamination of product
	Regular washing and sanitation of fishing vessels' holds and fish delivery conveyances
	Active supervision of fishermen's personal hygiene, including guidance and health checks
	Ongoing training of fishermen in basic food hygiene matters relevant to their tasks
Pathogen reduction measures	
	Starvation in clean water for >3 days of farmed fish before slaughter to reduce gut microbial content
	Depuration for >4 weeks of harvested contaminated bivalves in tanks or clean water ponds before sale
	Visual inspection of fish for *Anisakidae* larvae (alternatively freezing can be employed for control)
Aquaculture considerations	
	Net-bottomed floating cages are less botulinogenic in salmonid farming than earth ponds or raceways
	Automatic removal of biomass through funnel-shaped pond bottoms reduces *C. botulinum* contamination of fish
	Use of formulated feeds instead of 'trash fish' recommended in salmonid farming for reducing *C. botulinum* spore influx
	Increased growth of carp pond vegetation should be achieved with chemical fertilisers instead of using human sewage

material. To make informed statements on the latter a profound knowledge of the epidemiology (particularly the geographical spread) of the major pathogens is necessary.

Table 4. Prevalence surveys for C. botulinum *type E in aquatic environments and seafood.*

Country	*Sample type*	*Observed prevalence of* C. botulinum *type E (%)*	*Reference*
Denmark	Sea sediment	90	Huss,1980
	Marine fish	11	Huss and Pedersen, 1979
	Farmed fish	65	Huss *et al.*, 1974
Finland	Sea sediment	83	Ala-Huikku *et al.*, 1977
	Sea and freshwater sediment	71	Hielm *et al.*, 1998b
	Sea and freshwater sediment	*80*	*Hielm* et al., *1996*
	Marine and freshwater fish	*19*	*Hyytiä* et al., *1998*
	Farmed fish	7	Ala-Huikku *et al*, 1977
	Trout farms	68	Hielm *et al.*, 1998a
	Fish roe	6	Hyytiä *et al.*, 1998
France	Sea and freshwater sediment	4.5	Smith and Moryson, 1977
Germany	Freshwater fish	30	Hyytiä-Trees et al., 1999
Iceland	Freshwater sediment	1.7	Huss,1980
Norway	Marine fish	11	Tjaberg and Håstein, 1975
Poland	Soil	31	Zaleski *et al.*, 1978
	Marine fish	18	Zaleski *et al.*, 1978
Russia	Soil and water	62	Kravchenko and Shishulina, 1969
Sweden	Sea sediment	100	Johannsen, 1963
	Soil	71	Johannsen, 1963
	Marine fish	71	Johannsen, 1963
UK	Marine fish	7.2	Cann *et al.*, 1975
USA	Soil and sea sediment	22	Miller, 1975
	Marine fish	9	Craig *et al.*, 1968
	Marine fish	1.2	Houghtby and Kaysner, 1969
	Marine fish	5	Miller, 1975
	Freshwater fish	20	Bott *et al.*, 1966
	Freshwater fish	6	Chapman and Naylor, 1966
	Freshwater fish	17	Pace *et al.*, 1967
	Crab	12-87	Eklund and Poysky, 1966
	Crab and shellfish	22	Craig *et al.*, 1968
	Marine mammals	2.2	Miller, 1975
Canada	Sea and freshwater sediment	20	Laycock and Loring, 1972
Japan	River soil	60	Yamakawa and Nakamura, 1992
Thailand	Sea sediment	0.3	Tanasugarn, 1979
	Marine fish	0.2	Tanasugarn, 1979

5.1 Bacteria

Few bacterial species are both unequivocally pathogenic to man, and indigenous members of the normal marine microflora. The latter for instance applies to nonproteolytic (Group II) strains of *C. botulinum* and various *Vibrio* species. Other non-indigenous bacteria of fish (*Enterobacteriaceae, L. monocytogenes, S. aureus*, etc.) may contaminate seafood harvested from waters polluted by sewage from humans or other warm-blooded animals (ICMSF, 1998). The prevalence of such pathogens obviously reflects the gastrointestinal biota of surrounding communities, and will be enhanced by fish farming practices (WHO, 1999).

Clostridium botulinum
The causative agent of botulism, a severe neuroparalytic foodborne intoxication, is a ubiquitous. Its nonproteolytic strains, especially *C. botulinum* type E, can be isolated from sub-arctic and northern temperate waters on fishing grounds in the Baltic region, North America, Russia, and Japan (Table 4).

Extensive microbiological screening of the Baltic Sea over the past decades has substantiated that it probably represents the most polluted watermass in the world with regard to *C. botulinum* type E. Fish seem to be incidental carriers of *C. botulinum* spores, which can be found on the skin, gills, and especially in the intestines (Hielm, 1999). Passive spore contamination is probably the result of bottom dwelling, as pelagic species feeding on plankton as a rule tend to be less contaminated than bottom-feeders (Hyytiä-Trees, 1999). That the spore carriage in fish is incidental and not the result of active infectivity is in our opinion suggested by the broad diversity of *C. botulinum* type E strains isolated from fish and their surrounding environment. Molecular fingerprinting analyses such as pulsed-field gel electrophoresis (PFGE) typing have often shown the profiles of different isolates from one pond, one fish species, or even a single fish to be unrelated (Hielm et al., 1998a; Hyytiä et al., 1999a).

Botulism and especially spore contamination of fish products can become a big problem in the closed ecosystems encountered in salmonid fish farming (Huss et al., 1974a). In a study of marine net cage and freshwater pond trout farms in Finland, we found that contamination rates in both farm types were similar, although the environmental spore pressure is much higher in marine areas (Hielm et al., 1998a). The study indicated that trout having access to pond sediments are relatively easier contaminated with *C. botulinum* type E spores than are net cage trout having no access to sediments. However, the beneficial effects of marine farming seemed to be compounded by the sea farmers' habit of feeding such trout trash herring, severely contaminated with *C. botulinum* type E spores. In conclusion, we found that the best results were achieved using freshwater ponds equipped with an automatic bottom-suction device for the removal of biomass, originally developed for reducing the eutrophic fecal load of fish farms to the environment. In the Baltic, the 10-40 % presence of botulinal spores in wild and farmed fish must be taken into account in seafood processing (Hyytiä et al., 1998), although this contamination can be significantly reduced by hygienic gutting of fish or pre-slaughter starvation (Huss et al., 1974b). Activities such as vacuum-packaging, reduction of salt, banning of nitrite as an additive in fish, extension of sell-by dates and temperature abuse will all increase the risk of botulism from fish products (Hyytiä et al., 1999b). In tropical regions fishing grounds can occasionally be polluted by the terrestrial and mesophilic *C. botulinum* types A and B instead, which naturally dictate different safety requirements than does the psychrotrophic type E (Huss et al., 1995).

Vibrio spp.

At least 12 different species of *Vibrio* have been shown to cause infections in healthy adults. Of these, *V. cholerae* 01/0139, *V. parahaemolyticus* and *V. vulnifucus* have caused most concern (WHO, 1999). All vibrios are halotolerant mesophilic organisms, which are present in warm marine, brackish, or even fresh waters. Most *Vibrio* species except *V. cholerae* are indeed halophiles, and will not survive in fresh water. Several studies have shown the occurrence of vibrios to correlate with human wastewater, whereas it is known not to correlate with the numbers of faecal coliforms used as indicator organisms (Koh et al., 1994). Both the numbers of pathogenic vibrios as well as human infections correlate strongly with increased temperature, and a seasonal correlation is particularly marked for *V. parahaemolyticus* and *V. vulnifucus* (Veenstra et al., 1994; WHO, 1999). During cholera outbreaks, virulent strains of *V. cholerae* 01 capable of transmitting the disease can sometimes be isolated from fish; however, most environmental strains of this species lack virulence genes and are thus harmless to humans (Hackney and Garrett, 1985). *Vibrio* infections in Western countries are commonly associated with shellfish, while finfish are common vectors in Japan and other Asian countries (ICMSF, 1998). In general, vibrio's do not grow well below 15°C, so refrigeration can be relied upon for control - especially since the minimum infective dose (MID) needed to induce symptoms ranges from 10^5-10^8 viable cells and these numbers will only be reached if growth in the food is allowed to occur (Huss et al., 1997). Vibrios are, however, able to grow remarkably fast if subjected to temperature abuse conditions. For example, *V. parahaemolyticus* is a major cause of foodborne illness in Japan, where it has been responsible for more than 70% of bacterial gastroenteritis (Oliver and Kaper, 1997). Large numbers of *V. vulnifucus* retained or multiplied in raw oysters can cause septicaemia or wound infections in immunologically compromised individuals, and this organism (with a case-fatality rate of 0.39) is by far the leading agent (95%) of seafood-related deaths in the USA (Mead et al., 1999; Oliver and Kaper, 1997).

Listeria monocytogenes

This species presents a more complex problem, in that this halotolerant psychrotrophic pathogen is commonly present in waters polluted by agriculture or sewage (Manoj et al., 1991; Masuda et al., 1992; Ben Embarek, 1994; Markkula et al., 2000). Differences in salinity between brackish and ocean waters does not seem to influence the prevalence of *L. monocytogenes;* whereas isolation rates have been lower in tropical waters as compared to cold or temperate waters, possibly due to interference from a high background flora (Markkula et al., 2000). Thus most fish caught or grown close to human communities may be contaminated by *L. monocytogenes*, and it is the fish through which the pathogen is apparently carried into processing facilities. As *L. monocytogenes* is readily destroyed by heat and cannot compete in growth with the natural spoilage flora of fish, the biggest concern is that upon entry it will become part of a plant's in-house microflora, thereby gaining the possibility to contaminate high-risk seafood products prior to the final packaging stage. Products such as vacuum-packaged cold-smoked or 'gravad' salmon have been implicated in listeriosis outbreaks in the Nordic countries (Ericsson et al., 1997; Miettinen et al., 1999). Reduction of post-process contamination within an infected plant is difficult without extensive contamination analysis using molecular epidemiology aiming at exposing and eventually destroying virulent strains, perhaps hiding under biofilms inside complex machinery (Autio et al., 1999; Lundén et al., 2000). *L. monocytogenes* clearly illustrates the difficult task in differentiating between pathogens related to pre-harvest contamination on the one side and post-harvest contamination on the other.

5.2 Viruses

Pathogenic viruses of concern to humans are generally found on seafood originating from coastal waters polluted with human sewage. The major foods of concern are the molluscan bivalves that filter large amounts of water and thereby retain the virus particles. Finfish or shrimp will extremely rarely act as vectors of human viral diseases, and only if post-processing contamination occurs. Human pathogenic viruses are essentially host-specific enteric organisms unable to multiply extracellularly, and will therefore not increase in numbers upon harvesting. In general, enteric viruses are destroyed by similar heat treatments as most enteric bacteria (e.g., thorough cooking), while their infectivity is preserved at low temperatures (Lees, 2000). This is an obvious cause for concern in the seafood industry, which has largely based its standards for safety on a rigorous cold chain.

Caliciviruses

Norwalk-like and other small round-structured viruses (SRSVs) belong to the caliciviruses, which are roughly 30-35 nm in diameter and similar in appearance and size to its close relative, the hepatitis E virus (Pringle, 1998). They cannot be grown in any laboratory host, so diagnosis of caliciviruses is limited to electron microscopy and PCR amplification of species-specific RNA sequences in patient stools, to which the organism is shed in large numbers (10^8 g^{-1}). Mead et al. (1999) estimated that the Norwalk-like viruses are responsible for 23 million annual cases of gastroenteritis in the USA, representing 60% of all foodborne infections. While secondary person-to-person infections represent over half of the total caseload, it seems that a substantial amount of the initial cases are related to contaminated molluscan shellfish, primarily oysters and clams (Lees, 2000). To reduce the risk of disease in America, a classification scheme termed the *National Shellfish Sanitation Program* (NSSP) has been set up by the US Food and Drug Administration (FDA, 1993), where all waters where shellfish are harvested must meet certain standards. A four-step grading system divides the growth areas into approved, conditionally approved, restricted and prohibited zones; effectively dictating to producers when it is advisable to harvest and when it is not (Chen, 1995). Similar programs have also been put in place in European shellfish growing areas (European Community, 1991) and in countries exporting shellfish to the EU or the USA. In all control schemes the quality of the harvesting areas is chiefly sensored through indicator organisms such as faecal coliforms or *E. coli*, which do not correlate all that strongly with viral contamination. New indicators are now being investigated, such as F+ bacteriophages (Dore et al., 2000), *C. perfringens* spores, or adenoviruses (Jiang et al., 2001); which seem to mimic the spread of seafood-related viruses in water to a larger degree.

Hepatitis A

Hepatitis is a feco-orally transmitted liver disease caused by the hepatitis viruses A to E, of which only the A-type has been conclusively linked to the consumption of contaminated foods (Lees, 2000). However, only about 5% of the estimated annual cases (83,000) of hepatitis A in the US are attributable to foodborne transmission (Mead et al., 1999). As with the Norwalk-like viruses, hepatitis A is retained during filtration in molluscan bivalves if these are grown in waters polluted with human sewage. In several viral outbreaks the contamination of shellfish has resulted from gross "point" pollution (emptying of toilets overhead harvesting areas), rather than from dispersion of communal wastewaters. Consequently, bivalve shellfish that is eaten raw or inadequately heated are typical vectors of primary cases of hepatitis A. Hepatitis A is endemic in most developing countries, and most children have experienced mild infections and acquired active immunity by the age of six (Lees, 2000). However, in developed northern countries with improved sanitary conditions infections are rare and correspondingly severe and these are acquired in adulthood primarily through travel to endemic countries. As there is little or no immunity in developed countries, the prospect of a large-scale outbreak is frightening in as much as it is probable. Hepatitis A virus are shed some two weeks before symptoms of the disease occur, giving the disease ample time to be transmitted to others through unhygienic handling of foods. A vaccine giving protection against hepatitis A for at least 20 years is available, and due to their pivotal role in an epidemic setting the active immunization of food handlers against hepatitis A has been discussed in several countries. The first reported outbreak of shellfish-borne hepatitis A affected 629 patrons eating contaminated oysters in Sweden in 1955 (Roos, 1956), while the largest recorded outbreak involved almost 300,000 cases in Shanghai, China in 1988 and was attributed to clams (Halliday et al., 1991).

5.3 Helminthic parasites

A large number of finfish and shellfish species, both marine and freshwater, are potential sources of parasitic zoonoses of public health interest (WHO, 1999). Foodborne helminths share the common characteristic of requiring more than one host to complete their life cycles. Fish or crustaceans generally act as intermediate hosts, while the definite hosts can be found on the highest trophic level of the food chain (Hayunga, 1997). Adult forms of helminths are rarely host-specific, which means that the role of the definitive host can occasionally be transferred from aquatic predators to humans consuming raw or undercooked fish, resulting in disease. The often severe clinical manifestations of helminthic infections are by and large due to the invasive nature of adult worms, whereas propagation rarely occurs in the human host. Severity of helminthic disease is thus more dependent on the number of infective organisms consumed than in the case of microbial infections. Helminths can easily be destroyed by heating or freezing of the contaminated seafood, so infections are only prevalent in those communities around the world where eating raw or undercooked fish is a cultural habit.

Anisakiasis

Anisakis simplex or *Pseudoterranova decipiens*, also known as the sushi parasites, are transmitted through eating of raw finfish or squid. Anisakiasis is especially prevalent in

Japan (2000-3000 annual cases), but cases occur regularly in other developed countries as well (Sakanari and McKerrow, 1989; WHO, 1999). Symptoms range from "tingling throat syndrome" to the more severe gastrointestinal migration of larvae, for which the only effective treatment is surgical worm removal (Hayunga, 1997). Anisakidae larvae have been found in numerous fish species captured from temperate waters, and fish safety regulations state that susceptible fish must be visually inspected for worms or frozen before sale to eliminate the risk of infection. It has been reported that salmon farmed in Norway, Scotland and the USA do not harbour nematodes, probably due to formulated feeds, and is thus exempt from hygienic provisions requiring freezing (Reilly and Käferstein, 1997).

Diphyllobothriasis

Human tapeworm infections are caused by *Diphyllobothrium* species that mature in the small intestine, causing mild symptoms despite their relatively large size (up to 9 m in length). Human infections are reported from the Baltic countries, the Great Lakes region, Canada, and Japan, which result from the ingestion of raw freshwater fish containing infective *D. latum* larvae. The disease now occurs at a much lower frequency than before, largely due to communal sewage treatment and informed preparation practices such as freezing. Diphyllobothriases are still thought to be prevalent in Eastern Europe and Russia, and even in the USA the number of broad tapeworm carriers far outnumbers those suffering from anisakiasis (Reilly et al., 1997). The worldwide number of tapeworm carriers at present is not known, but in 1977 it was estimated to be 13 million (Crompton and Joiner, 1980).

Opisthorchiasis

Fishborne trematode infections are of great public health concern in various parts of the developing world, estimated to affect some 40 million people. Although seldom fatal in their own right, they can cause morbidity and complications leading to death. Infection is caused by the ingestion of viable metacercariae encysted in the flesh of freshwater fish that is served raw or otherwise minimally processed (WHO, 1999). In addition to humans, many farm animals can serve as trematode reservoirs, and the parasites' life-cycle is well maintained in the dense agri- and aquaculture settings of Southeast Asia. *Opisthorchis sinensis, O. viverrini* and *O. felineus* are the main pathogens of concern, causing chronic bile duct and liver inflammation, pancreatitis, and possibly cholangiocarcinoma in affected individuals. All *Opisthorchis* species are widespread through Southeast Asia, but *O. felineus* can also be found in Eastern Europe and parts of the former Soviet Union (Hayunga, 1997).

5.4 Marine toxins

In recent years, a substantial amount of the fishborne outbreaks of illness have been due to the ingestion of marine toxins that have accumulated in the flesh of finfish or shellfish. There is a wide range of different toxins and microalgal species producing them, and to date at least seven different types of intoxications are recognized: paralytic shellfish poisoning (PSP); diarrhetic shellfish poisoning (DSP); neurotoxic shellfish poisoning (NSP); amnesic shellfish poisoning (ASP); azaspiracid poisoning (AZP); ciguatera; and scombroid (histamine) fish poisoning (Aune, 1999). Most of these illnesses are caused by consumption of bivalve molluscs that have been feeding on toxic algae, but ciguatera is related to the accumulation of toxins in the flesh of large carnivorous fish that occupy the higher levels of the food chain. Formation of histamine (and other biogenic amines) is essentially a post-harvest problem, since it is the result of histidine decarboxylase activity in bacteria multiplying on dead fish after capture.

A wide range of environmental factors influence the production of toxins in the algae, e.g., temperature, salinity, pH, light, and nutrients like nitrogen and phosphorus (Aune, 1999). Toxic algal blooms and disease resulting from these have historically been connected to tropical waters, but these seem to have spread to temperate waters during the last decades. This may be the result of increased global shipping (Aune, 1999) or perhaps even global warming. The severity of illness from marine toxins depends on the ingested dose and the general condition of the patient, with children being more susceptible than adults. Death can on rare occasions result from PSP, less so with the other forms of intoxication (Gillespie et al., 1986). Prevention of diseases from marine toxins relies on active surveillance for suspect algae in fishing areas, as the toxins in questions are essentially tasteless and are not destroyed by routine cooking.

6. Conclusions

6.1 What has been achieved?

In more developed regions, sophisticated monitoring for indigenous seafood hazards, or indicators for these, is being carried out. This has reduced the incidence of hazards in commercial seafood products originating from both capture fishery and aquaculture. Sophisticated chilling equipment is today prevalent onboard large-scale fishing vessels, which effectively brings the start of the cold chain and termination of pathogen growth to the moment of harvest. Surveys for indigenous bacterial pathogens such as *C. botulinum* type E and vibrios have been carried out in many fishing waters, and today we have established a fairly comprehensive picture of the geographical spread and epidemiological features of these pathogens. Molecular diagnostic techniques such as PCR have been developed for the detection of many seafood pathogens, including caliciviruses and *C. botulinum,* which today enables more extensive and speedy surveillance.

6.2 What has been neglected?

Fish production intended for domestic use in developing countries is not yet properly regulated. The cold chain, if existing at all, does not commence directly upon harvesting which makes pathogenic bacterial growth and subsequent illness possible. Monitoring of waters for hazards is not proportional to the high population densities in the third world, resulting in unnecessarily high numbers of viral and bacterial gastrointestinal disease due to sewage pollution. Low-quality or even total lack of phytoplankton monitoring can also lead to large marine biotoxin outbreaks. Trematodiases of seafood origin are rampant in Southeast Asia, but also occur in South America and countries of the former Soviet Union; as these are spread through the eating of raw fish such dietary habits should be discouraged in the risk areas.

In developed countries, the move from traditional end-product testing to a HACCP-based food safety approach has not been as swift in the seafood industry as in other food industries. This reluctance to change is manifested through great product quality and safety variability, with low-grade products causing bacterial and viral seafood outbreaks in excess.

6.3 What needs to be done

In fighting foodborne disease, a good knowledge of food hygiene is paramount to other demands. The persons in most need of basic hygiene education are on one hand those workers that handle foods before consumption, and on the other hand the consumers. The need for education is evident in small-scale seafood production, where fishermen, fish farmers and processing employees should be offered government-sponsored tuition. Furthermore, the fish industries of developing countries should receive aid in building processing facilities compliant to the basic hygiene standards set out by the Codex Alimentarius Commission.

The transition to HACCP as the primary risk management tool in the fish industry of developed countries should be accelerated. Today's strong increase of international trade and changing dietary habits is likely to bring forth new seafood hazards as well as enhance some old ones; to combat these we need a program for healthy seafood based on the principles of risk analysis which, when needed, can react swiftly to novel scientific input.

References

Ala-Huikku, K., Nurmi, E., Pajulahti, H and Raevuori, M., 1977. The occurrence of *Clostridium botulinum* type E in Finnish trout farms and the prevention of toxin formation in fresh-salted vacuum-packed trout fillets. Nord. Vet. Med. 229, 386-391.

Aune, T., 1999. Detection of algal toxins and examples of marine food poisoning. Proc. Soc. Vet. Finlandiae, Fennovet, Helsinki, p 185-188.

Autio, T., Hielm, S., Miettinen, M., Sjöberg, A.-M., Aarnisalo, K., Björkroth, J., Mattila-Sandholm, T. and Korkeala, H., 1999. Sources of *Listeria monocytogenes* contamination in a cold-smoked rainbow trout processing plant detected by PFGE typing. Appl. Env. Microbiol. 65, 150-155.

Barile, L.E., Milla, A.D., Reilly, A. and Villadsen, A., 1985. Spoilage patterns of mackerel (*Rastrelliger faughni* Matsui). 1. Delays in icing. In: (Ed.) Reilly, A., Spoilage of Tropical Fish and Product Development, FAO Fisheries Report 371 (Suppl.), Rome, p 29-40.

Ben Embarek, P.K., 1994. Presence, detection and growth of *Listeria monocytogenes* in seafood. Int. J. Food Microbiol. 23, 17-34.

Bott, T.L., Deffner, J.S., McCoy, E. and Foster, E.M., 1966. *Clostridium botulinum* type E in fish from the Great Lakes. J. Bacteriol. 91, 919-924.

Bryan, F.L., 1987. Seafood-transmitted infections and intoxications in recent years. In: (Eds.) Kramer, D.E. and Liston, J., Seafood Quality Determination, Elsevier Science, Amsterdam, p 319-337.

Cann, D.C., Taylor, L.Y. and Hobbs, G., 1975. The incidence of *Clostridium botulinum* in farmed trout raised in Great Britain. J. Appl. Bact. 39, 331-336.

Chapman, H.M. and Naylor, H.B., 1966. Isolation of *Clostridium botulinum* type E from Cayuga Lake fish. Appl. Microbiol. 14, 301-302.

Chen, H-C., 1995. Seafood organisms and seafood safety. J. Food and Drug Analysis 3, 133-144.

Craig, J.M., Hayed, S. and Pilcher, K.S., 1968. Incidence of *Clostridium botulinum* type E in salmon and other marine fish in the Pacific Northwest. Appl. Microbiol. 16, 553-557.

Crompton, D.W.T. and Joiner, S.M., 1980. Parasitic Worms, Wyeham, London.

Disney, J.G., 1976. The spoilage of fish in the tropics. In: (Eds.) Cobb, B.F. and Stockton, A.B., Proc. 1st annual tropical and subtropical fisheries technol. Conf. Vol. 1, Texas A&M Univ. Sea Grant, College Station, p 23.

Dore, W.J., Henshilwood, K. and Lees, D.N., 2000. Evaluation of F-specific RNA bacteriophage as a candidate human enteric virus indicator for bivalve molluscan shellfish. Appl. Env. Microbiol. 66, 1280-1285.

Eklund, M. W. and Poysky, F., 1966. Incidence of *Cl. botulinum* type E from the Pacific Coast of the United States. In: (Eds.) Ingram, M. and Roberts, T.A., Botulism, Chapman and Hall, London, p 49-55.

Ericsson, H., Eklöw A., Danielsson-Tham, M-L., Loncarevic, S., Mentzing, L-O., Unnerstad, H., and Tham, W., 1997. An outbreak of listeriosis suspected to have been caused by rainbow trout. J. Clin. Microbiol. 35, 2904-2907.

European Community, 1991. Council Directive of 15th July 1991 laying down the health conditions for the production and placing on the market of live bivalve molluscs (91/492/EEC). Off. J. Eur. Communities.

FDA, 1993. National Shellfish Sanitation Program, Manual of operations, 1993 revision. USA Department of Health and Human Services, Public Health Service, Food and Drug Administration.

FAO, 1973. Code of practice for fresh fish. FAO Fisheries Circ. C318. Food and Agricult. Org. of the United Nations, Rome, Italy.

FAO, 2001. The state of world fisheries and aquaculture 2000. Food and Agricult. Org. of the United Nations, Rome, 119 p. Available at http://www.fao.org/docrep/003/x8002e/x8002e00.htm

Food Standards Agency, 2001. Agency probes nation on attitudes to food. Available at http://www.foodstandards.gov.uk/press_releases/uk_press/2001/pr10205.htm

Gillespie, N.J., Lewis, R.J., Pearn, J.H., Bourke, A.T.C., Holmes, M.J., Bourke, J.B. and Shields, W.J., 1986. Ciguatera in Australia: occurrence, clinical features, pathophysiology and management. Med. J. Australia 145, 584-590.

Hackney, C.R. and Garrett, E.S., 1985. Safety and quality of seafood. 38th Reciprocal Meat Conference Proceedings. American Meat Science Association, p 135-140.

Halliday, M.L., Kang, L.-Y. and Zhou, T.-K., 1991. epidemic of hepatitis A attributable to the ingestion of raw clams in Shanghai China. J. Infect. Dis. 164, 852-859.

Hatakka, M. and Halonen, H., 2000. Ruokamyrkytykset Suomessa vuonna 1999. National Food Adminstration No. 7/2000, Helsinki, 27 p.

Hayunga, E.C., 1997. Helminths acquired from finfish, shellfish, and other food sources. In: (Eds.) Doyle, M.P., Beuchat, L.R. and Montville, T.J., Food Microbiology - Fundamentals and Frontiers, ASM Press, Washington DC, p 463-477.

Hielm, S., 1999. Molecular detection, typing and epidemiology of *Clostridium botulinum*. PhD thesis, Helsinki University, 78 p.

Hielm, S., Björkroth, J., Hyytiä, E. and Korkeala, H., 1998a. Prevalence of *Clostridium botulinum* in Finnish trout farms: PFGE typing reveals extensive genetic diversity between type E isolates. Appl. Env. Microbiol. 64, 4161-4167.

Hielm, S, Hyytiä, E., Ridell, J. and Korkeala, H., 1996. Detection of *Clostridium botulinum* in fish and environmental samples using polymerase chain reaction. Int. J. Food Microbiol. 31, 357-365.

Hielm, S, Hyytiä, E., Andersin, A.-B. and Korkeala, H., 1998b. A high prevalence of *Clostridium botulinum* type E in Finnish freshwater and Baltic Sea sediment samples. J. Appl. Microbiol. 84: 133-137.

Hirn, J., Maijala, R. and Johansson, T., 1992. Foodborne disease outbreaks in Finland 1975-1990. Proc. 3rd World Congress Foodborne Infections and Intoxications, Berlin, p 33-36.

Houghtby, G.A. and Kaysner, C.A., 1969. Incidence of *Clostridium botulinum* type E in Alaskan salmon. Appl. Microbiol. 18, 950-951.

Huss, H.H., 1980. Distribution of *Clostridium botulinum*. Appl. Environ. Microbiol. 39, 764-769.

Huss, H.H., Ben Embarek, P.K. and Jeppesen, V.F., 1995. Control of biological hazards in cold smoked salmon production. Food Control 6, 335-340.

Huss, H.H., Dalgaard, P. and Gram, L., 1997. Microbiology of fish and fish products. In: (Eds.) Luten, J.B., Børresen, T. and Oehlenschläger, J., Seafood from Producer to Consumer, Integrated Approach to Quality, Elsevier Science, Amsterdam, p 413-430.

Huss, H.H., Pedersen, A. and Cann, D.C., 1974a. The incidence of *Clostridium botulinum* in Danish trout farms. I Distribution in fish and their environment. J. Food Technol. 9, 445-450.

Huss, H.H., Pedersen, A. and Cann, D.C., 1974b. The incidence of *Clostridium botulinum* in Danish trout farms. II. Measures to reduce contamination of the fish. J. Food Technol. 9, 451-458.

Huss, H.H. and Pedersen, A., 1979. *Clostridium botulinum* in fish. Nord. Vet. Med. 31, 214-221.

Hyytiä, E., Hielm, S., and Korkeala, H., 1998. Prevalence of *Clostridium botulinum* type E in Finnish fish and fishery products. Epidemiol. Infect. 120, 245-250.

Hyytiä, E., Hielm, S., Björkroth, J. and Korkeala, H., 1999a. Biodiversity of *Clostridium botulinum* type E isolated from fish and fishery products. Appl. Environ. Microbiol. 65, 2057-2064.

Hyytiä, E., Hielm, S., Mokkila, M., Kinnunen, A. and Korkeala, H., 1999b. Predicted and observed growth and toxigenesis by *Clostridium botulinum* type E in vacuum-packaged fishery product challenge tests. Int. J. Food Microbiol. 47, 161-169.

Hyytiä-Trees, E., 1999. Prevalence, molecular epidemiology and growth of *Clostridium botulinum* type E in fish and fishery products. PhD thesis, Helsinki University, 73 p.

Hyytiä-Trees, E., Lindström, M., Schalch, B., Stolle, A. and Korkeala, H., 1999. *Clostridium botulinum* type E in Bavarian fish. Arch. Lebensmittelhyg. 50, 79-82.

ICMSF, 1998. Fish and fish products. In: Micro-organisms in foods 6 - Microbial ecology of food commodities. Blackie Academic and Professional, London, p 130-150.

Jackson, T.C., Acuff, G.R. and Dickson, J.S., 1997. Meat, poultry, and seafood. In: (Eds.) Doyle, M.P., Beuchat, L.R. and Montville, T.J., Food Microbiology - Fundamentals and Frontiers, ASM Press, Washington DC, p 83-100.

Jiang, S., Noble, R. and Chu, W., 2001. Human adenoviruses and coliphages in urban runoff-impacted coastal waters of Southern California. Appl. Environ. Microbiol. 67, 179-184.

Johannsen, A., 1963. *Clostridium botulinum* in Sweden and the adjacent waters. J. Appl. Bacteriol. 26, 43-37.

Koh, E.G.L., Huyn, J.-H. and LaRock, P.A., 1994. Pertinence of indicator organisms and sampling variables to *Vibrio* concentrations. Appl. Env. Microbiol. 60, 3897-3900.

Kravchenko, A.T. and Shishulina, L.M., 1966. Distribution of *Cl. botulinum* in soil and water in the U.S.S.R. In: (Eds.) Ingram, M. and Roberts, T.A., Botulism, Chapman and Hall, London, p 13-19.

Laycock, R.A. and Loring, D.H., 1972. Distribution of *Clostridium botulinum* type E in the Gulf of St. Lawrence in relation to the physical environment. Can. J. Microbiol. 18, 763-773.

Lees, D., 2000. Viruses and bivalve shellfish. Int. J. Food Microbiol. 59, 81-116.

Lima dos Santos, C.A.M., 1981. The storage of tropical fish in ice - a review. Tropical Sci. 23, 97-127.

Lindqvist, R., 1999. Vi ser bara toppen av isberget - Få matförgiftningar rapporteras i Sverige. Vår Föda No. 5, 5-8.

Lundén, J.M., Miettinen, M.K., Autio, T.J. and Korkeala, H.J., 2000. Persistent *Listeria monocytogenes* strains show enhanced adherence to food contact surface after short contact times. J. Food Prot. 63, 1204-1207.

Manoj, Y.B., Rosalind, G.M., Karungasar, I. and Karungasar, I., 1991. *Listeria* spp. in fish and fish-handling areas, Mangalore, India. Asian Fish Sci. 4, 119-122.

Markkula, A., Autio, T. and Korkeala, H., 2000. *Listeria monocytogenes* in Finnish raw fish. Proc. 'Food Safety Assurance in the Pre-harvest Phase', University of Veterinary Medicine, Vienna, p 147-148.

Martin, R.E., Gray, R.J.H. and Pierson, M.D., 1978. Quality assessment of fresh fish and the role of the naturally occurring microflora. Food Technol. 32, 188.

Masuda, T., Iwaya, M., Miura, H., Kokubo, Y. and Maruyama, T., 1992. Occurence of *Listeria* species in fresh seafood. J. Food Hyg. Soc. Japan 33, 599-602.

Mead, P.S, Slutsker, L., Dietz, V., McCaig, L.F., Bresee, J.S., Shapiro, C., Griffin, P.M. and Tauxe, R.V., 1999. Food-related illness and death in the United States. Emerg. Inf. Dis. 5, 607-625.

Miettinen, M.K., Siitonen, A., Heiskanen, P., Haajanen, H., Björkroth, K.J. and Korkeala, H., 1999. Molecular epidemiology of an outbreak of febrile gastroenteritis caused by *Listeria monocytogenes* in cold-smoked rainbow trout. J. Clin. Microbiol. 37, 2358-2360.

Miller, L.G., 1975. Observations on the distribution and ecology of *Clostridium botulinum* type E in Alaska. Can. J. Microbiol. 21, 920-926.

Norling, B., 1994. Food poisoning in Sweden - results of a field survey. Report No 41/94. National Food Adminstration, Uppsala, Sweden.

Oliver, J.D. and Kaper, J.B., 1997. *Vibrio* species. In: (Eds.) Doyle, M.P., Beuchat, L.R. and Montville, T.J., Food Microbiology - Fundamentals and Frontiers, ASM Press, Washington DC, p 228-264.

Pace, P.J., Krumbiegel, E.R., Angelotti, R and Wisniewski, H.J., 1967. Demonstration and isolation of *Clostridium botulinum* types from whitefish chubs collected at fish smoking plants of the Milwaukee area. Appl. Microbiol. 15, 877-884.

Pringle, C.R., 1998. Virus taxonomy - San Diego 1998. Arch. Virol. 143, 1449-1459.

Reilly, A., Howgate, P. and Käferstein, F., 1997. Safety hazards and the application of HACCP in aquaculture. Proc. 2nd Int. Conf. Fish Insp. Quality Control: A Global Focus, Technomic Publishing, Lancaster,PA, p 353-373.

Reilly, A. and Käferstein, F., 1997. Food safety hazards and the application of the principles of the hazard analysis and critical control point (HACCP) system for their control in aquaculture production. Aquacult. Res. 28, 735-752.

Roos, R., 1956. Hepatitis epidemic conveyed by oysters. Svensk Läkartidning 53, 989-1003.

Sakanari, J.A. and McKerrow, J.H., 1989. Anisakiasis. Clin. Microbiol. Rev. 2, 278-284.

Shewan, J.M., 1949. Some bacteriological aspects of handling, processing and distribution of fish. J. Royal Sanit. Inst. 59, 394.

Shewan, J.M., 1961. The microbioogy of sea-water fish. In: (Ed.) Borgstrom, G., Fish as food, Vol. 1. Academic Press, New York, p 487.

Shewan, J.M., 1971. The microbiology of fish and fishery products - a progress report. J. Appl. Bacteriol. 34, 299-315.

Smith, G.R. and Moryson, C.J., 1977. The low prevalence of *Clostridium botulinum* in the lakes, marshes and waterways of the Camargue. J. Hyg., Camb. 78, 33-37.

Sumner, J.L., Gorczyca, E., Cohen, D. and Brady, P., 1984. Do fish from tropical waters spoil less rapidly in ice than fish from temperate waters? Food Technol. Australia 36, 328.

Tanasugarn, L., 1979. *Clostridium botulinum* in the Gulf of Thailand. Appl. Environ. Microbiol. 37, 194-197.

Tjaberg, T.B. and Håstein, T., 1975. Utbredelse av *Clostiridum botulinum* i norske fiskeoppdrettsanlage. Norsk Vet. Tidskr. 87, 718-720.

Veenstra, J., Rietra, P.J.G.M., Coster, J.M., Slaats, E. and Dirks-Go, S., 1994. Seasonal variations in the occurrence of *Vibrio vulnificus* along the Dutch coast. Epidemiol. Infect. 112, 285-290.

Ward, D.R. and Baj, N.J., 1988. Factors affecting microbiological quality of seafoods. Food Technol. 42, 85-89.

WHO, 1999. Food safety issues associated with products from aquaculture. World Health Organisation Technical Report Series No. 883, Geneva, p 9-18.

Yamakawa, K. and Nakamura, S., 1992. Prevalence of *Clostridium botulinum* type E and coexistence of *C. botulinum* nonproteolytc type B in the river soil of Japan. Microbiol. Immunol. 36, 583-591.

Zaleski, S., Daczkowska, E, Fik, A. and Jozwiak, A., 1978. Surveys on the occurrence of *Clostridium botulinum* in fresh Baltic herrings. Acta Alim. Polonica. 6, 159-161.

Synopses of other conference contributions

Preliminary results of *Salmonella* challenge experiments in broilers fed with fermented liquid feed

Lourens Heres
Department of Food Science, ID-Lelystad, The Netherlands.

Introduction

Salmonella is a major cause of food poisoning in Western Europe. Poultry is recognised as the major reservoir for human infections with *S. enteritidis* and *S. typhimurium*.
On the farm level hygienic measures, vaccination and the use of caecal cultures such as Broilact® can help to control salmonella in poultry flocks. More than once a protective effect of feeding fermented products on the prevalence of salmonella in pig husbandry was observed(1,2). Although the mechanisms behind this reduced risk for salmonella infections in pigs are not clarified, two challenge experiments in broilers were conducted with fermented feed. These experiments are carried in a framework in which the epidemiology of salmonella and campylobacter is studied more extensively.

Materials and Methods

A broiler feed, based on a commercial composition, gamma sterilised, was supplied to the control groups. The fermented liquid feed (FLF) was composed of 1.4 kg of water mixed with 1 kg of feed. Fermentation was done by adding *Lactobacillus plantarum* (10^7 cfu/g) and incubating the batches at 30(C. At the end of fermentation, after two days, a pH of 4.0 and 10^9-10^{10} cfu lactobacilli per gram liquid feed were reached. The salmonella strain that was used was a naladixic acid resistant mutant of *S. enteritidis* (SE-nal^r). In experiment A, 4 times 8 salmonella free broilers per group (control and FLF) of 1-week old were orally inoculated with 10^2, 10^3, 10^4, and 10^5 cfu SE-nal^r respectively. Inoculated chickens were housed in small pens, with one chicken per pen, in two compartments. After inoculation chickens were examined with cloaca swabs at 1, 2, 3 and 8 days after inoculation (p.i.).
At day 13 p.i. all animals were euthanised and salmonella was cultured out of the caeca.
In experiment B, 3 times 16 one-week chickens were inoculated with 10^5, 10^6 and 10^7 cfu SE-nal^{rr}. These chickens were all fed FLF. Inoculated chickens were housed in pairs in one compartment. Cloaca swabs were taken during 6 days post inoculation. At day 6 p.i. the presence of SE-nal^r of the caeca was estimated.
No contact between animals of different pens was possible. Chickens, feed and water equipment were handled with gloved hands and hands were disinfected with 70% alcohol to prevent spread between different pens.

Results and Discussion

Results of experiment A and B are summarised in table 1.
The 10^2 cfu inoculation dose was not high enough to infect any of the chickens in both groups during the first 3 days p.i.. At the 7th day p.i., that was the end of the experiment for this group, most caeca were still negative for salmonella, except for two control animals. The FLF counterparts stayed negative till the end of the experiment.
The FLF-fed group inoculated with 10^3 cfu stayed negative during the first days p.i.. At day 8 only two chicken was positive in the cloaca swab. At the end 6 of 8 chickens had salmonella positive caeca. In the control group 4 inoculated chickens shed salmonella already at day 2 after inoculation. At day eight, 6 out of 8 pens contained chickens that showed positive cloaca swabs. At the end all chickens in the control groups (10^3 infectious dose) were colonised with salmonella.
In the FLF group inoculated with 10^4 cfu SE one pen had an excreting animal from day 2 p.i. onwards. One other inoculated chicken was also positive in the cloaca swab at day 8. However, at the end of the experiment, 13 days p.i., all chickens were colonised in the caeca. In the control group of this inoculation level 3 of 8 cages stayed negative during the swabbing, and 6 of 8 cages were salmonella infected at the end of the experiment.
In the FLF group inoculated with 10^5 cfu non of the infected animals excreted SE for more than one day of the first 3 days after inoculation. Three chickens were found positive in the cloaca swabs only once. At day eight p.i. 5 out of 8 FLF-chickens were shedding SE. At the end of the experiment one cage housed a negative chicken. In the control groups all inoculated chickens were excreting salmonella within three days after inoculation.
In experiment B the inoculation dose of 10^5 was repeated for the FLF group. The number of excreting animals in this group tended to be a slightly lower than in experiment A. Inoculation with 10^6 and 10^7 cfu SE induced a higher number of excreting chickens, maximum 6 out of 16 and 11 out of 16 respectively. The maximum level of 11 out of 16 chickens in the 10^7 cfu group is lower than the 8 out of 8 in the dry feed control animals at an inoculation dose of 10^5 cfu in experiment 1.
During experiment A more and more chickens began to excrete salmonella, which can be seen in the rise of positive swabs at the 8 p.i. and the number of positive cultured caecal contents at day 13 of both feed groups. There might be at least two reasons for this increase in colonised chickens. 1) An aerogenic horizontal transmission within the department occurred. In our experiment this aerogenic transmission from cages with a high infectious pressure to pens with non colonised chickens can not be excluded. 2) Most chickens were already colonised shortly after inoculation but the detection level of the cloacal swab is too high to detect this colonisation.
The number of SE excreting chickens in the 10^7 FLF group are, with respect to the first days after inoculation, most comparable with the 10^3 control group. The 10.000 times difference in inoculation dose suggests an improved resistance against a single oral infection of chickens against SE when fed with fermented feed.

A mechanism that hypothetically explains the improved resistance against colonisation is the acidifying effect of the fermented feed in the first parts of the gastro-intestinal tract. The liquid feed contains a high concentration of lactate. This lactate in combination with

Table 1. Positive swabs after Salmonella *challenge in control and liquid fermented feed fed broilers.*

	inoculation dose										
	2		3		4		5		5	6	7
days p.i.	FLF	C	FLF	C	FLF	C	FLF	C	FLF	FLF	FLF
1	0/8	0/8	0/8	3/8	0/8	5/8	2/8	6/8	1/16	4/16	6/16
2	0/8	0/8	0/8	4/8	1/8	5/8	0/8	7/8	0/16	5/16	6/16
3	0/8	0/8	0/8	4/8	1/8	5/8	1/8	8/8	1/16	3/16	7/16
4									2/16	6/16	5/16
5									2/16	4/16	6/16
6									4/16	5/16	11/16
8			2/8	6/8	2/8	5/8	5/8	8/8			
FINAL	0/8	2/8	6/8	8/8	8/8	6/8	7/8	8/8	7/16	10/16	14/16
caecum	d 7	d 7	d 13	d 13	d 13	d 13	d 13	d 13	d 6	d 6	d 6

FLF = fermented liquid feed group C= control

the low pH in crop and gizzard will probably kill most of the salmonella. Thus the probability of reaching the caecum will be reduced.
It is also possible that FLF reduces the level of colonisation with salmonella in the caecum with a corresponding reduced excretion level. In other experiments with FLF we found reduced levels of enterobacteriaceae in the caecum. It is not yet clear whether these lower levels are caused by a reduction of transient bacteria or by competition with lactobacilli or their metabolites.

The results of experiment A and B clearly demonstrate a protective effect against a single oral infection with *S. enteritidis*.
Transmission experiments with salmonella are underway to verify whether the liquid feed not only protects against a single oral infection, but also prevents transmission between broilers. The challenge and transmission studies will be extended to campylobacter in the near future

References

Wolf PJ., Bongers JH., Elbers ARW., Franssen FMMC., Hunneman WA., Exsel ACA. et al. Salmonella infections in finishing pigs in The Netherlands: bacteriological herd prevalence, serogroup and antibiotic resistance of isolates and risk factors for infection. Veterinary Microbiology 1999;67:263-275.

Schie FWv. Some epidemiological and nutritional aspects of asymptomatic salmonella infection in pigs: Universiteit van Utrecht; 1987

Influence of different types of husbandry and feeding on lead contents of selected cattle tissues

Sonja Forster and A. Stolle
Institute for Hygiene and Technology of Food of Animal Origin, Veterinary Faculty, Ludwig-Maximilians-University, Munich

The toxicological significance of lead for human health is well known. In this respect it is important that food intake does not contribute to a considerable exposure to this element. The discussions about better ways of living based on ecological principles brought up the question if meat from animals reared on ecological farms is less contaminated with lead than meat originating from conventionally raised animals.

This study was intended to prove the suspected correlation between lead concentrations in cattle tissues and the type of farm the cattle were raised on. During 1997 and 1998 249 samples were taken and the respective lead concentrations were determined. For each of the three different kinds of farming examined (intensive, extensive and ecological) three farms were included in the investigations. From each farm tank milk, feed stuff and liquid manure samples were obtained. Three cows per farm were slaughtered, aged from four to 18 years. At the slaughterhouse muscle (M. tibialis cranialis, M. biventer), kidney, liver, blood and bone (tarsal joint) samples were taken.

The lead contents were determined via graphite furnance atomic absorption spectrometry. By parallel analyses of certified reference materials the quality of the examination was assured. Similar to other studies, kidneys and bones showed the highest contents of the animal samples while the lead levels on the other samples were comparatively low. The mean lead concentrations of the kidneys and bones were 0.240 mg/kg and 0.727 mg/kg wet weight, respectively. The lead levels of the feed stuff varied from 4.2 mg/kg (mineral food) to 0 mg/kg (mineral powder). For the liquid manure samples intermediate lead levels were observed.

No significant differences in the lead levels of samples of animal origin were noticed between the three different kinds of farming. Differences in the lead concentrations appeared when comparing individual farms. Because of significant correlations between the lead levels of animal tissues and corresponding liquid manure samples the investigation of liquid manure samples might serve as a indicator for the lead load of cattle.

Table. Mean of the lead concentrations (mg/kg) according to sample and kind of farms.

sample	ecological group	extensive group	intensive group
M. tibialis cran.	0,022	0,026	0,012
M. biventer	0,019	0,032	0,014
kidney	0,073	0,071	0,109
liver	0,037	0,044	0,078
blood	0,007	0,015	0,016
bone	0,780	0,734	0,776
milk1	<ng	0,011	<ng
milk 2	<ng	0,006	<ng
milk 3	<ng	0,007	<ng
gras	0,271	0,146	0,164
silage	0,275	0,172	0,252
hay	1,441	0,395	0,113
salt	0,676	0,176	0,943
cereals		0,078	0,510
mineral food	8,631	2,101	
spent hops	0,025	0,031	
mineral powder			<ng
cobs	0,355	1,834	
other concentr.feed	0,838		
vitamin feed	0,918		
liquid manure	0,092	0,110	0,237

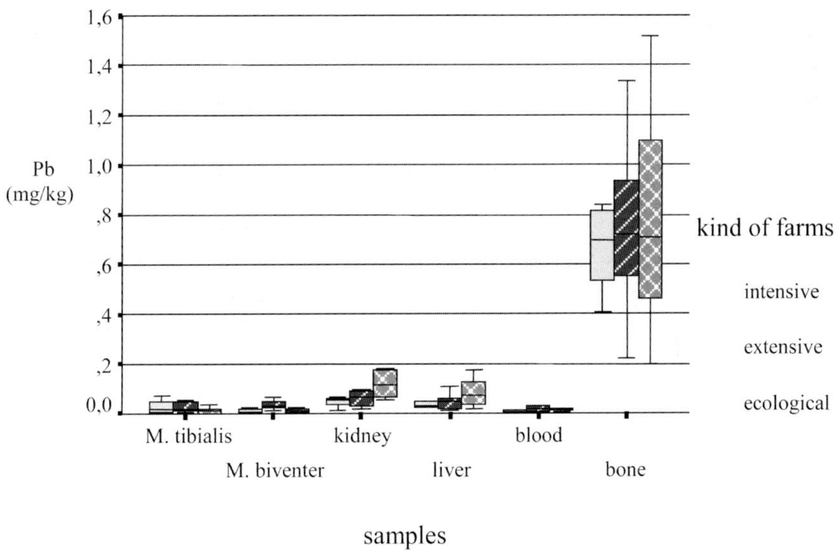

Figure. Pb-contents of animal tissues according to husbandry.

C. botulinum contamination in honey

Mari Nevas, Sebastian Hielm, Miia Lindström and Hannu Korkeala
Department of Food and Environmental Hygiene, Faculty of Veterinary Medicine, University of Helsinki, Helsinki, Finland

Clostridium botulinum is a pathogenic sporeforming organism that produces the most potent toxin known during its growth. The spores of *C. botulinum* are widely distributed in soil and may contaminate various food items via dust. Spores of *C. botulinum* are also found in honey. In the United States there have been evidence that 10-15% of honey would be contaminated with spores of *C. botulinum*. (Lilly *et al.* 1991). Honey is the only food item demonstrated to contain enough spores to act as a vehicle for infectional disease called infant botulism (Arnon *et al.* 1979).

Infant botulism is an infectious disease, caused by ingestion of spores of *Clostridium botulinum*. The first clinical cases of infant botulism were described in 1976 (Midura *et al.* 1976, Pickett *et al.*1976) and today it is the most common appearance of botulism in United States, where there are about 80 reported cases of infant botulism every year (CDC 1998). Immature infantile intestinal flora allows the ingested spores to germinate, multiply and to produce botulinal toxins (Midura et al. 1976). Children aged between 2 weeks and 6 months are the most susceptible to this infection (Arnon *et al.* 1981). Symptoms of infant botulism reflect flaccid paralysis which begins with the muscles of the head and neck, extending to involve the whole body. Prolonged constipation is often seen prior to the neurological signs. These symptoms are caused by botulinal toxin that irreversibly blocks the release of acethylcholine at peripheral synapses, most importantly in neuromuscular junction.

We have developed a PCR based method for detecting spores of *C. botulinum* in honey samples. Using this method we have performed a prevalence study of *C. botulinum* spores in Finnish honey. A total of 114 honey samples were examined and about 8 % of these samples were positive. The contamination routes of *C. botulinum* spores to the honey and the possible substrates for the spores to germinate and grow inside the bee hive are not yet explained.

As the origin of the positive Finnish honey samples have been identified we are now going to determine the possible contamination routes of *C. botulinum* to honey. The results gained from this study will provide us important knowledge of ecology of *C. botulinum*. We will be able to assess the role of hygienic processing methods in honey production in pre-harvest level. A proposal for improving the production methods to minimize contamination and growth of *C. botulinum* in honey may also be done according to the results.

References

Arnon S.S., Midura T.F. and Damus K., 1979. Honey and other environmental risk factors for infant botulism. J. Pediatr. 94: 331-336.

Arnon S.S., Damus K. and Chin J., 1981. Infant botulism: Epidemiology and relation to sudden infant death syndrome. Epidemiol. rev. 3: 45-66.

CDC, Centers for Disease Control and Prevention, 1998. Botulism in the United States, 1899-1996. Handbook for Epidemiologists, Clinicians and Laboratory workers, Atlanta, GE. Centers for Disease Control and Prevention.

Lilly, T., Rhodehamel, E. J., Kautter, D. A. and Solomon, H. M., 1991. *Clostridium botulinum* spores in corn syrup and other syrups. Journal of food protection. 54: 585 - 587.

Midura T. F. and Arnon S. S., 1976. Infant botulism: Identification of *Clostridium botulinum* and its toxin in faeces. Lancet ii: 934-936.

Pickett, J., Berg, B., Chaplin, E. and Brunstetter-Schafer, M., 1976. Syndrome of botulism in infancy: Clinical and electrophysiologic study. N. Engl. J. Med. 295: 770-772.

Honey: Relationships between heavy metals content and place of origin

S. Ghidini, G. Campanini and G. Delbono
Istituto di Scienza e Tecnologia degli Alimenti, Università Studi Parma

Introduction

Honey is usually considered a safe food and, because of this and of its eating quality and digestibility, it is eaten mostly by children and old people. The safety of this food, though, needs to be verified since it has been shown that in honey it is possible to detect both the residues of drugs used to prevent or to cure some diseases of the bees (e.g. sulphonamides) and environmental contaminants. Due to the very polar nature of honey non polar environmental contaminants are practically absent whereas heavy metals can still be present.
Bees pick up the pollen from a wide area (within 3 km) bringing in such way environmental contaminants to the apiaries and the honey. Bees can also be contaminated by the water they drink and bring contaminants on their body. For such reasons some scientists tried to use bees and honey to monitor environmental pollution (Accorti, Persano-oddo, 1986; Celli, 1983; Wallwork-Barber et al., 1982).

Materials and Methods

41 honey samples were bought directly from the producers to be sure of their origin. The samples were produced in the region Emilia Romagna and in particular in the provinces of Piacenza, Parma and Reggio Emilia in 1996, 1997 and 1998. They were classified on the basis of province of origin, on closeness to main roads, on predominant botanical essence and on the type of place of origin (mountain, hill, plain).
The samples were made homogeneous by heating to 60°C. Organic matrix was dissolved by humid mineralization and pressure heating in a microwave oven. Samples for the determination of lead were prepared by simple dilution in a solution of 0,5% HNO_3.
Analyses determinations were performed through Atomic Absorption Spectrophotometry (AAS). Flame atomization was used for the determination of copper, zinc and manganese, electrothermal atomization was used for lead and cadmium and cold vapour atomization was used for mercury. The data were statistically processed with the SPSS 9.0.1 program.

Results

In table 1 it is possible to see the results of the analytical determinations.

The mean value of lead content was found to be quite high. The result is due to a few samples with very high concentrations as opposed to several samples with low concentrations. There

Table 1. Mean values of heavy metals in honey. Concentrations are expressed as mg/kg.

	Zn	Cu	Mn	Pb	Cd	Hg
Mean	1,88	1,06	0,93	0,150	0,004	<0,010
Std. Dev.	1,67	0,93	0,63	0,091	0,003	
Max	11,10	4,01	2,79	0,424	0,013	
Min	0,47	0,20	0,34	0,018	<0,003	

are no legal limits set for lead in honey. Otto and Jekat (1977) suggested a threshold limit of 0,215 mg/kg. Nine honey samples are over this threshold. Some of the samples with high levels are from places near traffic congested roads but statistically relevant relationships were not found between these two factors. Some authors (Otto and Jekat, 1977; Stein and Umland, 1986; Celli et al., 1988) found this relationship, but they placed the apiaries in towns near very heavily traffic-congested roads and, indeed, in some cases they found concentrations up to 1 mg/kg. Furthermore these studies were made in a period in which leaded fuel was the only used, whereas nowadays unleaded petrol is the most common one. Oddi and Bertani (1987) found a relationship between lead content in honey and the prevalent botanical essence. They found higher lead concentrations in dark types of honey (in particular chestnut honey). No such relationship was found in the present investigation but an unexpected relationship was observed between lead concentration and place of origin. As shown also by the Sheffe post-hoc test (Table 2) the samples coming from mountains have higher concentrations and they belong to a completely separated whole. The same correlation was not found for the other metals we considered.

Table 2. Heavy metals content (mg/kg) of honey samples divided on the basis of the place of origin. Different superscripts stand for significant differences (Sheffe's test, P = 0,05).

Place of origin	Mn	Cd	Pb	Cu	Zn
Plain	$0{,}824^{a}$	$0{,}003^{a}$	$0{,}114^{a}$	$0{,}255^{a}$	$1{,}016^{a}$
Hill	$0{,}988^{a}$	0.004^{a}	$0{,}168^{a,\,b}$	$1{,}080^{a}$	$1{,}834^{a}$
Mountain	$1{,}319^{a}$	$0{,}003^{a}$	$0{,}289^{b}$	$1{,}122^{a}$	$2{,}009^{a}$

All samples had low concentrations of cadmium, only in two cases the concentrations of the metal exceeded 0,010 mg/kg.

Mercury concentrations were very low too and all samples were found to have concentrations lower than the detection limit (LOD= 0,010 mg/kg) of the method.

Very high concentrations of zinc were found in some cases (up to 11 mg/kg). This happened because some producers use zinc coated steel is sometimes used to build apiaries or containers to store the product. Honey is slightly acid and the leaching of zinc into honey is therefore possible.

Manganese and copper showed a similar behaviour through the different provinces. Their concentrations were significantly higher in the province of Piacenza, as it is possible to see from the post-hoc Sheffe test. In the case of copper the fact is easily explained because in that province tomatoes and grape are intensively cultivated and both cultures require copper treatments. The higher concentrations of manganese instead could be explained only by assuming a different nature of the soil.

Table 3. Heavy metals content (mg/kg) of honey samples divided on the basis of the province of origin. Different superscripts stand for significant differences (Sheffe's test, P = 0,05).

Province	Mn	Cd	Pb	Cu	Zn
Reggio Emilia	0,503 [a]	0,003 [a]	0,166 [a]	0,993 [a]	1,806 [a]
Parma	0,742 [a]	0.004 [a]	0,151 [a]	0,560 [a]	2,066 [a]
Piacenza	1,401 [b]	0.004 [a]	0,142 [a]	1,833 [b]	1,644 [a]

Comments

Zinc, manganese and copper are essential elements for human health and they are toxic only at very high concentrations. The concentrations here reported are exceptionally high in some cases, but honey is not eaten in great amounts so it is difficult that honey consumption could bring dietary introduction of such elements to toxic levels. However the use of zinc coated iron should be avoided for the apiaries and the containers.
Both cadmium and mercury are very toxic metals, but the concentrations observed were very low and absolutely safe. Other authors found cadmium levels comparable to ours (Oddi and Bertani, 1987; Galeno et al 1992). The data found in literature about mercury in honey are very few. Toporcak (1992) found concentrations ranging between 0,005 and 0,212 mg/kg in samples from polluted areas and concentrations between 0,001 and 0,003 mg/kg in samples from non polluted areas. Tong (1975) found concentrations lower than 0,1 mg/kg in all the samples examined.
The condition is different for lead contamination which in some cases seems considerable. There isn't a legal threshold limit for lead in honey, but the WHO fixed a Provisional Tolerable Weekly Intake (P.T.W.I.) of 3 mg a week for adults and ten times lower for infants. The Pb contents observed in some samples could contribute to reach this level. After the substitution of leaded petrol with the unleaded one the closeness to traffic-congested roads can't completely explain this contamination. Therefore the relationships with other sources (e.g. industrial productions) of this metal should be studied. In any case the proximity of the apiaries to traffic-congested roads should be avoided as much as possible because lead is a persistent polluting element and also because several vehicles still use leaded fuel. Further studies are required also to explain the higher contamination of samples coming from the mountains.

References

Accorti M. and Persano-Oddo L., 1986. Un servizio di monitoraggio ambientale urbano: "Apincittà". Informatore agrario, 42, 39-41.

Celli G., 1983. L'ape come insetto test della salute di un territorio. Proceedings of the XIII National Congress of Entomology, Sestriere, 27th June-1st July 1983, 637-644.

Wallwork-Barber M.K., Ferenbaugh R.W. and Gladney E.S., 1982. The use of honey bees as monitors of environmental pollution. Am. Bee J., 122, 770-772.

SPSS 9.0.1 for Windows (11/03/1999).

Oddi P. and Bertani P., 1987. Contaminanti nel miele. Nota 1: residui di Pb e Cd. Proceedings S.I.S.Vet., Vol. XLI, 998-1000.

Galeno N., Rocca L., Ferrari A., Molinelli E. and Arossa C., 1992. Il miele di produzione ligure. Caratteristiche organolettiche e residui di metalli pesanti. Proceedings S.I.S.Vet., Vol. XLVI, 657-660.

Otto K. and Jekat F., 1977. Experimentelle Untersuchungen über die Belastung eines Nahrungsmittels mit Rückständen von Blei, Zink, Cadmium. Ernährungs Umschau, 24(4), 107.

Stein K. and Umland F., 1986. Zeitschrift für Analytische Chemie, 323(2), 176-177.

Celli G., Porrini C., Balestra V. and Menozzi R., 1988. Monitoraggio di inquinanti atmosferici urbani mediante api. Workshop on health and environment, Cagliari, 27 th , 28 th of May 1988.

Tong S.S.C., Morse R.A., Bache C.A. and Lisk D.J., 1975. Elemental analysis of honey as an indicator of pollution. Arch. Environ. Health, Vol 30, July 1975.

Toporcak J., Legath J. and Kul'kova J., 1992. Levels of mercury in samples of bees and honey from areas with and without industrial contamination. Vet. Med. Praha, 37(7), 405-412, July 1992.

Prevalence of *Salmonella* along a meat pork production system

N. Korsak*[1], B. Groven[1], B. Jacob[1], G. Daube[1] and E. Flament[2]
[1]Microbiology Laboratory, Faculty of Veterinary Medicine, University of Liege, Belgium,
[2]Industrial partner

Introduction

In February 1999, a project called "Implementation of a *Salmonella*-free meat pork production system" started in Belgium. The first objective was to obtain a *Salmonella*-free pig production system and to maintain it. In November 1999 we begun a monitoring program on all the production chain.
This paper describes the results of this surveillance program and gives a diagnostic of this particular chain of production.

Material and Methods

Where?	When?	Which samples?
Feeds	Random sampling when loading the lorry	25 g of meal
Breeding	1x / month (every farrow should be inspected)	5 faeces samples pooled in 25 g Overshoes
Weaned pigs (→ 20 kg)	Once on every batch 8 days before day out/going out/departure?	5 faeces samples pooled in 25 g Overshoes
Fattening (20 - 110 kg)	Twice on every batch: (after 2 months and 4 months of fattening)	5 faeces samples pooled in 25 g Overshoes
Slaughterhouse	Once on every batch	5 samples of faecal contents pooled in 25 g
Carcasses	Once on every batch	Pool of 5 surface swabs (5 * 600 cm^2)
Cutting room	Random sampling	25 g of retail cuts
Mincing room	Random sampling	25 g of ground minced meat
Butcheries	Random sampling	25 g of ground minced meat

* PCR kit Probelia™ *Salmonella spp* (Biorad, Maasluis, NL)
* SP-VG-M002 method

Twenty five grams of faecal samples were analysed with the SP-VG-M002 method using Diassalm, a semi-solid enrichment media. The research of Salmonella in food samples as meal and meat was done with the AFNOR validated PCR kit Probelia™ *Salmonella spp* (Biorad, Maasluis, NL).

Results

Where ?	Feeds	Breeding	Weaned pigs	Fattening pigs	Slaughter-house	Carcasses	Cutting room	Mincing room	Butcheries
Positives	7	9	2	8	35	10	0	1	7
N	96	161	44	129	69	58	40	29	55
%	6,7%	5,6%	4,5%	6,2%	50,7%	17,2%	0%	3,4%	13%

We observe an important increase between faeces from farms and intestinal content in the slaughterhouse.
It seems that the stress due to transport and fasting period have a significant effect on competitive flora which lead to a higher level of detection.
Moreover, the hazard represented by *Salmonella* in the slaughterhouse seems to be relatively well controlled. 17,2 % of the carcass are contaminated with *Salmonella* when 51 % of the pigs are carriers of this bacteria.
This conclusion is also true for the cutting and mincing room seeing that none samples of cutted meat and only one sample of minced meat were found positive.

Discussion

Contamination rate is various along the food chain. The surveillance system described here allows to evaluate control measures. In order to prevent the main contamination sources, typing of isolates must also be done.

References

SP-VG - M002 - Official method from the Belgian Ministry of Public Health - site internet LMDA; http://mda04.fmv.ulg.ac.be .

Fedorka-Cray P.J., Gray J.T. and Wray C., 2000. *Salmonella* infections in pigs. In:Wray C. and Wray A. (eds), *Salmonella in domestic animals* ,CABI Publishing, pp 191-209.

Van der Zee Henk and Huis in't Veld Jos H.J., 2000. Methods for the rapid detection of *Salmonella*. In: Wray C. and Wray A. (eds), *Salmonella in domestic animals* ,CABI Publishing, pp 373-391.

This project is financially supported by DG6 from the Belgian federal ministry of agriculture and DGA from walloon regional ministry.

Epidemiology of *Cryptosporidium parvum* and *Giardia duodenalis* on a dairy farm

R.E.C. Huetink[1,2], J.W.B. van der Giessen[2], J.P.T.M. Noordhuizen[3] and H.W. Ploeger[1]
[1]Department of Parasitology and Tropical Veterinary Medicine, Faculty of Veterinary Medicine, University of Utrecht, P.O.Box 80.165, 3508 TD Utrecht, The Netherlands, [2]National Institute of Public Health and the Environment, Microbiological Labatory for Health Protection, P.O.Box 1, 3720 BA Bilthoven, The Netherlands, [3]Ruminant Health Unit, Department of Farm Animal Health, Faculty of Veterinary Medicine, University of Utrecht, P.O.Box 80.151, 3508 TD Utrecht, The Netherlands

Introduction

Cryptosporidium parvum and *Giardia duodenalis* are intestinal parasites identified in many mammals causing diarrhoea. Outbreaks of cryptosporidiosis and giardiosis in humans have been attributed to contamination of drinking water by dairy pasture run-off (Anderson, 1998*)*. *C. parvum* is commonly found in young cattle, being a major cause of diarrhoea in calves of 2-3 weeks old, but in adult cows cryptosporidiosis is rarely reported. *G. duodenalis* has been found in high prevalences in dairy calves and is assumed to be a potential pathogen in cattle.
In this study prevalences of *C. parvum* and *G. duodenalis* in relation to age and season, associations with diarrhoea, transmission routes and genotypes of *C. parvum* and *G. duodenalis* were researched on a dairy farm.

Materials and Methods

The study was done on a Dutch top dairy farm. The whole herd was sampled 5 times, calves <3 months were sampled every 2-3 weeks. An immunofluorescence test was used to investigate faecal samples. Associations with diarrhoea and potential transmission routes of *C. parvum* were investigated and *C. parvum/G. duodenalis* positive samples were identified to genotype level.

Results

Shedding of *C. parvum* was found in all age categories but highest in calves 1-3 weeks old (39.1%) (Fig.1).

Herd prevalence of shedding for *C. parvum* varied from 2.4% in June to 22.2% in Dec. (Table 1). Shedding of *G. duodenalis* was found in all age categories but peaked in animals 4-5 months old (54.5%) (Fig.1).

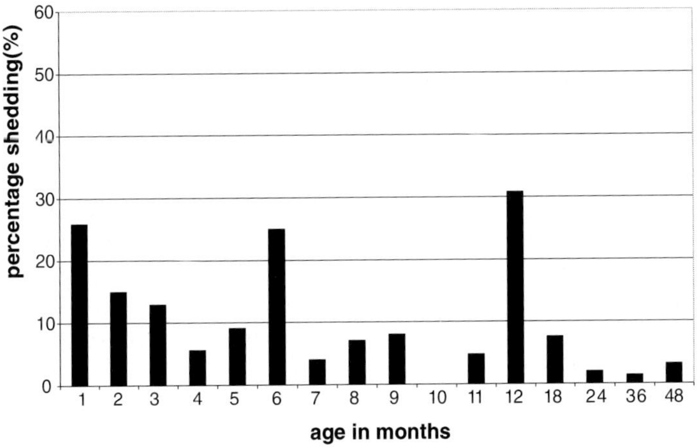

Figure 1. Percentage of animals of different ages shedding Cryptosporidium parvum *oocysts throughout the year.*

Table 1. Herd prevalences of Cryptosporidium parvum *oocyst and* Giardia duodenalis *cyst shedding and variations in time(%).*

Month	*C. parvum*	*G. duodenalis*
Oct. 99	8.7(11/127)	11.0(14/127)
Dec. 99	22.2(28/126)	14.3(18/126)
Feb. 00	4.3(6/126)	15.1(19/126)
Apr. 00	4.8(6/124)	4.0(5/124)
Jun. 00	2.4(3/125)	0.8(1/125)

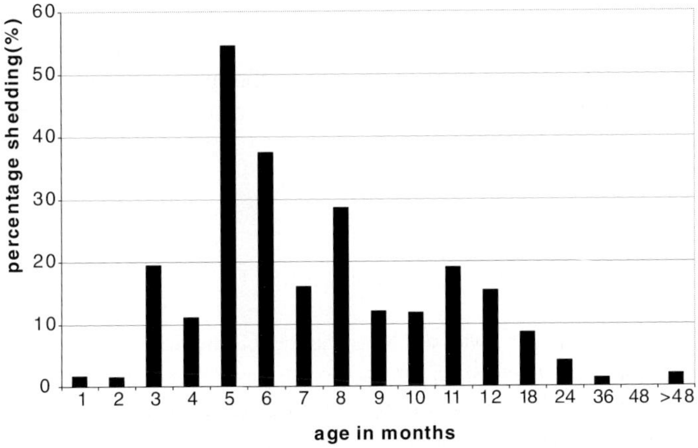

Figure 2. Percentage of animals of different ages shedding Giardia duodenalis *cysts throughout a year.*

Herd prevalence of shedding for *G. duodenalis* varied from 0.8% in June to 15.5% in Feb. (Table 1).

C. parvum, G. duodenalis and rotavirus were not significantly associated with diarrhoea in individual hutch calves, neither in older animals. If looked at simultaneous presence of diarrhoea and *C. parvum* oocysts in faecal samples as such, there was a significant association between both.

Of the *C. parvum* oocyst shedding hutch calves 24.0% was probably infected in the maternity pen, 51.5% had a shedding neighbour and 10.3% had a shedding predecessor. No oocysts were found in the scrapings of empty hutches and in samples of cats on the farm. In samples of swans sitting around the farm *C. parvum* and *G. duodenalis* were detected in Dec. The 2 strains of *C. parvum* found here (C1 and C3) were both (zoonotic) calf types. The identified strains of *G. duodenalis* were of an unknown type and their zoonotic potential is not known, but 3 samples had a type closely related to the human type.

Discussion

The pattern of *C. parvum* shedding in calves was similar to previous research (O'Handley et al., 1999). The relatively high prevalences found in 6 and 12 months old calves were completely due to the Dec. sampling with high prevalences in all animals <2 years. The seasonal differences in *C. parvum* shedding found here can possibly be explained by peaks in newborn calves (increasing infection pressure) and a peak in milk production in Dec. reducing immunity to infection.

The relatively high age at which *G. duodenalis* cysts were first found might be a consequence of management. Until 2-3 months calves didn't have contact with each other, reducing the chance of transmitting infection. A possible explanation for the peak of shedding *G. duodenalis* during the winter months, in contrast with previous research, is that in this period animals 2-6 months old suffered from serious respiratory problems, rendering them more susceptible for infections such as *Giardia*.

C. parvum was associated with diarrhoea on this farm, but there was no strong association as in previous research (de Graaf et al., 1999). Since samples were taken every 2-3 weeks and *C. parvum* oocyst shedding periods and diarrhoea episodes usually last <1 week, both could easily have been missed in individual calves.

Some calves probably were infected with *C. parvum* in the maternity pen and transported oocysts to the hutches where they infected others. Direct calf-to-calf transmission couldn't occur, but for instance care takers or cats might have acted as (mechanical) vectors. Hutches were not an important factor in transmission.

References

Anderson, B.C., 1998. Cryptosporidiosis in bovine and human health. J. Dairy. Sci. 81, 3036-3041.

de Graaf, D.C., Vanopdenbosch, E., Ortega-Mora, L.M., Abbassi, H. and Peeters, J.E., 1999. A review of the importance of cryptosporidiosis in farm animals. Int. J. Parasitol., 29, 1269-1287.

O'Handley, R.M., Cockwill, C., McAllister, T.A., Jelinski, M., Morck, D.W. and Olson, M.E., 1999. Duration of naturally acquired giardiosis and cryptosporidiosis in dairy calves and their association with diarrhea. J. Am. Vet. Med. Assoc. 214, 391-396.

The influence of the immune status of slaughter pigs on the endogenous contamination of muscle tissue

Th. Alter and K. Fehlhaber
Institute of Food Hygiene, Faculty of Veterinary Medicine, Leipzig University, Germany

Slaughter animals are subject to different stressors on their way from the farmyard to the abattoir.
Those intense stress-situations (e.g. transport, handling) are not only an ethical problem, they do not only diminish meat quality but increase the risk of endogenous contamination of meat by various bacteria.
The latter is caused by the stress-induced rise of the permeability of the gut wall and the diminished efficiency of certain parts of the immune system (Figure 1).

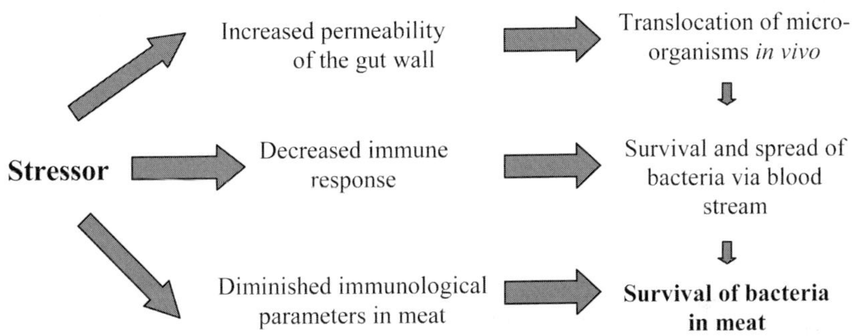

Figure 1. Impact of stress-situations on the degree of endogenous contamination (summary).

The first line of defence against intruding bacteria consists of non-specific immunological factors.
The complement activity (CH-50, measured by a haemolytic complement assay) and the serum bactericidal activity (SBA) were compared under the influence of typical stress situations for pigs right before slaughter.
Our findings revealed immune modulations after the impact of these stressors.
Whereas short term transport or heat only reduced bactericidal activities slightly or not at all, massive losses occurred after simulating extreme transport conditions (combined effect of heat, high humidity and handling - the most potent stressor) (Figure 2).
Survival rates of the tested micro-organisms increased massively in serum of stressed animals, i.e. parts of the immune system were restricted by those stressful events.
(Haemolytic complement assays carried out on the same serum samples supported these results.)

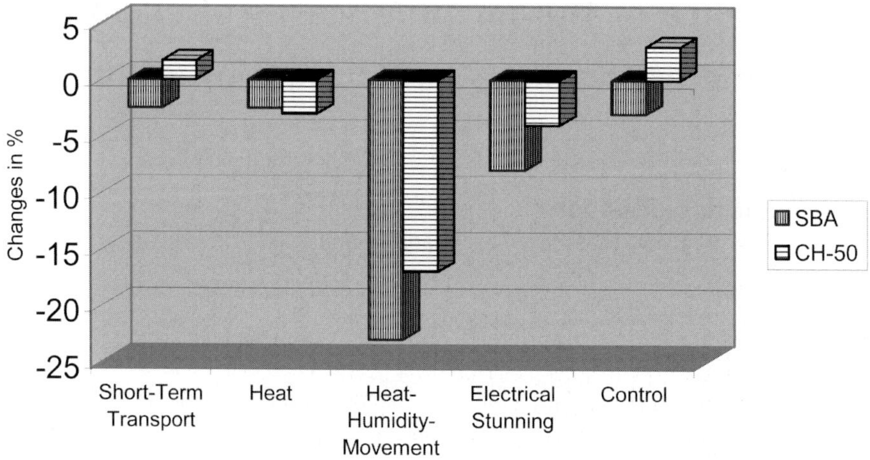

Figure 2. Comparison of CH-50/SBA under different stress conditions.

Resting time before slaughter could not compensate stress-induced losses of bactericidal and complement activity.
By diminishing the immunological competence of the animal, increased survival and spread of micro-organisms throughout the organism is to be expected. On the other hand - by improving the immunological status of the animal it is possible to minimise the spread of bacteria into muscle tissue.

But can immune parameters still be found post mortem in muscle tissue? Do they have the ability and strength to halt growth of intrinsic bacteria?
Complement (C3c)-concentration was established in muscle tissue right after slaughter of stressed and non-stressed animals and trailed for 72 hrs in meat samples.
To cover even more immunological factors in muscle tissue post-mortem the remaining bactericidal activity of meat juice samples was established. Conclusions could be drawn from the survival rate/growth rate of the bacterial cells after incubation in meat juice and linked with the remaining immunocompetence of the corresponding meat sample.

It could be observed that immunological factors still exist right after slaughter in muscle tissue for several hours. Those factors were able to limit the growth of different bacterial strains for a short period.
Complement concentration decreased rapidly within 10 to 20 hrs p.m. The degradation of complement depends on storage temperature and the quality of the immune system of the animal before slaughter. Ante-mortem stress situations had an impact on the post-mortem complement concentration as well (Figure 3).

The question to be solved now is, if it is possible - by making the immune system fitter in its fight against intruding micro-organisms - to make the „meat fitter in its fight against these micro-organisms" as well.

The usage of probiotics or paramunity inducers can not replace the application of antibiotics but could support and stabilize the animal defence system in stressful situations and consequently reduce the endogenous contamination of meat.

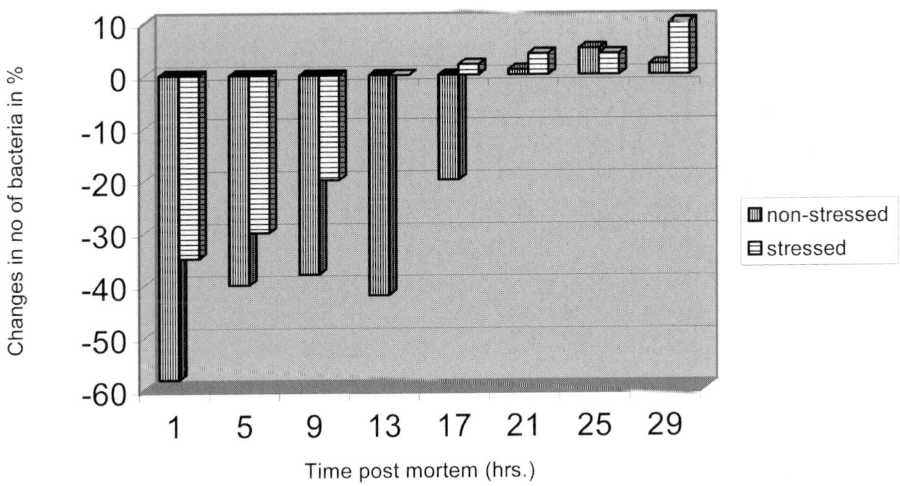

Figure 3. Bactericidal effect of meat-juice on E. coli *K12 (comparison of stressed and non-stressed animals).*

Yersiniosis: Foodborne disease and underestimated diagnostical problem

H. Neubauer
Institut für Mikrobiologie, Sanitätsakademie der Bundeswehr, Neuherbergstr. 11, 80937 München, Germany

Introduction

Yersiniosis in humans and animals is caused by the gram negative bacterium *Yersinia enterocolitica*. Cardinal symptoms in humans are diarrhea due to acute gastro-enteritis, stomach ache caused by mesenterial lymphadenitis, fever, nausea and sometimes a sore throat. The reservoir for human infection is believed to be the pig. Contamination of pork and pork products ocurrs during the slaughtering process or manufacturing. A current sero-prevalence study shows that up to 40% of the healthy German population has anti-*Yersinia* antibodies (8). Statistically yersiniosis is the third most important bacterial food borne disease apart from salmonellosis and campylobacteriosis in Germany. Being aware of this threat, the 'Infektionschutzgesetz' implemented in July 2000 orders that isolations of or indirect proofs for *Y. enterocolitica* connected with acute human disease have to be notified to the authorities. As no current epidemiological data on yersiniosis in different animal species are available, veterinary public health has not only to clarify chains of infection but also to identify ways to control them, as to contribute to the efforts made by human public health to protect the consumer. The first step in this process has to be the evaluation of the currently available diagnostical tools or the development of sensitive and specific identification systems if those are not available.

Material and Methods

More than 400 biochemically characterised strains of all eleven species of the genus *Yersinia* were included in the various investigations foccusingespecially on apathogenic and pathogenic
Y. enterocolitica and *Y. pseudotuberculosis* strains. Strains were maintained and propagated using standard procedures. Of the biochemical identifcation systems the 'gold standard' API 20E, the API rapid 32 IDE, (both BioMerieux, Nürtingen, Germany), the GNI card of the Vitek system (bioMerieux Vitek Inc, Hazelwood, USA) and the Micronaut E (Merlin, Germany) were evaluated. Commercially available molecular identification systems for members of the species *Y. enterocolitica* included in the studies were the *Yersinia enterocolitca* Amplification Set (Kreatech, NL), based on a PCR assay, and the Gene Trak *Yersinia*-Assay (Gene-Trak, USA), a dip-stick probe test. Testing was performed according to the recommendations of the manufacturers.
DNA preperation, PCR, sequencing alignment studies and relatedness calculations were done using standard procedures. Development of a newly composed semiautomated

biochemical identification system was done using the configuration and the research software supplied by Merlin (Germany).

Results and Discussion

Biochemical Identification Systems

The 'gold standard' API 20E had a sensitivity of 79%, the API rapid 32E of only 42% and the Micronaut E of 72% (1). The Vitek system showed a middle range sensitivity of 57% (2). It has to be stressed that most systems can identify an isolate at the genus level with a high sensitivity above 90%. All systems identified *Yersinia* virulence plasmid harbouring - presumptative enteropathogenic *Y. enterocolitica* and *Y. pseudotuberculosis* - strains with sufficient accuracy, however, apathogenic *Y. enterocolitica* strains and isolates of apathogenic *Yersinia* species were regularly misidentified. These strains are often found in milk but also in stool samples of patients. Possible reasons are: a short incubation time, no updated identification indices and most importantly configurations of reactions which do not include the key reactions used for differentiating the species of the genus *Yersinia*. We therefore had to conclude that currently available test systems are not useful for the identification of *Yersinia* isolates. In an effort to overcome these shortcomings we rearranged the layout of the Micronaut E identification system in which 20 key reactions for the differentiation of the species of the genus but also for the identification of the biovars of the species *Y. enterocolitica* were included. The substrates / enzymes are lyophilised in the wells of a 96 well plastic plate, reading is done by a scanner and calculation of the results is faciliated by a computer program to minimise bias and systemic errors. The prototype revealed an overall sensitivity of 98% (9). The results are obtained within 24 hours. A sensitive tool for monitoring *Yersinia* contaminants in food and feeds but also for HACCP in the pre-harvest status is hence available now.

Molecular Identification Systems:

All commercially available molecular identification systems for the species *Y. enterocolitica* are based on sequences of the 16S rRNA gene. The Gene Trak dip stick probe assay proved to identify only strains which have a 16S rRNA sequence signature of American strains. It did not identify European pathogenic or apathogenic isolates (3, 4). On the other hand the PCR assay was not able to identify strains of the American *Y. enterocolitica* subspecies (3) and had only a sensitivity of 85% (1). These assays can therefore not be recommended for the use in a routine laboratory. In an own investigation we could demonstrate that the 16S rRNA gene sequence is only useful to identify *Yersinia* isolates which have already been typed to be a member of the genus by means of biochemical analyses (5). Sequencing as the method of choice is too expensive and time consuming to be used in veterinary hygiene control where a great number of samples is to be expected. The number of PCR assays used to identify members of the pathogenic species *Y. enterocolitica* and *Y. pseudotuberculosis* in research is enormous (review: 6). But an assay used in routine diagnosis should correlate with a simple phenotypic pathogenicity test and the gene product of the underlying sequence ought to be used in the serological assay which is in use for surveillance in the living animal. We compared various already established and newly designed assays based on genes, which might fullfill the above described requirements. However, it was not even possible to correlate the data of the PCR assays to the standard pathogenicity test for enteropathogenic *Yersinia*, the autoagglutination test (7, 10). The immunogenicity of various proteins involved in virulence has not been investigated in the reservoir animals pig and cattle although their gene sequences have been known for a long time. A sensible target gene for diagnostic molecular and serological assays cannot be named in the moment. In the opinion of the author, at present, no PCR assay which can identifiy pathogenic *Y. enterocolitica* or *Y. pseudotuberculosis* strains has been evaluated for routine use. The question how to prepare bacterial DNA for the use in amplification techniques from the various matrices connected with rigorous hygiene control in stables and during food or feeds production has not been solved yet.

References

1. Neubauer, H., Sauer, T., Becker, H., Aleksic, S. and Meyer, H., 1998. Comparison of systems for the identification and differentiation of the species within the genus *Yersinia*. J. Clin. Microbiol., 36 (11): 3366-68
2. Linde, H.-J., Neubauer, H., Meyer, H., Aleksic, S. and Lehn, N., 1999. Identification of *Yersinia* Species by the Vitek GNI card. J. Clin. Microbiol., 37 (1): 211-214
3. Neubauer, H., Reischl U., Köstler J., Aleksic, S. and Meyer, H., 1999. Variations in the 16S rRNA gene sequence of *Yersinia enterocolitica* isolates influence the specificity of molecular identification systems. Zentralbl. Bakt., 289: 329-337
4. Neubauer, H., Hensel, A., Aleksic, S., Finke, E. J. and Meyer, H., 2000. *Yersinia enterocolitica* 16S rRNA gene types belong to the same genospecies but form three different homology clusters. Int. J. Med. Microbiol., 290: 61-64
5. Neubauer, H., Hensel, A., Aleksic, S. and Meyer, H., 2000. Identification of *Yersinia enterocolitica* within the genus *Yersinia*. System. Appl. Microbiol., 23: (1) 58-62
6. Neubauer, H., Sprague, L.D., Scholz, H. and Hensel, A., 2000. Die Diagnostik der *Yersinia enterocolitica*-Infektionen: Eine Übersicht über klassische Nachweistechniken und neue molekularbiologische Methoden. BMTW, 113: 1-7
7. Neubauer, H., Hensel, A., Aleksic, S. and Meyer, H., 2000. Evaluation of a *Yersinia* adhesion gene (*yadA*) specific PCR for the identification of enteropathogenic *Yersinia enterocolitica*. Int. J. Food Microbiol., 57:225-227
8. Neubauer, H., Rasolomaharo, M., Brooks, T. and Chanteau, S. and Splettstösser, W.D. IgG / IgM ELISA and immunoblot for the serodiagnosis of human plague. Epidemiology and Infection, accepted
9. Neubauer, H., Molitor, M., Rahalison L., Aleksic, S., Bakes, H., Chanteau, S. and Meyer, H. A semiautomated system for identification of *Yersinia* species within the genus *Yersinia*. Clin. Lab., accepted
10. Neubauer, H., Sprague, L.D., Hensel, A., Aleksic, S. and Meyer, H.. Specific detection of plasmid bearing *Yersinia* isolates by PCR. Clin. Lab., accepted

Prevalence of thermophilic *Campylobacter* spp. in Danish turkeys at slaughter

B. Borck, K. Pedersen and M. Madsen
Danish Veterinary Laboratory, Aarhus, Denmark

Campylobacter are reported to be the most common cause of human acute bacterial diarrhoea in most industrialised countries, outnumbering the cases caused by Salmonella (Skirrow, 1994). In Denmark, the number of Campylobacter infections has steadily increased since 1992 and has more than tripled within the last five years reaching the highest incidence ever registered in 1999 (78 cases/100 000 inhabitants) (Anonymous, 2000).
Several investigations have consistently indicated that consumption of poultry is one of the dominant sources of human infections (Friedmann et al., 2000). Much research has already been carried out in broilers, however little data is available on Campylobacter in turkeys. In 1999, the Danish Veterinary Laboratory initiated a three-year project with the purpose of investigating the occurrence of Campylobacter throughout the production of Danish turkeys; from stable to table. Approximately one million turkeys are slaughtered in Denmark annually. The turkeys are raised on 32 different farms and slaughtered at the only large turkey slaughterhouse in Denmark. The project was initiated by carrying out a surveillance study for thermophilic *Campylobacter* spp. comprising all batches of turkeys at slaughter.
Faecal swab samples were systematically collected from all turkey flocks; ten cloacal swabs from each batch of turkeys at the time of slaughter. The majority of turkey flocks were slaughtered over a period of 2-4 days and most flocks were therefore sampled more than once.
On arrival to the laboratory, the ten samples were pooled to one sample and left in a tube with 3 ml of sterile water for approximately two minutes to release bacteria. Culture for thermophilic *Campylobacter* spp. was carried out in accordance with the recommendations given by the Nordic Committee on Food Analysis (Anonymous, 2000). Briefly, 10 µl of the solution was streaked onto modified CCDA plates (Campylobacter blood-free selective agar base [Oxoid CM739] with CCDA selective supplement [Oxoid SR155], cefaperazone 32 (g/ml and amphotericin B 10 µg/ml). The plates were incubated in a microaerophilic atmosphere at 37°C for 48 hours. Campylobacter-like colonies were subcultured one or more times until monocultures were obtained. One colony was selected, subcultured and identified to the species level by reactions in indoxyl acetate, catalase, and hippurate tests and by their susceptibility to nalidixic acid and cephalotin.
The results from the first 11 months of the surveillance study (September 1999 - July 2000) are presented below.

A total of 551batches (332 flocks) were tested for the presence of Campylobacter and an average of 49% (56% flocks) were found positive. *C. jejuni* was isolated from 91% of the positive samples, whereas *C. coli, C. lari* and *C. hyointestinalis* were isolated from only 6%, 1% and 1% of the positive samples, respectively. During the 11-month period the

Table 1. Number of positive batches/flocks in the turkey production from Sept. 1999 to July 2000.

	Sept.	Oct.	Nov.	Dec.	Jan.	Feb.	Mar.	Apr.	May	June	July
Pos. Batches	32	22	27	24	34	39	9	2	28	27	27
	(71%)	(50%)	(52%)	(67%)	(71%)	(75%)	(15%)	(5%)	(47%)	(48%)	(48%)
Pos. Flocks	24	19	21	17	23	20	7	1	18	16	19
	(83%)	(56%)	(66%)	(74%)	(70%)	(67%)	(22%)	(4%)	(62%)	(48%)	(59%)

percentage of *Campylobacter* positive batches fluctuated between 50% and 75% from September 1999 to February 2000, but the percentage dropped dramatically to 15 % Campylobacter positive batches in March and 5% in April. During the months of May, June and July the percentage of *Campylobacter* positive batches once again rose to almost 50%. This variation in prevalence does not correspond to the distinct seasonal distribution, with a well-defined summer peak, seen in broilers in Denmark (Wedderkopp et al., 2000). The explanation for the difference between broilers and turkeys is unknown and further investigations are required. However, factors such as age at slaughter and differences in farm management may be contributing factors.
In conclusion, more than 50% of the turkey flocks slaughtered in Denmark, from September 1999 to July 2000, were positive for thermophilic *Campylobacter* spp., *C. jejuni* being by far the most dominant species. The results presented above emphasise the need for qualitative as well as quantitative investigations of Campylobacter in turkeys. Quantitative studies are already being carried out on broilers, at the farm level, to provide data for risk assessment in the poultry industry. Similar studies for turkeys will be undertaken in the near future.

References

Anonymous, 1990. *Campylobacter jejuni/coli*. Detection in foods. Nordic Committee on Food Analysis 2nd.ed. Method No. 119.

Anonymous, 2000. Annual Report on Zoonoses in Denmark 1999. Ministry of Food, Agriculture and Fisheries.

Friedman, C.R., Neimann, J., Wegener, H.C., and Tauxe, R.V., 2000. Epidemiology of *Campylobacter jejuni* infections in the United States and other industrialized nations. In 'Campylobacter'. (Eds. I. Nachamkin and M.J. Blaser.) pp. 121-38. ASM Press: Washington, D.C.

Skirrow, M.B., 1994. Diseases due to *Campylobacter, Helicobacter* and related bacteria. *Journal of Comparative Pathology* **111**, 113-149.

Wedderkopp, A., Rattenborg, E., Madsen, M., 2000. National surveillance of *Campylobacter* in broilers in Denmark in 1998. *Avian Diseases* (in press).

European follow-up study in *Salmonella* seronegative and seropositive pig herds

D.M.A. Lo Fo Wong[1,2], A. Wingstrand[2], J. Dahl[3], P.J. van der Wolf[4], B.M. Thorberg[5] and A. von Altrock[6]

[1]Royal Veterinary and Agricultural University, Groennegaardsvej 8, DK-1870 Frederiksberg C, Denmark; [2]Danish Zoonosis Centre, Danish Veterinary Laboratory, 27 Bülowsvej, DK-1790 Copenhagen V, Denmark; [3]Danish Bacon and Meat Council, 3 Axeltorv, DK-1609 Copenhagen V, Denmark; [4]Animal Health Service, P.O. Box 4, NL-5280 AA Boxtel, The Netherlands; [5]National Veterinary Institute, P.O. Box 7073, Ultuna, S-75007 Uppsala, Sweden; [6]Free University of Berlin, 69 Koenigsweg, D-14163, Berlin, Germany

Introduction

Surveillance and control is an important aspect of food safety assurance strategies. In contrast to clinical salmonellosis, sub-clinical *Salmonella* infections in pigs are not uncommon in most pig producing countries. Since pigs can carry *Salmonella* without obvious symptoms of disease, there is a need for tools that can be used to identify herds housing infected animals, which ultimately can lead to contaminated end products.

When considering a surveillance and control programme, it is important to have knowledge on the dynamics and epidemiology of *Salmonella* infections in pig herds. Pig production is a dynamic process with intensive transfer of animals and feedstuffs. Each introduction into a herd is a potential source of infection and may thus lead to a change of *Salmonella* status. Questions regarding herd incidence and the persistence of (or freedom of) infection need to be answered. In addition, the tools used to assess the *Salmonella* status of a herd need to be evaluated in a practical setting over a period of time. Therefore, this study set out to investigate the stability of herd *Salmonella* status over time and the relation between serological and bacteriological herd classification.

Materials and Methods

For this study, slaughter pig herds with a seropositive and seronegative *Salmonella* status were selected. A herd was assigned a seropositive *Salmonella* status if more than 4% of the blood samples from finishers tested seropositive. Four herds participated in Germany, 10 in Denmark, 13 in The Netherlands and 5 in Sweden. Thus, a total of 32 slaughter pig herds participated in the follow-up study, 17 of which were initially classified as seropositive and 15 as seronegative.

The selected herds were followed by 7 subsequent serological testing rounds. At each testing, 25 blood samples from slaughter pigs close to slaughter and 25 blood samples from growers (with a minimum weight of 20 kg) were taken. In addition, 10 pooled faecal pen samples (5 from grower pens, 5 from finisher pens) of approximately 25 grams were taken per visit. Testing was done approximately 3 months apart, so that each herd was followed for approx. 2 years. German, Danish and Swedish blood samples were analysed with the Danish mix-ELISA and the Dutch blood samples with the Dutch mix-ELISA. Samples with an Optical Density Percentage (OD%) larger than 10 were considered positive. Both tests were inter-calibrated so that test results could be compared.

Results

In order to investigate the stability of an assigned *Salmonella* status, the number of times a herd changed status during the observation period was recorded. The number of status shifts during the study period varied from 0 to 3 (see Table 1). Overall, 12 out of 32 herds (38%) continued to have the same *Salmonella* status as appointed at the start of the study during the observation period.

Table 1. Frequency distribution of the number of herds shifting from one Salmonella *status to the other during the observation period. Herds were assigned a seropositive status if more than 4% of samples > 10 OD%.*

Initial *Salmonella* status	No. of status shifts	No. of herds	% of herds
Seropositive herds	0	6	35%
	1	8	47%
	2	2	12%
	3	1	6%
	Total	17	100%
Seronegative herds	0	6	40%
	1	2	13%
	2	6	40%
	3	1	7%
	Total	15	100%

Among the 6 herds which remained negative, 3 herds were sampled over the entire 2 year period, while sampling was stopped prematurely for the remaining 3 herds (after 12-18 months). Of the 17 seropositive herds, 6 herds (35%) remained positive. However, 2 of the positive herds were not sampled over the entire 2-year period (respectively 6 and 15 months), and 3 herds were missing 3 to 5 sampling rounds.

The period of time it took for 8 seropositive herds to become seronegative during the observation period varied from 2 to 22 months. There was no significant difference in the mean duration of the initial status between seropositive and seronegative herds (p=0.6780).On average, herds that were classified as being seronegative in the initial sampling round, remained negative for 381 days, ranging from 49 to 695 days. Seropositive herds stayed positive for 419 days on average, ranging from 33 to 861 days.

A positive correlation between bacteriological results and a positive serological status in the same sampling round was found (r=0.32; p=0.0001). After taking between-country variation into account, the odds for culturing *Salmonella* in pen samples were 10 times higher if the herd was classified as seropositive at the time of sampling (i.e. more than 4% seropositive samples) compared to a seronegative status (OR=10.2; p=0.0001). There was no significant difference between growers and finishers in this respect.

Discussion

The stability of an initial *Salmonella* status, was found to vary noticeably. Sixty-two per cent of herds in this study shifted from their initial status at least once during the observation period, which likely would have been more had all herds been followed for the entire two year period. Steady-status periods varied from a month to more than two years, with no apparent difference between initially positive and initially negative herds. There are many factors that can lead to a loss of status, such as introduction of feed and animals or a change in feed or management strategy. However, herds that are consistently negative over longer periods of time than in this study are not uncommon in most pig producing countries. It is difficult to find a general practice for these herds which could explain why they remain negative, especially since management strategies, commonly accepted as risky with respect to *Salmonella* infections do not seem to play a role.

Traditionally, researchers have relied on bacteriological results to demonstrate a *Salmonella* contamination or infection. Bacteriological testing methods are often not practically and economically feasible in a situation where regular testing of many samples is involved, particularly at the pre-harvest level. Therefore, serological testing could be a practical alternative, especially since latent carriers or intermittent shedders may be detected by this method. This study showed a general good correlation between bacteriological and serological classification of slaughter pig herds. In conclusion it could be stated that a *Salmonella* herd-status should not be based on a single sampling round. Regular testing is necessary for correct classification of pig herds and to enable producers, advisors and authorities to react to a sudden increase in the *Salmonella* prevalence on a herd- or national level.

The influence of type of production on the occurrence of *Salmonella* in the Danish table-egg production

T. Hald[1], S. Kabell[2] and M. Madsen[2]
[1]Danish Zoonosis Centre, Danish Veterinary Laboratory, 27 Bülowsvej, DK-1790 Copenhagen V, Denmark; [2]Danish Veterinary Laboratory, 2 Hangoevej, DK-8200 Aarhus N, Denmark

Introduction

During recent years, Denmark has experienced an increasing consumer demand for both animal welfare and food safety. As a consequence of the concern for animal welfare, the number of layer flocks producing table-eggs under some kind of non-conventional concept like free-range or organic production is expected to increase. Common to most of the non-conventional systems is that the animals have more space including access to outdoor areas. However, the knowledge about the occurrence of zoonotic bacteria in such systems is limited, since most research so far has been directed towards the conventional production types, where the establishment of bio-security measures has been one of the cornerstones in the prevention and control of zoonotic diseases. In production systems with outdoor facilities, implementation of the traditional hygiene barriers is more difficult and the risk of introducing zoonotic diseases from the environment may be increased. Furthermore, the obvious difficulties in applying the recommended cleaning and disinfecting procedures after an infection may increase the risk of survival of infective agents and thereby give rise to re-infections. In the study presented here, results from the Danish *Salmonella* Surveillance and Control Programme in the table-egg production were used to investigate the influence of production type on the occurrence of *Salmonella* in layer flocks.

Materials and Methods

In March 1998, an improved plan for elimination of *Salmonella* in table-egg layers was implemented in Denmark. Accordingly, all layer flocks were monitored for *Salmonella* by serological testing of egg yolk and bacteriological examination of faeces (Feld et al.). Commercial layer flocks producing eggs for authorised egg-packing centres were sampled every nine weeks, whereas flocks producing eggs for barn-yard sale were sampled twice a year. At each sampling, 60 eggs and a pooled sample of 60 fresh faecal droppings or 5x2 so-called "sock samples" were collected (Skov et al., 1999). The latter samples consisted of 15 cm pieces of tube gauze mounted on the footwear during inspection in floor producing units. Seropositive flocks, defined as flocks with two or more birds (eggs) with an calculated OD% > 40, or flocks with *Salmonella* positive faecal samples were suspected of being infected (Feld et al.). Suspected flocks were re-tested in order to confirm the infection. If the second set of samples was positive, the flocks were declared infected.

Eggs from suspected and infected flocks were diverted to pasteurisation. The reduction of *Salmonella* in layer flocks has been achieved primarily by eradication of infected breeder and rearing flocks, but also by increased bio-security measures at hatcheries and in layer farms.

Based on the serological test results collected from January 1998 through March 2000, the proportion of seropositive flocks in each production type was determined. Further, in order to assess whether the within flock seroprevalence was varying between production types, the proportion of individual seroreactors per flock was calculated.

Results

In the 1st quarter of 1998, a considerable difference in the proportion of seropositive flocks between types of production was noticed: battery: 37%, deep litter: 24%, free-range: 18%, organic: 9% (Figure 1). Flocks producing eggs for barn-yard sale were not included in the surveillance until the 3rd quarter of 1998, where 11% of the flocks were seropositive. As a consequence of the control programme, the proportion of seropositive flocks within all production systems was reduced to 0-5% at the end of the study period.

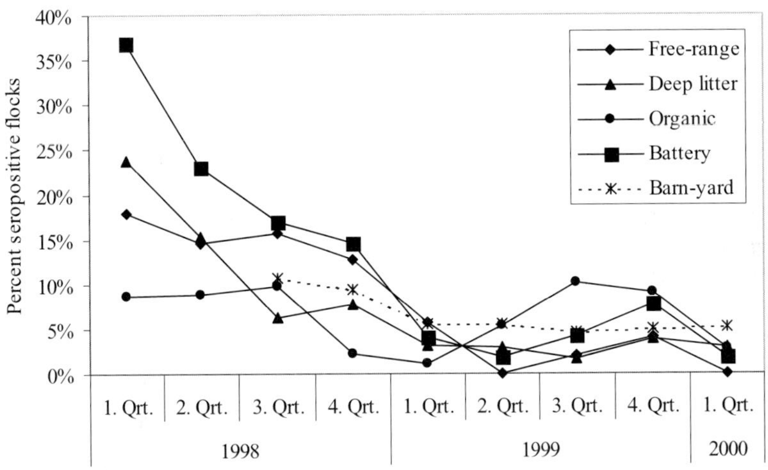

Figure 1. Percent seropositive table-egg producing flocks by type of production, 1998 - 1st quarter 2000.

In 1998, the highest proportion of flocks with many seroreactors (within flock seroprevalence >10%) was found among the cage layers (16%), whereas the lowest proportion occurred in organic flocks (1.3%). In 1999, the proportion of flocks with a seroprevalence above 10% was reduced to almost the same level within all types of production ranging from 2.5% in the battery production to 4.5% in the barn-yard and deep litter production.

Discussion

The descriptive results from this study suggest that the type of production does not influence the occurrence of *Salmonella* in table-egg layer flocks. The difference in the proportion of seropositive flocks observed in 1998, was not considered to be due to the type of production, but rather depending on the hatchery, since one hatchery that faced *Salmonella* problems in 1997 supplied a large part of the conventional production. Considering that most infections in Danish layer flocks so far have been caused by *S.* Enteritidis, which is often vertically transmitted (Humphrey, 1994), this is believed to be a reasonable conclusion. Unfortunately, it was not possible to make a statistical evaluation of the effect of the hatchery since this information was not routinely entered in the surveillance database. In the broiler production, however, studies have shown that both the hatchery and the *Salmonella* status of the parent flock is major risk factors for infection (Chriél et al., 1990; Davies et al., 1994).

References

Feld NC, Ekeroth L, Gradel KO, Kabell S, Madsen M. Evaluation of a serological *Salmonella* mix-ELISA for poultry used in a national surveillance programme. Accepted for publication in Epidemiology and Infection.

Skov MN, Carstensen B, Tornøe N, Madsen M. Evaluation of sampling methods for detection of salmonella in broiler flocks. Journal of Applied Microbiology 1999; 86: 695-700.

Humphrey TJ. Contamination of egg shell and contents with *Salmonella enteritidis*: a review. International Journal of Food Microbiology 1994; 21: 31-40.

Chriél M, Stryhn H, Dauphin G. Generalised linear mixed models analysis of risk factors for contamination of Danish broiler flocks with *Salmonella typhimurium*. Preventive Veterinary Medicine 1990; 40: 1-17.

Davies RH, Wray C. An approach to reduction of salmonella infection in broiler chicken flocks through intensive sampling and identification of cross-contamination hazards in commercial hatcheries. International Journal of Food Microbiology 1994; 24: 147-160.

Detecting *Cryptosporidium parvum* and *Giardia intestinalis* from surface water by polymerase chain reaction

Ruska Rimhanen-Finne and Marja-Liisa Hänninen
Department of Food and Environmental Hygiene, Faculty of Veterinary Medicine, University of Helsinki, Helsinki, Finland

The aim of this study is to find out the prevalence of Cryptosporidia and Giardia in Finnish surface waters. Waterborne outbreaks of cryptosporidiosis and giardiasis have been reported. Outbreaks assosiated with contaminated drinking water have been reported especially from USA,UK and Sweden. Infections are usually caused by fecally (either animal or human origin) contaminated water. It has been thought that in Finland these organisms are greatly under-diagnosed. In Finland the prevalence of Cryptosporidium oocysts and Giardia cysts in source water as well as in reservoir animals is so far unknown. There has not been any outbreaks caused by these organisms reported either.

Protozoans Cryptosporidia and Giardia are causes of human diarrheal diseases both in developed and undeveloped countries. Cryptosporidia are highly prevalent in young calves worldwide and calves do present diarrhea. Older cattle with abomasal cryptosporidiosis have no diarrhea, although they can excret oocysts for several months. Infection route is feco-oral. The severity and duration of the illness both in humans and cattle is influenced by the immune response. Immunocompromised persons present with resolving and recurring diarrhea, with *Cryptosporidium parvum* the ilness may become persisting and life threatening.

In this study we developed a protocol for detecting Cryptosporidium oocysts and Giardia cysts from surface water. Conventional methods for the detection of Cryptosporidia and Giardia from water are labor-intensive and time-consuming. Various techniques (f.ex. flow cytometry,enzyme-linked immunosorbent assay) have been tried to provide better detection of Cryptosporidia and Giardia. Polymerase chain reaction (PCR) is thought to be more sensitive and specific than the conventional methods. Our method consists of development of a water pretreatment including filtration, immuno magnetic capture of oocysts and cysts and silica-based DNA extraction and is based on PCR reaction. The sensitivity of the method is 50 cysts/oocysts for both Cryptosporidium and Giardia per 2 litres sample. As far only 7 environmental watersamples has been investigated by this newly developed method.

This study is a part of my Ph.D. work which will evaluate the prevalence of Cryptosporidium parvum and Giardia intestinalis in Finnish environment (surface waters, waste water, sewage) and in animal reservoirs (cattle, beavers, reindeer) and as risks for human health from contaminated water or food.

References

Fayer R. Cryptosporidium and Cryptosporidiosis. CRC Press, 1997.

Rochelle P.A., De Leon R.,Stewart M., Wolfe R.L. Comparison of primers and optimization of PCR conditions for detection of *Cryptosporidium parvum* and *Giardia lamblia* in water. Applied and Environmental Microbiology, Jan. 1997, p. 106-114.

Epidemiology of *Yersinia pseudotuberculosis* in Finland

Taina T. Niskanen
Department of Food and Environmental Hygiene, Faculty of Veterinary Medicine, P.O. Box 57, 00014 University of Helsinki, Finland, Fax: +358 9 19149 718, E-mail: taina.niskanen@helsinki.fi

The aim of my Ph.D. research is to study the prevalence and pathogenicity of *Y. pseudotuberculosis* and potential contamination routes of this bacteria. The epidemiology of *Y. pseudotuberculosis* is poorly studied and the reservoirs are not well known. In Finland yersiniosis is the fourth common bacterial enteric infection with approximately 700 diagnosed cases annually, which are caused by *Yersinia enterocolitica* and *Yersinia pseudotuberculosis* 93% and 7%, respectively (Anon. 2000). From 1990 to 1999 few *Y. pseudotuberculosis* outbreaks have reported in Finland. There has been indirect evidence that food is a source of human infections, but in most of the cases the source of the infection or the vehicle were unknown. No *Y. pseudotuberculosis* has been isolated from food, but *Y. pseudotuberculosis* serotype O:3 has been isolated from human faeces in most outbreaks in Finland. In 1998 first time in the world, a food vehicle was identified in a case-control study in *Y. pseudotuberculosis* outbreak in Finland (Anon. 1999).

Y. pseudotuberculosis is rarely isolated from food or clinical samples due to considerable difficulties associated with detecting it in any natural samples. *Y. pseudotuberculosis* can be detected during clinical infections in domestic and wild animals (Weber 1988, Fukushima and Gomyoda 1991) Some reports can be find about the findings of *Y. pseudotuberculosis* in the tonsils of clinically healthy pigs (Narucha and Westendoorp 1977, Weber and Knapp 1981, Fukushima et al. 1989). However, the possible significance of *Y. pseudotuberculosis* in human infections is unclear and needs further investigation. I'm researching domestic and wild animals in order to find out their importance as a reservoirs of *Y. pseudotuberculosis*. Enviromental and food samples are studied as possible source of infections. The pathogenicity of the isolated strains is determined using both phenotypical and genotypical methods. We have established the pulsed-field gel electrophoresis (PFGE) technique for characterisation of *Y. pseudotuberculosis* and to get more information about the genetical diversity. Important part of my work is to try to improve the sensitivity of isolation of *Y. pseudotuberculosis* by developing conventional culture methods and PCR-technique for enviromental, food and animal origin samples.

In my first and present study, I have determined the prevalence of *Y. pseudotuberculosis* in fattening pigs using the conventional culture method. A total of 426 pig tonsil samples were collected from seven different abattoirs in Finland. Different isolation methods, including cold enrichment were used in order to find out the most efficient and productive alternative to isolate *Y. pseudotuberculosis*. The pathogenicity of the strains was confirmed by the PCR (polymerase chain reaction) targetting the *virF*- gene on the virulence associated plasmid of *Y. pseudotuberculosis*. Isolated *Y. pseudotuberculosis* strains were characterised using both fenotyping and genotyping methods. The genetic diversity of the isolates were studied by PFGE (pulsed-field gel electrophoresis).

References

Anonymous, 1999. KTL, National Public Health Institute, 6/1999, Helsinki, Finland.

Anonymous, 2000. Infectious diseases in Finland 1995-1999. National Public Health Institute, KTL B4 / 2000, National Public Health Institute. Helsinki, Finland.

Fukushima, H., Maryama, K. Omori, I., Ito, K. & Iorihara, M., 1989. Role of the contaminated skin of pigs in fecal Yersinia contamination of pig carcasses at slaughter. Fleischwirtsch. 69:369-372.

Fukushima, H., Gomyoda, M., 1991. Intestinal carriage of *Yersinia pseudotuberculosis* by wild birds and mammals in Japan. Appl. Environ. Microbiol. 57: 1152-1155.

Narucha, U. , Westendoorp, J. F., 1977. Studies for the presence of *Yersinia enterocolitica* and *Yersinia pseudotubeculosis* in clinically normal pigs. Tijdschr Diergeneeskd. 1: 299-303.

Weber, A., Knapp, W., 1981. Über die jahreszeitlich Abhängigkeit des Nachweises von *Yersinia enterocolitica* und *Yersinia pseudotuberculosis* in Tonsillen gesunder Schlachtschweine. Zentralbl. Bacteriol. Hyg. 250: 78-83.

Weber, A., 1988. Serotyping of *Yersinia pseudotuberculosis* strains isolated from various animals species in the Federal Republic of Germany. Tierärtztliche-Umschau. 43: 262, 265-267.

Prevalence of *Clostridium botulinum* type B, E and F in faecal samples from Swedish pigs

Maria Dahlenborg [1,2], Peter Rådström[1] and Elisabeth Borch[2]
[1]Applied Microbiology, Lund Institute of Technology, Lund University, P.O. Box 124, SE-221 00 Lund, Sweden; [2]Swedish Meats R&D, P.O. Box 504, SE-222 40 Kävlinge, Sweden, Fax number: +46-46-73 61 37; E-mail: maria.dahlenborg@smrd.com

Introduction

The consumer's demand for ready-to-eat foods that require minimal preparation time, are of high quality and contain low concentrations of preservatives has increased. This has resulted in the development of chilled ready-to-eat foods with an extended durability and *sous-vide* foods. Non-proteolytic strains of *Clostridium botulinum* is considered as a hazard in chilled ready-to-eat foods with an extended shelf life. This is due to lower heat treatment than, for example canned foods and the anaerobic atmosphere provided by the processing of the foods. These conditions favour germination of spores and toxin formation from non-proteolytic *Cl. botulinum,* which can multiply and produce neurotoxin of types B, E or F at temperatures as low as +3.0 °C.
Cl. botulinum can be found naturally in the environment for example in soil and in sea sediment (Hauschild, 1993). Only a limited number of studies have been made in Europe to survey the distribution of *Cl. botulinum* in the environment and in the raw material for foodstuffs. In Sweden, the latest survey was made by Johannsen *et al.*1963 (Johannsen, 1963) in soil and coast sediment. Both type B and E were found.

Objective

The objective in this work is to study the occurrence and distribution of *Cl. botulinum* type B, E and F in faecal samples from both outdoor and indoor pigs in Sweden.

Materials and Methods

Faecal samples from pigs (both outdoors and indoors) have been collected in slaughterhouses in different parts of Sweden. The samples have been collected over two periods of time during the year (June-September and December-March). The samples are collected at the slaughter line.

In a first step, 78 faecal samples have been analysed as composite samples with faeces from 3 pigs in each samples. This was to increase the sample size, due to predicted low numbers of spores in the faeces. In a second step, positive composite samples by PCR, will be divided and faeces from each pig and analysed separately.

A PCR-based method has been developed for detection of endospores of *Cl. botulinum* type B, E and F in faecal samples from pigs. Two sets of primers were designed, one specific for part of the type B neurotoxin gene and the other specific for part of both the type E and F neurotoxin genes.

Due to predicted low number of *Cl. botulinum* spores in the environment and a high concentration of non-target cells in the faecal samples (10^{10} bacteria cells per gram from pigs (Melin et al., 1997), an enrichment procedure pre-PCR were optimised. The aim of this enrichment procedure is to select for and activate spores of *Cl. botulinum* to germinate and finally cultivate the bacteria to detectable levels. Before PCR the faecal samples are heat treated at 70°C for 10 min and enriched in TPGY medium at 30°C for 18 h.

Results and Conclusion

The detection limit for the enrichment PCR procedure is, with a detection probability $\geq$ 95 %, 10 spores per gram faecal sample for type B and corresponding value is 10^3 spores per gram faecal sample for type E and F.

Results from the analysed composite samples showed an occurrence of *Cl. botulinum* type B between 14-42 % in faecal samples from pigs in Sweden. No *Cl. botulinum* type E or F has so far been detected.

Acknowledgement

This work is supported by a national, industry-oriented Swedish programme for research and PhD education, LiFT - Future Technologies for Food production.

References

Hauschild, A.H.W., 1993. *Clostridium botulinum* in the environment . In: Hauschild, A.H.W. and Dodds, K. L. (eds.) *Clostridium botulinum* - Ecology and Control in Foods. Marcel Dekker, New York, pp. 21-51.

Johannsen, A., 1963. *Clostridium botulinum* in Sweden and the adjacent waters. J. Appl. Bacteriol. 26:43.

Melin, L., Jensen-Waern, M., Johannisson, A., Ederoth, M., Katouli, M., Wallgren, P., 1997. Development of selected faecal microfloras and of phagocytic and killing capacity of neutrophils in young pigs. Veterinary Microbiology. 54:287-300

Epidemiology and evolution of *enterobacteriaceae* infections in humans and domestic animals Peri- and post-harvest risk identification and quantification of foodborne transmission

D.J. Mellor[1], S.A. McEwen[2], W.J. Reilly[3] and S.W.J. Reid[1]
[1]Comparative Epidemiology and Informatics, Departments of: Veterinary Clinical Studies, University of Glasgow Veterinary School, Bearsden Road, Glasgow G61 1QH, UK, &, Statistics and Modelling Science, University of Strathclyde, 26 Richmond Street, Glasgow, G1 1XH, UK; [2]Department of population Medicine, Ontario Veterinary College, University of Guelph, Ontario, Canada; [3]Scottish Centre for infection and Environmental Health, Clifton House, Clifton Place, Glasgow, G3 7LN, UK.

There is evidence that the infection status of production animals and their products can change during events that occur from the point of departure from farm to final consumption of meat. Current tenets are that risk reduction interventions must be placed at intervals along the entire food chain if the public is to be adequately protected. Knowledge of the dynamics of meat contamination is needed if we are to estimate the impact of interventions such as prevalence reduction in herds, hide cleanliness of slaughter animals, testing programs for animals or meat product, and segregated slaughter. Traditional experiments and observational studies have produced qualitative information in these areas, but new approaches are needed to improve understanding of the interrelationships among the variables affecting contamination of meat, to assemble existing data and to deal with uncertainties that exist.

This study seeks to assess quantitatively factors of importance in the contamination of meat products with EHEC and drug resistant *Salmonella* and *E. coli*, from the point of departure of cattle from farms, through transport, mixing, slaughter and processing, distribution and finally to consumption. The study will test a number of hypotheses concerning risk associated with the animal-human transmission, viz: (I) that the prevalence of infection in cattle at slaughter has direct bearing on the contamination of food product; (ii) that the source of animals is an important risk factor for the production of contaminated product; (iii) that human cases are associated geographically and temporally with specific cattle sources. These will be addressed in collaboration with other members of the WIPRAVE partnership under focused questions such as (1) Does the prevalence of faecal shedding on the farm have a direct bearing on meat contamination and human illness? (2) What is the effect on carcass contamination of mixing infected with uninfected animals during transport, in market yards or in lairage prior to slaughter? (3) What is the relative contribution of faecal shedding and hide contamination to the presence of bacteria on carcasses? (4) What events during harvest are the most important to meat contamination? (5) Would reducing meat contamination affect the incidence of sporadic disease in humans?

Sampling of animals in Scotland has begun and will be conducted at intervals throughout the chain to obtain data for stochastic risk models and to provide a basis for the trace-forward/back studies. The sampling strategy has been formulated in collaboration with the Scottish Agricultural Colleges and is based on data from their recently published study of the prevalence of *E. coli* 0157 on Scottish farms, funded by the Scottish Executive. Samples (faeces, hides, carcasses, meat, equipment) will be collected at each of the following steps: departure from farm; arrival at slaughter; immediately *post mortem;* carcasses and equipment on line; and in chill.

Throughout the study, identified human cases within Scotland will be subjected to rigorous trace-back. This will build upon the current (separately funded) *E. coli* 0157:H7 case-control study being carried out by the Scottish Centre for infection and Environmental Health (SCIEH) and will involve the development of molecular markers. Information will be available from source farm, processing plant and human case population, enabling us to test the hypothesis that human disease is directly associated with animal.

Data from these sources and the literature will be used to build stochastic risk models of meat contamination and human infection. Stochastic simulation modelling has utility where the system under study is complex and when it is highly desirable to allow for variability and uncertainty in input variables. Risk models will form the basis for peri- and post-harvest Hazard Analysis and Critical Control Point (HACCP) strategies and Statistical Process Control (SPC) designs by incorporating interventions and their associated critical limits. Interventions are needed throughout the food chain to reduce risks of foodborne pathogens and these should be adaptable to HACCP systems. In many of the peri- and post-harvest points in the food chain, however, more knowledge is needed of the likely critical control points and their critical limits (e.g. Pathogen concentrations) and the risk models developed in this study would be used to specify the criteria for improving the microbial safety of meat.

Potential for the spread of food-borne pathogens in the cattle lairage environment

A. Small, C.-A. Reid, S. Avery and S. Buncic
Division of Food Animal Science, Department of Clinical Veterinary Science, University of Bristol, Langford, Bristol BS40 5DU

The major food-borne bacterial pathogens connected with contamination of red meat are *E. coli* 0157, *Salmonella* spp. and *Campylobacter* spp. These organisms are asymptomatically carried in the intestines of cattle and other animals and are shed in faecal material. This faecal material (and hence bacterial pathogens) is carried on the animals hide and hooves, and spread within the environment and between animals. Certain factors are known to increase the shedding of food-borne pathogens in faecal material, and these include diet, stress, feed withdrawal and seasonal variation i.e. higher shedding of *E. coli* 0157 within the summer months (Hancock *et al*, 1997; Harmon *et al*, 1999).
In terms of food poisoning, the highest risk of meat contamination is during slaughter and processing, where any pathogens present on the animal have the potential to be transferred to the carcass at various stages. In addition to animal hides being contaminated on the farm, they may also be freshly contaminated during transport to the slaughterhouse, whilst resting within the lairage, within the area leading to the stunning box, or within the stunning box itself. Any contamination of the hide can lead to meat contamination during the dressing and processing of the carcass, which may lead to infection in humans (Bell, 1997). Pathogen transfer within the lairage may arise from animal-to-animal or environment-to-animal contact.

The aim of this pilot study was to evaluate the prevalence of the major foodborne bacterial pathogens *E. coli* 0157, *Salmonella* and *Campylobacter* spp., within the lairage environment during the summer months. This study forms part of an on-going three year project on the pre-slaughter epidemiology and control of pathogens in abattoirs.

Methods

A total of 90 samples were taken from the lairage of three abattoirs within the South-West of England during the months of July - August, using a one-pass swabbing technique. Samples were taken from a 10 x 10cm area using a metal template and a moistened swab (A 10 x 11cm sponge); three from each of ten sites within the lairage environment. Sample sites included :-

- Stunning box floor
- Walls of holding pens
- Crush
- Water trough
- Ramp at unloading area
- Gates of stunning box/race
- Walls of funnel into race
- Post on turning into lairage
- Floor of waiting area
- Corner from unloading to lairage area

Samples were held on ice for transportation back to the laboratory. Samples were processed with maximum recovery diluent (MRD) and volumes alloquated to Buffered peptone water (BPW), Park Sanders broth, and modified Tryptone Soya broth (mTSB) for the enrichment of *Salmonella* spp., *Campylobacter* spp., and *E. coli* 0157 respectively. Following enrichment standard accredited methods were used for the isolation of the above pathogens, with biochemical and confirmatory tests used as appropriate for identification.

Results

Table 1. Number of positive isolates identified in 30 samples from each lairage.

	Abattoir 1	Abattoir 2	Abattoir 3
E. coli 0157	3	N.D*	4
Salmonella spp.	3	N.D*	8
Campylobacter spp.	N.D*	N.D*	N.D.*

*N.D. = not detected; preliminary identification using latex- and poly-0- agglutination

Discussion

Preliminary examination of the data collected has indicated that, on average, *E. coli* 0157 was detected in 7.7% of all lairage samples with detection from individual abattoirs ranging between 0-13.3%; *Salmonella* spp. were detected in 12.2% of samples on average, ranging between 0-27% within individual abattoirs, and *Campylobacter* spp. were not detected in any of the 90 samples. Previous prevalence studies have indicated that *E. coli* 0157 can be isolated from the faeces of cattle herds at a rate of 3-5% during the summer months (Hancock *et al*, 1997; Brown *et al*, 1997). Therefore, the detection levels of this food-borne pathogen within the lairage environments examined was higher than the levels previously detected in cattle herds. The lairage areas where most isolates of *E. coli* 0157 were detected were the pen floors (2 out of 3 abattoirs positive) and the race leading to the crush (2 out of 3 samples positive for abattoir 1).

The prevalence of *Salmonella* spp. within cattle faeces is reported to be 5.5%, with 3.5-7.4% isolated from pen floors of shedding animals (Fedorka-Cray, 1998). Within this study, 2 out of the 9 positive *Salmonella* spp. samples (22.2%) were detected from the pen floors, indicating a much higher prevalence within the abattoirs studied. *Salmonella* spp. were also detected from the water troughs (2 out of 3 abattoirs) and the stunning box (2 positive isolates from abattoir 3).

This indicates that the pen floors, water troughs (cattle are known to splash water around), stunning box and area leading into the box are the main problem areas within the lairage as a source of fresh hide contamination, particularly as all animals will pass through, and have direct contact with, these contaminated areas.

This initial study indicates that the major bacterial food-borne pathogens contaminating beef carcasses can be present within the lairage of some abattoirs at a relatively high overall frequency (up to 27%). As these pathogens were isolated from areas within the lairage which all animals to be slaughtered must come into contact (such as the race and stunning box), the risk of intra-lairage contamination of the hide, and consequently the carcass can be high. Further investigations will be aimed at identifying lairage designs and/or practices which contribute to this type of contamination.

References

Bell, R.G., 1997. J. Food Protection 82, 292-300.

Brown, C.A., Harmon, B.G., Zhao, T. and Doyle, M.P., 1997. Appl. Environ. Microbiol. 63, 27-32.

Fedorka-Cray, P.J., Dargatz, D.A., Thomas, L.A. and Gray, J.T., 1998. J. Food Prot. 61, 525-530.

Hancock, D.D., Besser, T.E., Rice, D.H., Herriott, D.E. and Tarr, P.I., 1997. Epidemiol.Infection 118, 193-195.

Harmon, B.G., Brown, C.A., Tkalcic, S., Mueller, P.O.E., Parks, A., Jain, A.V., Zhao, T. and Doyle, M.P., 1999. J. of Food Protection 62, 574-579.

The use of plasmid profile analysis to study the epidemiology of salmonellas in broiler chicken flocks

C.J. Crowley[1], J.E.L. Corry[1], E. Liebana[2], and R.R.H. Davies[2]
[1]University of Bristol, [2]Veterinary Laboratory Agency, New Haw, UK.

Introduction

Broiler chicken flocks are an important reservoir for human salmonellosis. By eliminating *Salmonella* sources in broiler flocks, the number of cases of human infection should be reduced. In order to achieve this, it is important to identify potential sources of contamination so the necessary control measures can be implemented (ICMSF, 1998).

Epidemiological investigations can identify sources of contamination. In such studies it is essential to further subdivide isolates by phenotypic and/or genotypic (molecular) typing (Threlfall et al., 1990; Farber, 1996). The molecular typing method of plasmid profile analysis has been successfully used by various authors, including Limawongpranne et al.(1999). This study was undertaken as part of a research project with two major UK integrated broiler production companies to identify sources of contamination using various molecular typing methods. This poster will cover the results obtained from the plasmid profiling analysis.

Method

Selected sites: hatchery, feed mill, broiler farm and abattoir within two broiler production companies (A & B), were sampled on several occasions in order to follow flocks through production over a three year period. Samples were cultured individually using sensitive culture techniques and all suspect *Salmonella* cultures were serotyped. Cultures of *S. typhimurium* and *S. enteritidis* were phage typed and all isolates were plasmid profiled using a modification of the method described by Kado and Liu (1981).

Results and Discussion

Low numbers of salmonella (<5% of carcasses at the end of production) were isolated from both companies. However, eight potentially significant *Salmonella* serotypes were detected: *S. typhimurium* (DT104); *S. enteritidis* (PT4); *Salmonella*. 4,12:d:-; *S. kedougou; S. binza; S. senftenberg; S. ohio* and *S. agama*.

Table 1. Salmonella *isolates from Company A.*

Serotype	No. isolates Profiled	No. plasmid	No. PPTs	Most common PPT (%)	Plasmid size
S. enteritidis (PT4)	23	all	6	2 (60.9)	≈63.63kb
Salmonella 4,12:d:-	264	40	6	1 (77.5)	plasmid-free
S. kedougou	17	all	5	2 (47.1)	2067-2972bp
				1 (29.4)	plasmid-free

Company A

S. enteritidis - Type 2 was the most common plasmid profile type (PPT), containing only the sero-specific plasmid described by Chart et al. (1989). This PPT was isolated from the hatchery, feed mill, broiler farm and abattoir throughout over the course of the study. It was first isolated from the hatchery.
Salmonella 4,12;d;- - Type 1 was the most common PPT. This PPT was isolated from the feed mill, broiler farm and abattoir, being first isolated from the broiler farm.
S.kedougou - Type 2 was the most common PPT, followed by type 1. PPT 2 was isolated from the feed mill, broiler farm and abattoir; it was first isolated from the broiler farm. PPT 1 was isolated on more visits but from fewer samples, it was isolated from the feed mill and broiler farm only, and was first isolated from the broiler farm.

Table 2. Salmonella *isolates from Company B.*

Serotype	No. isolates Profiled	No. plasmid	No. PPTs	Most common PPT/s (%)	Plasmids size
S. typhimurium (DT104)	74	all (64)*	8	2 (78.7)	63.63-148.48kb 2067-2972bp
S. enteritidis (PT4)	44	all	2	2 (93.2)	≈63.63kb
S. binza	264	40	6	5 (52.8)	6030-7045bp ≈2067bp ≈2067bp
				3	
S. senftenberg	17	all (62)*	5	4B (35.5)	16.6-36.21kb, >148.48kb ≈3990bp, 2067-2972bp
				4A (16.1)	>148.48kb ≈3990bp, 2067-2972bp
S. agama	5	all	1	1 (100)	plasmid-free

*indicates number of stable reproducible profiles

Company B

S. typhimurium (DT104) - Type 2 was the most common PPT. This PPT was isolated from the chick delivery box liners, feed mill, broiler farm and abattoir, continously over the course of the study. It was first isolated from the feed mill. Its prevalence fluctuated towards the beginning of production, but was isolated on 1/2 abattoir visits.
S. enteritidis (PT4) -Type 2 was the most common PPT, containing only the sero-specific plasmid described by Chart et al. (1989). This PPT was isolated repeatedly, from the hatchery, feed mill, broiler farm and abattoir. It was first isolated from the hatchery.
S. binza - Type 5 was the most common PPT. This PPT was isolated from all sites with different frequencies over the three year study. It was first isolated from the hatchery.
S. senftenberg - Type 4B was the most common PPT, followed by type 4A. These were isolated from the hatchery only, on two visits.
S. agama - Only five isolates were obtained, all were plasmid-free. This PPT was isolated from the feed mill and broiler farm only.
S. ohio - Fifty-seven isolated were obtained, however 36 isolates were profiled. These produced very unstable profiles and therefore the results were not analysed.

Conclusions

- A range of PPTs were obtained for all serotypes except *S. agama*, indicating more than one contamination source or genetic variation occuring in salmonellas originating from persistent sources.
- The results indicate that major contamination sources are presistent contamination of equipment in the feed mill and hatchery.
- Serious limitations of plasmid profiling have been demonstrated
  - The majority of *S. enteritidis* strains isolated in the study, just carried the serovar-specific virulence plasmid, in agreement with other studies (Chart et al., 1989).
  - Plasmid profile analysis is only useful for some serotypes. Unstable plasmid profiles were obtained for all *S. ohio* isolates, also some *S. typhimurium and S. senftenberg* isolates
  - Other molecular methods have also been applied as part of this research project.

Acknowledgements

This work was funded by the UK Food Standards Agency.

References

Chart, H. E. et al., 1989. Micro. Letters. 58:299-304.
Farber, J.C., 1996. J. Food Protect. 59:1091-1101.
ICMSF, 1998. Micro-organisms in Food.. Ecology of food commodities. Blackie, London, pp75-129.
Kado. C.I. and Liu. S.T., 1981. J. Bacteriol. 145:1356-1373.
Limawongpranee, S. et al.,1999. J. Vet. Med. Sci. 61:255-259.
Threfall, E. J. et al., 1990. Epidiol. Infect. 104:243-251.

An investigation into the changed physiological state of *Vibrio* organisms as a survival mechanism in response to nutrient limitation at low temperatures

Moira Johston
Unilever Research, GB, Mk44 1LQ Beds

Among the *Vibrio* organisms are important human pathogens. Infection by *Vibrio vulnificus* was the leading cause of death in the U.S.A. from the consumption of seafood in 1999 (Bryan *et al.*, 1999). *Vibrio parahaemolyticus* also causes gastrointestinal disease (Jiang and Chai, 1996), whilst *Vibrio cholerae* has long been of concern as the etiologic agent of cholera (Massad and Oliver, 1987). These organisms have all been found to exist in marine and estuarine environments (Massad and Oliver, 1987; Oliver *et al.*, 1991; Jiang and Chai, 1996). There is a general inability to isolate estuarine Vibrios during the winter months or from colder waters (Wolf and Oliver, 1992). In the past this was believed to be due to the sensitivity of these bacteria to cold temperatures (Oliver, J.D., 1989). More recent studies have presented evidence that the bacteria, although unable to be cultured, have entered into what has been termed the "viable but non-culturable" form in response to the reduced temperatures (Oliver *et al.*, 1991; Wolf and Oliver, 1992; Huq and Colwell, 1995; Jiang and Chai, 1996; Bryan *et al.*, 1999).

In this study, *V. vulnificus*, *V. parahaemolyticus* and *V. cholerae* were all grown to late logarithmic phase, before being washed and resuspended in Artificial Sea Water (ASW). Organisms were kept in the ASW at 20°C as well as at 4°C and were monitored for changes in physiology and any decrease in culturability over 60 days. Direct Total Counts were carried out by light microscopy, using a Thoma slide, whilst culturability was monitored by plating onto non-selective as well as selective media. The non-selective agar was Heart Infusion Agar (HIA) and the selective agar was Thiosulphate Citrate Bile Salts Agar (TCBS) which is routinely used for the isolation of pathogenic *Vibrio* organisms (Nicholls, *et al.*, 1976; West *et al.*, 1982).

The *V. vulnificus* in ASW at 20°C remained culturable on HIA, giving counts of $>10^7$ per ml after 60 days. Counts on TCBS were lower, dropping from 10^9 per ml to approximately 10^5 per ml after 3 days, with 10^4 per ml being recoverable on this media after 60 days. The *V. vulnificus* in ASW at 4°C, however, became rapidly non-culturable on TCBS, with counts of <10 per ml after just 3 days and remaining non-culturable throughout the 60 day period. The counts on HIA also dropped rapidly and the organism could not be detected (counts of <10 per ml) after 21 days. Although these organisms had become non-culturable, the Direct Total Count showed $>10^9$ per ml for all samples, with the non-culturable samples showing a change in cell morphology from rods to coccoid cells. Electron microscopy showed the change in shape more clearly and that these coccoid cells all demonstrated formation of blebs in the membrane.

The Direct Total Count does not distinguish between live and dead bacteria, so cells were stained using acridine orange by a method developed by Kogure *et al.* (1979) for counting living marine bacteria. The method is based on giving the cells a nutrient source, in the form of yeast extract, along with nalidixic acid which prevents the cells from dividing. Other synthetic pathways continue to operate and due to the abundance of cellular RNA, acridine orange stained bacteria growing at higher rates fluoresce red orange under the fluorescent microscope. In contrast, inactive bacteria which are poor in cellular RNA fluoresce greenish white. The cells of *V. vulnificus*, which had become non-culturable at 4°C in ASW were stained and found to be living cells, of a coccoid shape.

A study was carried out on these cells to see whether they could be resuscitated by removing the stress of the low temperature. A sample of the non-culturable cells in ASW at 4°C was removed and transferred to a temperature of 20°C. Within 24 hours numbers as high as 10^7 per ml were observed on agar plates, reaching 10^8 per ml within 48 hours (on HIA and, surprisingly, also on TCBS). This was not believed to be due to growth of a few survivors, because there were no nutrients added to the sea water to promote growth. Furthermore, microscopy revealed a reversion from the coccoid form back to a rod shaped organism as the cells became culturable again. This first resuscitation study was carried out on cells which had been held in ASW at 4°C for 42 days. When repeated at 63 days, the samples showed no resuscitation over 48 hours. These samples were not stained, so it is not certain whether they may have died within this time, or gone so far into the non-culturable form that they could not be made culturable again purely by the upshift in temperature.

The chilled samples demonstrated that *V. vulnificus* enters the VBNC state quickly (taking only 21 days for the first batch of cells and, on repeating the study, 8 days for a second batch). The culturability of *V. cholerae* at 4°C decreased steadily with it becoming non-culturable on TCBS after 42 days, but still remaining culturable after 60 days on HIA, although at a very low number (30 cells per ml). The chilled samples of *V. parahaemolyticus* took longer still to become non-culturable on TCBS (57 days) and remained culturable on HIA at 6 x 10^2 per ml after 60 days. The change in morphology from rod shaped to coccoid cells was also evident for these species of Vibrio, as the organisms became non-culturable. The *V. cholerae* and *V. parahaemolyticus* samples in sea water at 20°C remained culturable on HIA and TCBS at approximately 10^7 per ml after 60 days.

Having demonstrated that these organisms can exist in a changed physiological state, it is important that they can be isolated by our *Vibrio* detection methodology. As an upshift in temperature appears to render the cells culturable again, it may be that the pre-enrichment stage in current methodology will be sufficient to do this. However, TCBS does not appear to be suitable for the detection of *V. vulnificus*, even for full recovery of normal cells. The times and temperatures required to optimise resuscitation also need to be investigated further. Another important fact to establish is whether the changed physiology cells have similar resistance to heating and freezing as the normal cells. Initial studies suggest that the changed physiology form may be more heat resistant, but this needs further investigation.

It is only by carrying out these studies on persistence, detection and enumeration, that we will be able to establish the risks of harvesting fish from temperate and warm compared with colder waters.

References

Bryan, P.J., Steffan, R.J., DePaola, A., Foster, J.W. and Bej, A.K, 1999. Adaptive response to cold temperatures in *Vibrio vulnificus*. Current Microbiology 38, 168-175.

Huq, A. and Colwell, R.R. ,1995. A microbiological paradox: viable but nonculturable bacteria with special reference to *Vibrio cholerae*. Journal of Food Protection 59, 96-101.

Jiang, X and Chai, T-J., 1996. Survival of *Vibrio parahaemolyticus* at low temperatures under starvation conditions and subsequent resuscitation of viable, nonculturable cells. Applied and Environmental Microbiology Apr, 1300-1305.

Kogure, K., Simidu, U. and Taga, N., 1979. A tentative direct microscopic method for counting living marine bacteria. Canadian Journal of Microbiology 25, 415-420.

Massad, G. and Oliver, J.D., 1987. New selective and differential medium for *Vibrio cholerae* and *Vibrio vulnificus*. Applied and Environmental Microbiology Sept, 2262-2264.

Nicholls, K.M., Lee, J.V. and Donovan, T.J., 1976. An evaluation of commercial thiosulphate citrate bile salt sucrose agar. Journal of Applied Bacteriology 41, 265-269.

Oliver, J.D., 1989. *Vibrio vulnificus*. In Foodbourne bacterial pathogens ed. Doyle, M pp. 569-599. Marcel Dekker. Inc, New York.

Oliver, J.D., Nilsson, L. and Kjelleberg, S., 1991. Formation of nonculturable *Vibrio vulnificus* cells and its relationship to the starvation state. Applied and Environmental Microbiology Sept, 2640-2644.

West, P.A., Russek, E., Brayton, P.R. and Colwell, R.R., 1982. Statistical evaluation of a quality control method for isolation of pathogenic Vibrio species on selected thiosulphate citrate bile salts sucrose agars. Journal of Clinical Microbiology Dec, 1110-1116.

Wolf, P.W and Oliver, J.D., 1992. Temperature effects on the viable but non-culturable state of *Vibrio vulnificus*. FEMS Microbiology Ecology 101, 33-39.

Development of methods to detect *Cryptosporidium parvum* and *Giardia lamblia* in biosolids applied to land

Jamie Massanet-Nicolau, Andrew Blackwell and Ashley Jonas
Thames Water Plc., Registered Office 14, Cavendish Place, London W1M 0NU.

The application of biosolids to agricultural land combined with an increased awareness of the health and commercial implications of a biosolid related contamination incident, has stimulated considerable interest in the routes of transmission of many pathogens including *Giardia* and *Cryptosporidium*. The ability to test for the presence of *Cryptosporidium* and *Giardia* (oo)cysts in biosolid matrices is becoming increasingly important and the development of accurate and reliable methods for such testing is now the focus of considerable research effort. At present no standard methods exist for determining the presence of *Cryptosporidium* and *Giardia* (oo)cysts in biosolids samples and little is known about the ability of biosolid treatment processes to reduce parasite loading and minimise the risk of contamination or infection.

In this study several different biosolid matrices have been investigated; sewage sludges, (comprising of raw untreated sludge and products from various sludge treatment processes including anaerobic digestion and thermal hydrolysis), animal manures, (pig, horse, cattle and poultry) and abattoir wastes, (blood, manure (including rumen contents) and processed effluent from the abattoir).

Several isolation techniques were investigated:

- Immuno Magnetic Separation (IMS). This technique involves conjugating the (oo)cysts with magnetic beads and using magnets to isolate the bead/oocyst complex from the rest of the matrix.
- Ether clarification. In this technique ether is added to a dilute sample of biosolid and after centrifugation a layer containing ether and unwanted organic mater is formed that can be discarded.
- Sucrose clarification. The sample is underlain with sucrose and centrifuged. Denser particles of debris are pelleted but lighter particles including (oo)cysts collect at the boundary between the sucrose and biosolid layers.
- Dual density sucrose clarification. The sample is underlain with two layers of sucrose of different densities. After centrifugation denser particles of debris are pelleted and lighter particles are trapped on the boundary between the lighter sucrose and the biosolid layers. Debris of intermediate density including (oo)cysts are trapped at the boundary between the two sucrose layers.
- Sedimentation. The sample is mixed with a surfactant such as tween 20 and left to settle for 1 hour. Larger particles of debris fall out of suspension and the supernatant including the (oo)cysts is collected.

The recovery efficiency of each technique varied greatly depending on the biosolid matrix and mass of sample analysed. However it was found that sucrose clarification or sedimentation followed by IMS had the best associated recovery efficiencies (30-50%) and that sucrose clarification, in particular dual density clarification, was most effective at removing debris from the samples prior to processing using IMS.

Two techniques for determining the viability of (oo)cysts were also evaluated, these were:

- DAPI/PI staining. This technique involved staining with nucleogenic dyes one of which is excluded from live (oo)cysts.
- In vitro excystation. Isolated (oo)cysts were placed in a culture media conducive to excystation, the percentage excysted oocyst was taken as a measure of viability.

Both assays of viability could be used in conjunction with the isolation techniques described above. Further more both gave a similar measure of viability when applied to oocysts isolated from a biosolid matrix. However, in vitro excystation, due to its reliance on light microscopy proved impractical when analysing samples containing a high level of debris.

The sucrose clarification and DAPI/PI staining techniques were used to characterise *Cryptosporidium* numbers and viability in samples taken from a series of bench scale anaerobic digesters over a period of days. The results of this study are summarised in Figure 1 and 2.

In conclusion the recovery efficiency for isolation and purification of Cryptosporidium oocysts and Giardia cysts varied greatly depending on the biosolid matrix and mass of sample analysed. Two techniques were demonstrated as suitable for a range of biosolids, sucrose filtration and sedimentation both followed by IMS. These techniques were compatible with viability assays.

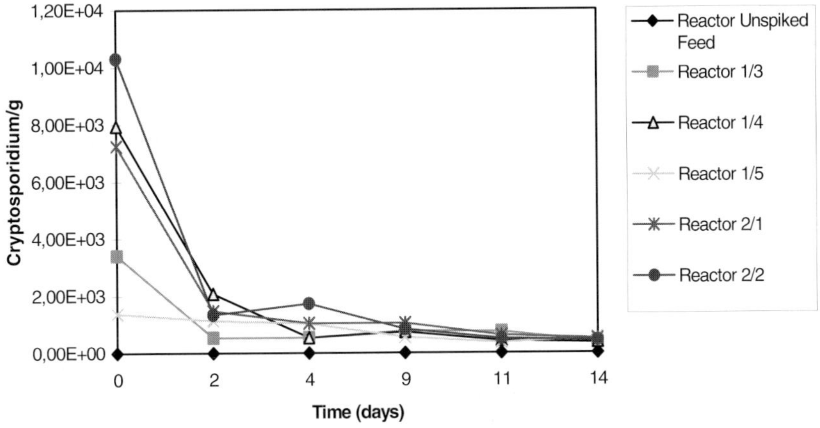

Figure 1. Effect of anaerobic digestion on Cryptosporidium *numbers.*

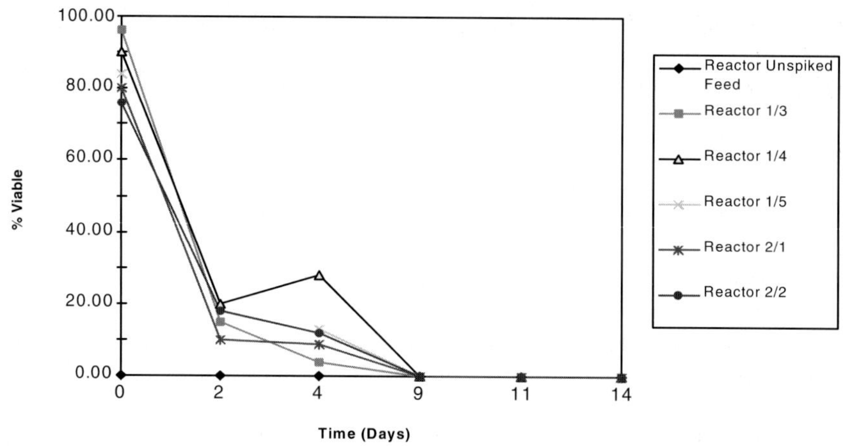

Figure 2. Effect of anaerobic digestion on Cryptosporidium *viability.*

On-farm hazard analysis critical control points (HACCP) for improving the safety of lamb and beef

David Edwards
Royal Veterinary College, London, UK

Introduction

The HACCP system has become the recognised standard for food safety quality assurance worldwide. Farms not only need to improve their role in food safety but provide documentation that verifies what steps are taken and what controls are in place. For such purposes the implementation of the HACCP system on the farm has real potential to improve public health and animal health and welfare. The international use of the HACCP system by food manufacturers and producers is seen as a logical progression enabling harmonisation of international food safety regulations and the removal of non-tariff barriers to trade arising from food safety (Caswell and Hooker, 1996).

This is a report of three farms, two cattle and one sheep, that were audited so that HACCP plans could be drawn up for the prevention of chemical residues, microbiological foodborne pathogens (salmonellae) and parasites resulting in gross lesions (e.g. *C. bovis* and *E. granulosus*). These farms serve as examples of commercial sheep and cattle farm environments distinct from intensive poultry or pig production systems, although more extensive system exist elsewhere in the UK. The aim of this study was to give practical examples of the strengths and weaknesses of the HACCP system for providing on-farm food safety assurance for beef and lamb production in the UK.

Materials and Method

Three farms that produced livestock destined for eventual slaughter were visited and a HACCP assessment carried out. The farmer/manager was interviewed and the farm inspected during a farm visit lasting approximately two hours. With the information gathered from the visit a flow diagram of the farm and a HACCP decision tree with control points (CP) and critical control points (CCP) was produced for salmonellae, meat residues and parasitic cysts. Then at a later date the HACCP team was consulted to verify the HACCP flow diagram and decision tree, checking for accuracy and further comment. With all three farms changes were made to the HACCP plan on the second visit. For each farm a HACCP team reviewed the farm's practices, identified potential farm level hazards to food safety and considered what CPs and CCPs were in place or should be introduced. The HACCP team consisted of the auditor (D. Edwards), the farm manager or farmer and on two farms a senior member of staff. The main reason for the small size of the HACCP teams was the low number of staff on the farms, which ranged from one to three.

Results

Farm 1 Hertfordshire Dairy Herd Producing Beef Cross Calves

This was an 80 cow dairy herd in Hertfordshire that produced beef breed cross calves for sale at market or to calf dealers in addition to Friesian heifers for herd replacements. The farm had a new large portal framed building with straw yards for the milking herd in winter, purpose built calving pens, a covered handling area with cattle crush, the milking parlour, bulk tank and office. The calves were housed in a section of older building in calf pens adjacent to workshops and a feed store. The farm had permanent pasture and arable rotations with temporary grass leys and maize for silage. This farm included sheep, beef cattle and pig enterprises that were managed separately from the dairy herd. The cows in the dairy herd were generally in good health with milk production averaging over 7,000 litres per cow. The dairy herd calved all year round. There was no history of salmonellosis on Farm 1.

Farm 2 Lowland Sheep Flock in Wiltshire

This was a lowland sheep farm in Wiltshire with a flock of Scottish Blackface cross Rouge ewes producing Suffolk cross lambs for direct sale to abattoir. The flock produced lambs in March and April with March born lambs marketed in the summer and some of the April born lambs being stored over winter. Farm 2 did not have a history of *Salmonella* and the ewe flock was not affected by infectious abortion. The farm was on a tenanted holding with buildings and grazing land in several disconnected parts each several miles from the other. The lambing area was in a large portal framed building surrounded by permanent pasture. Other grazing land included marshy flood meadow and temporary leys over chalky soil. Purpose built sheep handling facilities on concrete were next to the best grazing land adjacent to neighbouring farms silage clamps and a large portal framed building, which housed goats.

Farm 3 Lowland Suckler Beef Farm in Hertfordshire

This farm had a 120-cow suckler beef herd and finished the beef cattle intensively on a cereal-based diet. The breed of cows was Devon or Devon cross. The slaughter cattle were sent directly to an abattoir from the farm. The farm was a Farm Assured British Beef and Lamb (FABBL) member. The main farmyard included a series of portal-framed buildings and old brick cattle barns used for housing the cattle. Sealed feed bins were adjacent to the largest portal-framed building. Other buildings were used for the farm's arable enterprise. Approximately three miles from the main farmyard was a large old portal-framed building and an old brick barn that were also used for cattle housing. The cattle grazed on temporary leys that were part of an arable rotation. Apart from some poultry that had free range of the farmyard there were no other livestock on the farm. A major motorway ran through the centre of the farm under which there was a large access tunnel. The Herd calved in the autumn. There was no history of salmonellosis on Farm 3.

All three farms would need to make changes in order to implement control points or critical control points as part of the HACCP system.

- Netting over the buildings to prevent access by birds. This would require installing one of the systems available for lifting the netting to enable machinery entry into the building.
- Use of professional pest control to control rats and mice. Only Farm 1 used a professional pest controller on a routine basis.
- Replacement of unsuitable buildings on Farm 3. A highly expensive proposition.
- Farm 3 needed to rest the buildings, after mucking out, before the introduction of new livestock.
- Use of a livestock medicine book. Farm3 only obtained one at the time of the HACCP audit in order to qualify for FABBL status.
- On farms 2 and 3 buildings did not have a floor surface that could be cleaned effectively. Concrete flooring would be easier to clean, however, concrete can lead to problems such as poor drainage and can be slippery for livestock if improperly surfaced.
- The use of disinfectant boot dips and provision of clean protective clothing is something that all three farms neglected to do. The farmer on Farm 2 was enthusiastic about the use of disinfectant boot dips and protective clothing due to his concern over the introduction of organisms capable of causing abortion in the ewe flock which was abortion free. With farm 1 the large number of visitors to the dairy unit would appear to make the provision of at least disinfectant boot dips a prudent measure against disease introduction.
- Water troughs on farms 1 and 2 were visibly contaminated with silt and debris. On Farm 1 calves feed and water buckets were visibly fouled or contaminated. Both of these farms needed to pay more attention to the cleanliness of water troughs and buckets otherwise they will remain a likely source of infection with pathogenic microorganisms and residues.
- Farms 1and 2 needed to improve their feed storage practices. With Farm 2 requiring a dedicated feed store and Farm 1 needing to ensure that the calf feed was not left near the calf pens and that the calf feed store was no longer used for storing seed or other agricultural material.

Conclusions

It was clear from this study that the implementation of true CCPs to control microbial pathogens on these beef and sheep farms was largely impractical and, with many of the controls points, more quantitative data is needed. However in the absence of such quantitative data farm level HACCP does provide a verifiable means of ensuring a high standard of hygiene, enabling proactive management to tackle food safety. For the prevention and control of chemical residues and zoonotic parasite infections farm level HACCP has far more potential due to the readily identifiable CCPs that can be easily monitored and corrective action taken if critical limits are exceeded.

Caswell JA and Hooker N H., 1996. HACCP as an international trade standard. *American Journal of Agricultural Economics*, 78: 775-779.

The disclosure of tuberculous lesions at slaughter in cattle from attested herds in CO. Monaghan in 1995

Paul White
Vet. College Dublin, Tubercul.Invest.Unit, Dublin, IRL

Introduction

In all animals where a suspected lesion is detected at slaughter in an animal from an attested herd, (i.e. a herd not subject to movement restrictions for tuberculosis) that lesion is submitted to the Department of Agriculture, Food and Forestry's Veterinary Research Laboratory for histological and/or microbiological examination. Where a tuberculous lesion is confirmed, the herd of origin is traced, and listed for a factory lesion retest.

During the period 1988 to 1995, the confirmed lesion rate (as a percentage of the kill) for animals slaughtered at export meat plants increased from 0.146% to 0.202% (Costello et al 1995). Over the same period, this has resulted in an increase in the annual number of factory lesion re-tests nationally from 1,138 to 2,172. In County Monaghan the figure for the corresponding period increased from 48 to 155.

Materials and Methods

A questionnaire was used for systematic recording of data from herd files in Co. Monaghan for herds that underwent a factory lesion test in 1995Where a factory lesion retest resulted from the disclosure of more than one animal with confirmed lesion(s), a separate questionnaire was completed for each animal. Where animals had been purchased form another herd within Co. Monaghan since the last test, a trace-back was carried out. The results were computerised using the Microsoft Access™ Version 2.0 data analysis package.

Results

Table 1. Summary of Herds.

Description	Value
Number of herds with a factory lesion retest in 1995	155
Number of herds recorded and analysed	143 (92% of herds)
Number of animals with confirmed lesions	198 (1.38 per herd)

Of the 198 animals with a confirmed factory lesion, 174 (88%) had an identity card at the time of slaughter, indicating that the animal and herd in question had an attested status. The other 24 animals were slaughtered under movement permit.

Table 2. The number of confirmed factory lesion animals sent on either permit or identity card in 1995 which were inconclusive to severe interpretation on the previous tuberculin test.

Movement Status	Animals Inconclusive	Animals Tested	Percentage	Total Animals
Identity Card	2	114	2%	174
Permit	7	21	33%	24
Total	9	135	4.5%	198

*Animals are recorded as inconclusive to the severe interpretation of the tuberculin test when the bovine reaction is 2mm smaller, 1mm smaller or equal to the avian reaction, and there are no clinical signs at the bovine site.

Information about the tuberculin test preceding the slaughter of an animal was not available for 88% of purchased animals. Of the 135 animals with a result available, 2 (2%) of the identity card animals and 21 (33%) of permit animals were inconclusive to the *severe interpretation of the test (Table 4). Under Department policy, animals inconclusive to standard interpretation were all taken as reactors in 1995 and so these herds were omitted from this study.

Table 3. The movement history of animals with confirmed factory lesions.

Movement History	Number of animals	Percentage
Purchased since previous test	50	29
Total (I.D. card available)	174	

Table 4. The number of 'identity card' lesion animals that were purchased since the previous herd test grouped by animal class.

Animal Class	Animals Purchased	Total	Percentage
Cow	8	58	14
Heifer	9	37	24
Steer	32	73	44
Bull	1	2	50
Not specified	0	4	0
All classes	50	174	29

Table 5. The association between the proportion of herds which showed reactors at subsequent tests and the purchase of one or more (identity card) animals that gave rise to confirmed lesions.

	Outcome of subsequent testing			
Animal movement history	Herds with reactors	Herds with reactors reactors	Total Herds	% of herds showing reactors
Animal(s) purchased	6	40	46	13
No animals purchased	50	105	155	32
Total Herds	56	145	201	28

Discussion

The finding of tuberculous lesion(s) at post-mortem examination in cattle slaughtered from attested herds is retrospective evidence of infection with *Mycobacterium bovis*. In many cases it is difficult to determine when an animal became infected. While the sensitivity of the test may be as high as 95% when carried out under ideal conditions (Monaghan, 1994), the test as applied in the field is subject to a number of constraints which may result in a lower sensitivity. It should be noted that the severe interpretation of the tuberculin test was used in all tests carried out during 1994 and 1995 in Co. Monaghan. Under these circumstances fewer animals would have been falsely classified as negative to the test resulting in a higher sensitivity than if the standard mode of interpretation had been used.

Of the confirmed factory lesion animals, 88% were free to trade (i.e. they were accompanied by an identity card). Of these, 29% had been purchased since the previous test. The percentage of animals purchased varied according to animal class. Of the animals showing confirmed factory lesions, a higher than average percentage of steers (44%) and a lower than average percentage of cows (14%) had been purchased since the last herd test.

Where no confirmed factory lesion animals had been purchased since the last test, the risk of the herd showing subsequent test reactors was higher ($P<0.05$) than where the animal(s) had been purchased. (Table 5)

These findings call for further investigations into the possible sources of bovine tuberculosis in breakdown herds particularly where steers have been purchased and no reactors occur at factory lesion retest. Further case control studies are ongoing in this regard.

References

Monaghan M.L. Doherty M.L. Collins J.D. Kazda J.F. Quinn P.J., 1994. The tuberculin test. Veterinary Microbiology, 40, 111-124

Costello E. Dolan L.A. and Quigley F., 1995. Confirmed Tuberculous lesions in non-tuberulin-reactor cattle slaughtered at export licenced meat plants in the Republic of Ireland (1988-1995) Selected Papers of the Tuberculosis Investigation Unit. University College Dublin 1995

Animal health & pre-harvest food safety in organic dairy farming

C. Micheál and M.V.B. O'Mahony
Faculty of Veterinary Medicine, University College Dublin, Ireland, e-Mail:-Micheal.OMahony@ucd.ie

Background

Organic farming is a large and expanding sector of European agriculture. The movement is driven by increasing consumer concerns over food safety, animal welfare, and environmental issues.

Rational scientific investigations of organic production are scarce in the literature. Some studies exist on cow disease incidences (Weller and Cooper 2000, Hovi and Roderick 2000), and production levels (Resken and others, 2000) on organic farms, but extrapolation to food related aspects has been minimal (Busato and others 1999, Weller and Davies 1998). Parasite control strategies have been described by Thamsburg (1999).

Organic production is a pre-existing example of proposed vertically integrated food quality assurance schemes. The putative superiority assigned by consumers to organic produce, manifest in the supplement paid at retail level, underline consumer perceptions of deficiencies in conventional food production. Any scientific basis for this notional superiority has, at best, not been established. All of the existing pre-harvest food safety legislation, such as minimal withdrawal periods, have a sound scientific basis and any deviations on the side of caution under organic regulations would appear arbitrary. Antimicrobial resistance may prove different, as current legislation pertains to chemical residues rather than to residual effects on microbial populations, Also, any effort to minimise antimicrobial usage in the pre-harvest phase is commendable. Some practices, justifiable by their quasi-therapeutic nature and economic viability, such as blanket long-acting intramammary antimicrobial administration or frequent anthelmintic administration to young grazers, have attained widespread penetration into conventional agriculture. Real comsumer concerns exist about this xenobiotic dependent food production.

Aims

1. To establish baseline data on disease incidence on Irish organic dairy farms
2. To compare these data with data from matched conventional Irish dairy herds.
3. To measure the economic cost of disease and analyze the cost/benefit of any intervention.
4. Using the newly established institutional organic dairy farm, to design and implement treatment protocols for individual diseased animals in organic herds.
5. Using the newly established institutional organic dairy farm, to design and implement disease prevention protocols for organic dairy herds.
6. To assess indicators of animal welfare in organic farming.

7. To study the selected immune responses of animals at the institutional organic farm
8. To study antimicrobial resistance patterns in organic dairy farms
9. To investigate implications of manure composting on animal health and pre-harvest food safety in organic farms.
10. To critically examine regulations governing organic farming, their implementation, and implications from animal health and pre-harvest food safety perspectives.

Materials and Methods

A 40 cow organic herd has been established at Johnstown Castle, Wexford Ireland. A conventional herd already exists at this location and will serve as a control for our project. The computerised herd management program dairyCHAMP has been installed at both of these locations. Every animal health event is recorded on this program with detailed information concerning diagnosis, treatment, duration of illness, outcome, and withdrawal periods. Production data such as milk output and fertility events will also be gathered. Inputs, such as feeds will also be monitored.

Fundamental differences between conventional and organic production have been identified and will be actively investigated. Parasite burdens of first and second season grazers will be analysed frequently. Individual cow milk somatic cell counting and microbiological milk analysis will be performed at drying-off. Micronutrient levels and metabolic indices will be analysed from a sample of cows at critical points in the production cycle.

To remove any bias of institutional herds practices, and to increase sample size, the study will be broadened to encompass many of the commercial Irish organic dairy farms. These farms will be contacted and matched to conventional herds in their area. Animal health data will be gathered from these herds.

Antimicrobial resistance of pathogenic bacteria collected from diseased animals throughout the study will be investigated. In vitro investigation of selected animals' immune response will be undertaken. The effects of manure composting on survival, and hence annual recycling, of faecal microorganisms, e.g. verocytotoxigenic *E. coli* (Fukushima and others 1999)will also be studied.

Progress

A comprehensive literature review has been performed. Critical examination of organic regulations as they pertain to animal health and pre-harvest food safety has been performed. Some of the hypothesis generated for analysis are listed in the following table.

Organic Standard	Pre-Harvest Implication
Minimal conventional antimicrobial input	Antimicrobial resistance transfer risk reduced
Non-conventional therapy embraced	Absence of scientific residue data
Managemental parasite control	Residue risk reduced
Reduced non-forage feed input	Food closely reflects soil constituents
Absence of artificial fertiliser	Natural ecological cycles prosper
Absence of blanket dry cow therapy	Bulk milk somatic cell counts higher
Manure composted	Annual pathogen recycling reduced

The data collection system has been in place since January 2000 and a full production cycle will be allowed to elapse before data analysis. Commercial organic producers have been contacted and the suitability of their records will be ascertained.

References

Busato, A., Ttraschel, P., Schalibaum, M. and Blum, J.W., 2000. Udder health and risk factors for subclinical mastitis in organic dairy farms in Switzerland. Preventative Veterinary Medicine 44, 205-220

Fukushima, H., Hoshina, K. and Gomoyda, M., 1999. Long term survival of Shiga-toxin producing *Escherichia coli* 026,0111 and 0157 in bovine faeces. Applied and Environmental Microbiology 65, 5177-5181

Hovi, M. & Roderick, S., 2000. Mastitis and mastitis control strategies in organic milk. Cattle Practice 8, 259-264

Resken, O., Tverdal, A and Ropstad, E., 1999. A comparative study of reproductive performance in organic and conventional husbandry. Journal of Dairy Science 82, 2605-2610

Thamsburg, S.M., Roepstorff, A. and Larsen, M., 1999. Integrated and biological control of parasites in organic and conventional production systems. Veterinary Parasitology 84, 169-186

Weller, R.F. and Cooper, A., 2000. Health status of dairy herds in organic dairy farming. Veterinary Record 146, 80-81

Weller, R.F. and Davies, D.W.R., 1998. Somatic cell counts and incidence of clinical mastitis in organic milk production. Veterinary Record 143, 365-366

The biocontrol of *E. coli* 0157:H7 at farm level

P. McGee[1], J.J. Sheridan[1], B. Earley[2] and N. Leonard[3]
[1]Teagasc, The National Food Centre, Castleknock, Dublin 15. [2]Teagasc, Grange Research Centre, Dunsany, Co. Meath and [3]Faculty of Veterinary Medicine, University College Dublin, Ballsbridge, Dublin 4.

Introduction

Since its recognition as a human pathogen in 1982, *Escherichia coli* 0157:H7 has emerged as a considerable challenge to the cattle industry. The organism can colonise the gut of ruminants, as a commensal, and is then excreted in the faeces. A diversity in prevalence of *E. coli* 0157:H7 has been demonstrated between and within herds, suggesting that farm management variables may play a role in the occurrence of the pathogen. Studies investigating the persistence of the organism in animal drinking water, suggest that water may play an important role in the infection, and resultant horizontal transmission of the organism across a herd.

McGee *et al.* (1999) suggested the possible role of bacterial inhibitors as a form of biocontrol of *E. coli* 0157:H7 in farm water. Considerable success has been achieved using certain bacteria, particularly fluorescent *Pseudomonas* species in the biocontrol of important fungal plant diseases. However the possible on-farm biocontrol of important bacterial pathogens such as *E. coli* 0157:H7, has received little attention.

The objective of the present study was to investigate the putative antimicrobial activity of bacteria isolated from both soil and vegetation, against *E. coli* 0157:H7. The effect of such isolates in a water environment under controlled laboratory conditions was also examined.

Materials and methods

From an area of grassland at the National Food Centre, vegetation and soil samples were removed. Samples were then stomached in a solution of maximum recovery diluent (MRD). From a dilution series aliquots were plated onto Plate count agar (PCA) and incubated at 10°C for 72h and 25°C for 48h. Following incubation isolates were chosen based on morphological differences, and transferred twice in 30ml of Tyrptone soya broth (TSB). Antimicrobial activity was determined by spot inoculating 10 µl of culture at three equidistant points on Mueller Hinton agar (MH) and Potato Dextrose agar (PDA) and reincubating at 10°C for 72h and 25°C for 48h. MH agar is used for the detection of siderophores, while PDA is used for the detection of antibiotic-type substances. Plates were then overlayed with a lawn of sloppy agar containing 1ml of 4 $\log_{10}$ *E. coli* 0157:H7 suspension, and incubated at 37°C for 48h. If inhibitory substances were produced, an inhibition zone appeared on the indicator lawn. The *E. coli* 0157:H7 used, a non toxigenic strain (NCTC 12900), was made resistant to streptomycin sulphate and naldixic acid.

The antimicrobial activity of a *Pseudomonas fluorescens* strain isolated from soil, which showed inhibition on both agar types, was investigated in an untreated farm water source. *E. coli* 0157:H7 and the *Pseudomonas* strain were both added at a level of 3 $\log_{10}$ cfu/ml, to 500ml volumes of the water, and stored at 25°C in the laboratory. Samples were removed and filtered through a 0.2 (m-pore-size filter. The filter was then suspended in MRD and serial dilutions performed. Aliquots from the dilution series were plated on Sorbitol MacConkey Agar (SMAC) and also onto Tryptone Soya Agar (TSA), overpoured with SMAC. The latter process is for the recovery of stressed or injured cells.

Results

From a total of 302 isolates chosen, 20 exhibited antimicrobial activity towards *E. coli* 0157:H7. The majority of the strains isolated were *Pseudomonas fluorescens*. The inhibition from seven of these strains was iron regulated indicating that it was probably due to siderophore production. All other strains exhibited activity on PDA agar, indicating the role of antibiotics. When the activity of one of the strains was examined in farm water at 25°C, it was shown to eliminate *E. coli* 0157:H7 within three days, whereas in the control treatment, a 0.5 log increase in *E. coli* 0157:H7 numbers was observed (Fig. 1).

Conclusion

The results of this study indicate that organisms, mainly *Pseudomonas* species which possess antimicrobial properties towards *E. coli* 0157:H7 exist as part of the natural flora in soil and vegetative environments. The elimination of *E. coli* 0157:H7 in a farm water source within three days, by one of the organisms isolated, is very promising. However, work with these organisms is at a preliminary stage. Further work will address the influential role of the global regulator GacA/GacS system which controls secondary metabolite production.

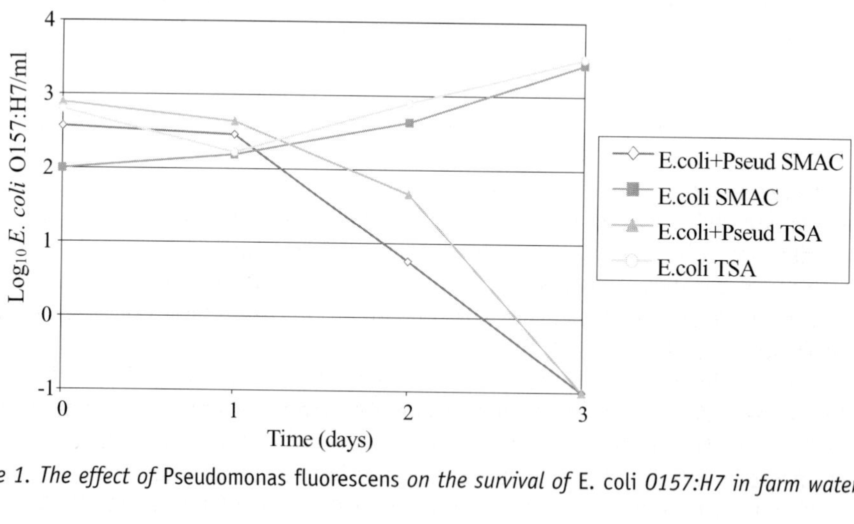

Figure 1. The effect of Pseudomonas fluorescens *on the survival of* E. coli *0157:H7 in farm water.*

Alteration of intestinal cell monolayers by pathogenic *Escherichia coli* : importance of the tracer used to measure paracellular permeability

A.B. Blanc-Potard, S. Etenna and A.L. Servin
INSERM U 510, Faculté de Pharmacie Paris XI, 92296 Châtenay-Malabry, France.

Food-borne pathogens EPEC and EHEC alter the permeability of intestinal epitheliums by increasing the paracellular permeability. This increased paracellular permeability, which is due to the opening of the tight junctions of polarized intestinal cells, can be measured by using radioactive probes. We demonstrate that ^{14}C mannitol should be used instead of ^{3}H mannitol, which can give false positive results. Our results are important for studies that would measure paracellular permeability to investigate the extent of virulence of food-borne pathogens.

The tight junction of intestinal cells forms a regulated barrier in the paracellular space, restricting the movement of molecules as small as ions across cell sheets. Several enteric bacterial pathogens, including *Vibrio cholerae, Helicobacter pylori*, pathogenic *Escherichia coli* (enteropathogenic or EPEC, enterohemorragic or EHEC and diffusely adhering or DAEC), cause an increase in the paracellular permeability of the intestinal epithelium by altering the function of the tight junctions (Papini et al., 1998; Fasano et al., 1991; Philpott, 1996; Philpott, 1998; Peiffer et al., 2000). Alteration of the tight junctions can be monitored by measurement of the transepithelial resistance (TER). In addition, the passage of radioactive probes from the apical to the basal side of cell monolayers has been widely used to measure the paracellular permeability. With this method, the extent of the alteration can be estimated by using probes of different molecular weight. Radioactive mannitol (either ^{3}H labeled or ^{14}C labeled) has been used as a probe to measure paracellular permeability after infection by pathogenic *E. coli*.

EPEC strain JNP15/pMar7 (Jerse et al., 1990) and DAEC strain C1845 (Bilge et al., 1980) cause an increase in the paracellular permeability to radioactive mannitol [2-^{3}H] with polarized intestinal Caco-2/TC7 cells grown as monolayers on filters (Table 1). No increase is observed with the laboratory strain HB101, which is usually used as control in studies on the effect of pathogenic *E. coli* on paracellular permeability. Surprisingly, we showed that other non pathogenic laboratory *E. coli* strains, DH5□ and MG1655, induced an increased paracellular permeability, to a similar extent than the one observed with pathogenic strains (Table 1).
The only strain which does not increase the paracellular permeability, HB101, harbors a mutation in the *mtlA* gene, which encodes an uptake system for extracellular mannitol. We hypothesized that the metabolism of the radioactive mannitol by bacteria that are retained on the filters might interfere with our assay. To test this hypothesis, we have complemented the HB101 strain with a plasmid encoding the *mtlA* gene (pGJ9). As shown in Table 1, the complemented strain induced an increase in the paracellular permeability

to mannitol [2-^{3}H] similar to the one observed with pathogenic *E. coli* strains. Taken together, these data suggest that false-positive results are obtained due to the metabolism of the mannitol [2-^{3}H] probe by bacteria.

Table 1. Paracellular permeability to mannitol [2-^{3}H] of Caco-2/TC7 monolayers.

Strain	Percentage of radioactivity in the basal compartment
Non infected	0.4%
JNP15/pMar7 (EPEC)	38%
C1845 (DAEC)	35%
HB101	0.6%
DH5□	30%
MG1655	28%
HB101/pGJ9 (*mtlA*$^{+}$)	36%

The radioactive hydrogen of the [2-^{3}H] mannitol can be lost during the conversion of mannitol into fructose-6-P. In contrast, the radioactive carbone of [1-^{14}C] mannitol is retained. Therefore, we have postulated that ^{14}C mannitol should be used rather than ^{3}H mannitol. After infection of intestinal monolayers by a DAEC strain, we have studied by thin layer chromatography the radioactive molecule present in the basal compartment. We have shown that a radioactive molecule of the size of the mannitol is present when [1-^{14}C] mannitol is used but not when [2-^{3}H] mannitol is used. This result demonstrates that [1-^{14}C] mannitol should be preferentially used in such studies.

Conclusion

One of the virulence traits of EPEC and EHEC strains is to increase paracellular permeability of intestinal cell monolayers. This trait can be used to investigate the extent of virulence of clinical isolates or food contaminants. We showed that [2-^{3}H] mannitol should not be used for this purpose because it leads to false-positive results. Our data indicate that [1-^{14}C] mannitol should be used instead.

References

Bilge, S.S., Clausen, C.R., Lau, W. and Moseley, S.L., 1989. Molecular characterization of a fimbrial adhesin, F1845, mediating diffuse adherence of diarrhea-associated *Escherichia coli* to HEp-2 cells. J. Bacteriol. 171, 4281-9.

Fasano, A., Baudry, B., Pumplin, D.W., Wasserman, S.S., Tall, B.D., Ketley, J.M. and Kaper, J.B., 1991. *Vibrio cholerae* produces a second enterotoxin, which affects intestinal tight junctions. Proc Natl Acad Sci U S A. 88, 5242-6.

Jerse, A.E., Yu, J., Tall, B.D. and Kaper, J.B., 1990. A genetic locus of enteropathogenic *Escherichia coli* necessary for the production of attaching and effacing lesions on tissue culture cells. Proc Natl Acad Sci U S A. 87, 7839-43.

Papini, E., Satin, B., Norais, N., de Bernard, M., Telford, J.L., Rappuoli, R. and Montecucco, C., 1998. Selective increase of the permeability of polarized epithelial cell monolayers by *Helicobacter pylori* vacuolating toxin. J Clin Invest. 102, 813-20

Peiffer, I., Blanc-Potard, A.B., Bernet-Camard, M.F., Guignot, J., Barbat, A. and Servin, A.L., 2000. Afa/Dr diffusely adhering *Escherichia coli* C1845 infection promotes selective injuries in the junctional domain of polarized human intestinal Caco-2/TC7 cells. Infect Immun. 68, 3431-42.

Philpott, D.J., McKay, D.M., Sherman, P.M. and Perdue, M.H., 1996. Infection of T84 cells with enteropathogenic *Escherichia coli* alters barrier and transport functions. Am J Physiol. 270, G634-45.

Philpott, D.J., McKay, D.M., Mak, W., Perdue, M.H. and Sherman, P.M., 1998. Signal transduction pathways involved in enterohemorrhagic *Escherichia coli*-induced alterations in T84 epithelial permeability. Infect Immun. 66, 1680-7.

Epidemiologic study on post mortem findings recorded in an Italian abattoir for sheep

D. Ranucci and S. Serra
Fac.Vet.Med.,Univ.degli Studi di Perugia

Introduction

The post mortem inspection of slaughter animals needs to be modernised and practices which are irrelevant to guarantee safe meat should be reduced or eliminated. An approach based on the risk analysis could be of great interest. The availability of sufficient information about the most common diseases which can be detected in a specific animal species slaughtered in a well defined context is of utmost importance in order to adopt a correct post mortem inspection procedure.
The results of a survey carried out in an abattoir for sheep are reported in this study.

Materials and Methods

The post-mortem inspection findings of 23,395 lambs and 2887 sheep were collected during a period of 3 months in an industrial abattoir for sheep. A large number of the animals were imported from various European countries (Table 1).
The distribution of the pathological findings detected in the respiratory system and liver was compared with the category and the country of origin of the animals.

Table 1. Numbers of slaughtered animals.

Country		Total examined animals
SPAIN	Lamb	8058
	Sheep	2500
U.K.	Lamb	758
	Sheep	50
FRANCE	Lamb	1217
	Sheep	1889
HUNGARY	Lamb	10,229
	Sheep	0
ITALY	Lamb	9753
	Sheep	518

Results

The results of the observed findings are shown in figures 1,2,3,4, for the respiratory system and the liver, respectively. Lesions observed in other organs were so rare and statistically irrelevant that they were not considered in this study. The enzootic pneumonia is the most frequently detected finding in lambs while the parasitic pneumonia and distomatosis are the most frequently observed findings in the adults. Our results agree with the data reported by other authors(Green et al, 1995; Builljihad et al., 1995; Severini et al., 1995). Echinococcus cysts were present in a large number of sheep coming from the Mediterranean region.

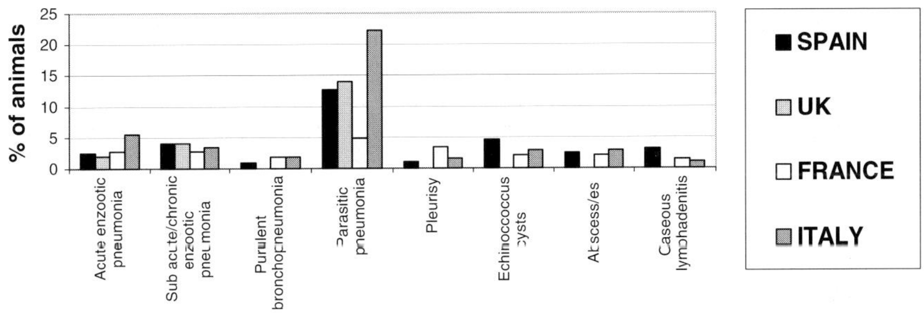

Figure 1. Respiratory system findings in sheep derived from different countries (% of animal with lesions for each Country).

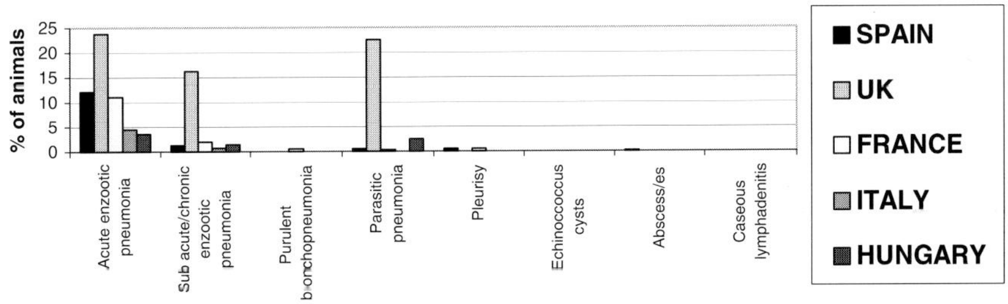

Figure 2. Respiratory system findings in lambs derived from different countries (% animals with lesions for each Country).

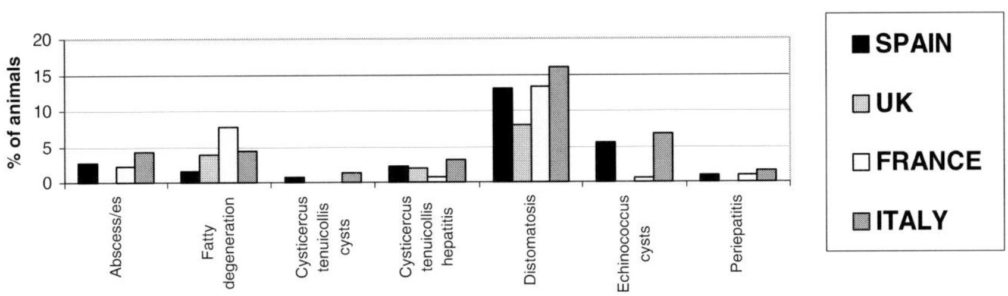

Figure 3. Liver findings in sheep derived from different countries (% animals with lesions for each Country).

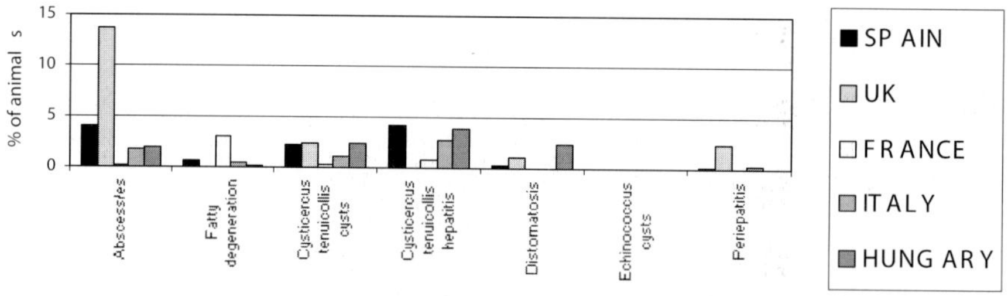

Figure 4. Liver findings in lambs from different countries (% animals with lesions for each Country).

Conclusions

The majority of the findings which have been observed are of limited importance for human health even though the involved organs are not suitable for human consumption. However, several lesions must be carefully checked and put under control (caseous lymphadenitis, echinococcosis, puruent lesions) because of their importance to the assessment of public health. When acute pneumonia is detected in animals which have been stressed it must be accurately evaluated, in order to exclude that bacteriemia took place. Recording data at post mortem inspection of sheep seems of great interest to identify the risk in relation to the age and origin of the animals. This could allow the Italian slaughter industry and the veterinary authority to improve the epidemiologic information about the health status of domestic and imported animals, to evaluate the risk for the human health, and to plan proper inspection measures at the abattoir. In this context, post mortem inspection should be a very useful tool in order to assure safe meat and also to record data about the pre-harvest health status of the slaughter animals.

References

Builljihad, M., Berrag, B. and Leipold, H.W., 1995. Journal of Veterinary Medicine. Series B, 42:9, 513-521.

Green, L.E., Berriatua, E., Cripps, P.J. and Morgan K.L., 1995. Preventive Veterinary Medicine, 22:1, 115-126.

Severini, M., Loschi, A.R., Aglietti, I. and Roncella, R., 1995. Atti S.I.S.Vet., 49,405-406.

Salmonella prevention in styrian parent flocks

H. Gruber and J. Köfer
Department Of Veterinary Administration i Styria, Zimmerplatzgasse 15, A-8010 Graz, Email: Post@Favw.Stmk.Gv.At

Salmonella prevention in Styrian parent flocks is based on extensive *Salmonella* monitoring through regular bacteriological examinations of drag and wipe swabs and on vaccination against *S. enteritidis*. During the period under investigation (January 1995 - December 1999) the number of Salmonella-positive flocks decreased from 16 in 1995 to 4 in 1997, 2 in 1998 and 1 in 1999. Since vaccination of all parent flocks against *S. enteritidis*, this pathogen has not been detected in any of the flocks monitored. The number of other *Salmonella* isolates was also reduced considerably.

Introduction

Salmonella-induced infections in humans are almost always associated with the consumption of products originating from infected poultry flocks. Human *Salmonella* infection, in particular those caused by *S. enteritidis*, have experienced a steep increase in all industrialised countries since 1985. In 1997 the incidence of human salmonelloses amounted to 94.6 cases per 100,000 inhabitants (Thiel, 1998). The Styrian Salmonella monitoring and control programme comprises all stages of production, from parent flocks to the slaughterhouse, and is based on WHO recommendations (Wray and Davies, 1994).

Material and Methods

Monitoring programme

Approximately 120,000 animals in 22 parent flocks participate in the programme. The structural and hygienic conditions of the parent flock establishment are assessed on the basis of a check list. Samples are taken prior to the introduction of new chickens, on the day the chicks are delivered to the establishment, during the raising period at 7 days, 4, 7 and 16 weeks of age. Thereafter the owner of the animals and the veterinarian take wipe and drag swabs for examination at monthly intervals. Samples taken from the floor and equipment on the premises are pooled and analysed. 10 x 10 cm Regal swabs of the company Johnson & Johnson are used as drag and wipe swabs. The samples are pre-enriched in a peptone solution at 37°C for 24 hours. Subsequently 1 ml of the solution is transferred to a Müller-Kauffmann bouillon for selective enrichment and incubated at 43°C for 24 hours. After 24 hours of incubation at 43°C and another 24 hours of incubation at 37°C the bouillon is spread onto XLD agar.

Vaccination programme

The *Salmonella* vaccination programme for parent flocks was initiated in 1996. The parent animals are vaccinated with *Salmonella* vac T® and Talovac 109 SE® of co. TAD. The peroral administration of *Salmonella* vac T® with drinking water is performed on the 1st day and during the 7th week, respectively. Talovac 109 SE® is administered by subcutaneous injection into the neck during the 16th week after birth.

Results

In the period under investigation from January 1995 to December 1999 a total of 3,117 samples from parent flocks (drag and wipe swabs, chicken paper) were subjected to bacteriological examinations for *Salmonella*. The percentage of positive results decreased from 7.5% (950 samples examined) in 1995 to 3.9% (483 samples) in 1997, to 2.5% (558 samples) in 1998 and finally to 0.66% (610 samples) in 1999.

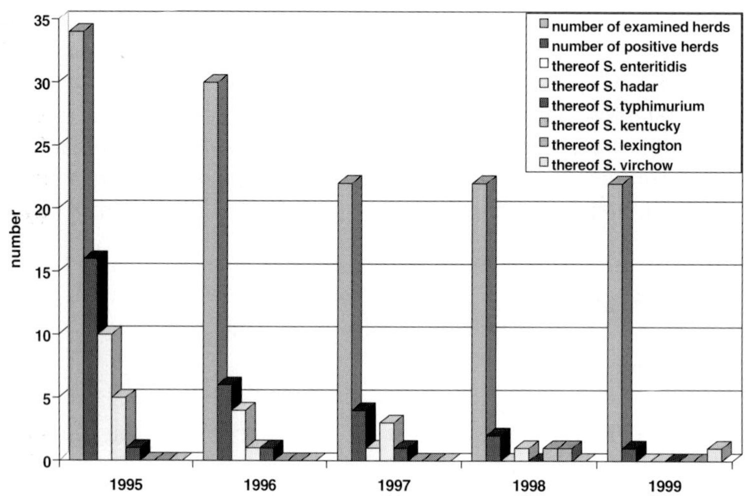

Figure 1. Salmonella *investigations in parent animal flocks.*

Discussion

The Styrian *Salmonella* monitoring programme involved 34 parent flock establishments in 1995, 30 establishments in 1996, and 22 in 1997, 1998 and 1999. In this time period the number of flocks in which bacteriological examinations rendered *Salmonella*-positive results, decreased from 16 in 1995 to 2 in 1998 and 1 in 1999, while the number of flocks that tested positive for *S. enteritidis* decreased from 10 in 1995 to 1 flock in 1997. In 1998 and 1999 none of the flocks was positive for *S. enteritidis* (Fig. 1). In May 1996 the first newly introduced parent flocks were vaccinated against *S. enteritidis*. The wide-spread sceptical attitude towards the vaccination, which was due to negative reports on vaccination-induced injuries, kept some farms from participating in the programme from the beginning. The *Salmonella* infections of 1996 and 1997 were restricted to unvaccinated flocks. From 1997 onwards all parent flocks are vaccinated against *S. enteritidis*. Since that time no incidence of *S. enteritidis* has been observed in any of the flocks monitored. In addition to this, the number of other *Salmonella* isolates has decreased considerably. In 1997 *S. hadar* was identified in three flocks, and more than once in only one flock. The two *S. typhimurium* isolates identified in 1996 and 1997 concerned the vaccine strain and were diagnosed only once in both cases. In 1998 one flock had been found to be infected with *S. hadar*. In another flock *S. kentucky* and *S. lexington* were identified once. In 1999 one flock had been found with *S. virchow*. Successful control of *Salmonella* in poultry requires consistent monitoring and control measures in the breeder flocks, which also includes the culling of positive flocks. The risk of having to cull an entire flock without compensation has prompted many operators of parent flock establishments to give up this line of business.

References

Thiel, W., 1998. Pers. Mitteilung.

Wray, C. and Davies, R. H., 1994. Guidelines on detection and monitoring of *Salmonella* infected poultry flocks with particular reference to *Salmonella enteritidis*. WHO, ZOON., 173.

Salmonella monitoring of styrian swine herds

J. Köfer, P. Pless and K. Fuchs
Department Of Veterinary Administration i Styria, Zimmerplatzgasse 15, A-8010 Graz, Email: Post@Favw.Stmk.Gv.At

Styria has implemented a Salmonella surveillance programme in pork production based on the serological monitoring of swine herds. 10,000 blood samples of slaughter pigs from 1,000 pig farms were investigated in 1999 using the SALMOTYPE® Meat Juice ELISA. 95.5 % of the herds tested negative, 4.2 % were suspected to be positive and 0.3 % produced positive results.

Introduction

The implementation of Directive 92/117/EEC ("Zoonoses Directive"), which requires measures to be taken for the control of "food-borne diseases", resulted in the setting-up of a Salmonella surveillance programme for pork production in Styria (Köfer et al., 2000). The Styrian system is similar to the "Danish Salmonella surveillance and control programme" (Mousing et al., 1997) and includes breeding and fattening farms, slaughterhouses and meat processing plants. As a first step, a comprehensive study was carried out in order to gain an overview of the Salmonella situation in Styrian slaughter pig herds. According to a two-stage stratified sample model, the prevalence of Salmonella in pork was estimated in the initial phase. 2.9% of animals from 4.3% of different farms were identified as latent carriers of Salmonella. By extrapolation from the sample to the population of Styrian slaughter pigs the estimated percentage $\hat{p}$ of Salmonella-positive pigs was calculated at 2.9 % [α = 5 %; CI(1.9 % $\leq$ p $\leq$ 4.2 %)] (Köfer et al., 1999). After bacteriological Salmonella screening, comprising determination of serovars and resistance patterns, a serological monitoring programme was established.

Material and methods

Sampling

Based on a representative sampling plan, a total of 1,000 stocks with at least 10,000 animals were tested in our monitoring programme. To calculate the number of stocks we used the formula proposed by Cannon and Roe (1982); $n \sim [1 - (1-\alpha)^{1/d}] * (N - ^{d/2}) +1$, with n = stock sample size, N = total number of stocks, d/N = expected prevalence, (= type I-error.
For N $\approx$ 40000, d/N = 0.3% and β = 5% we get a sample size of 1000 stock. To calculate the number of animals to be tested per stock we used the formula proposed by Cochran (1972):

$$\beta \le \frac{\binom{N-M}{n_S}}{\binom{N}{n_S}}$$

with n_S = sample size per stock, N = number of animals per stock, M = expected number of infected animals per stock; β = type II-error: For N > 50, M = 20 % and b < 10 % we get a sample size per stock of n_S=10. So we tested at least 10,000 animals in our screening programme.

Serological examination

The herd monitoring scheme makes use of the SALMOTYPE®-ELISA, a test kit based on the so-called Danish mix ELISA test (LPS-antigens 0:1, 4, 5, 6, 7, 12), licensed and commercially available from Labor-Diagnostik (Leipzig). The "cut-off value" (more than 40 % serum titre of antibodies against Salmonella for positive and >20≤40 % for doubtful samples) was specified according to the guidelines for a programme aimed at reducing Salmonella transmission into meat production by way of slaughter pigs as defined by the German Federal Ministry for Food, Agriculture and Forestry.

Results and discussion

About 10,000 pigs from 1,000 fattening farms were examined in the course of the serological meat juice monitoring programme performed in 1999. 95.5 % of the herds tested negative (level 1), 4.2 % were suspected to be positive (level 2) and 0.3 % (= 3 farms) produced positive results (level 3). According to these test results, herds are classified into three levels (1, 2 or 3). When a herd is classified as level 2 or 3 (moderate or high sero-prevalence) a follow-up programme with epidemiological investigations is carried out.

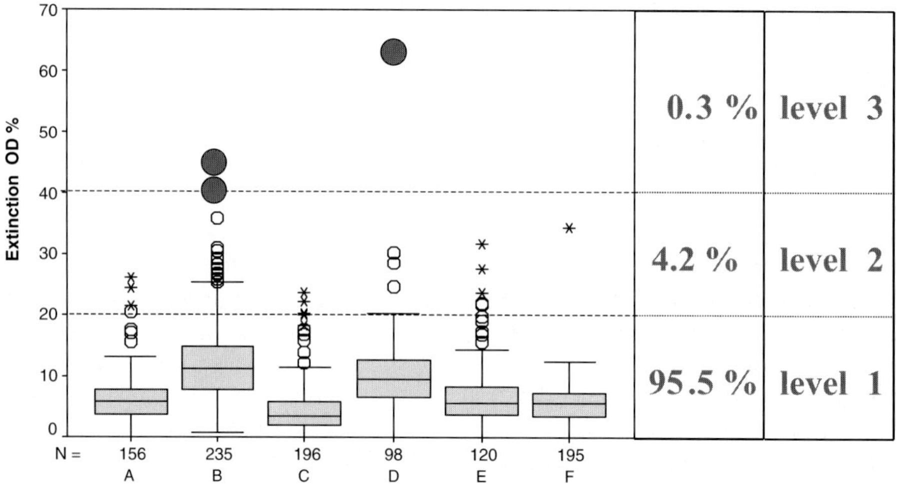

Figure 1. Salmonella *herd serology - meat juice ELISA monitoring 1999.*
Legend: N = number of farms investigated
A - F = slaughterhouses

References

Cannon, R.M. and R.T. Roe, 1982. Livestock Disease Surveys: A Flied Manual for Veterinarians. Australian Bureau of Animal Health. Canberra.

Cochran W.G., 1972. Stichprobenverfahren. Walter de Gruyter, Berlin, New York.

Official Journal of the European Communities. Council Directive 92/117/EEC of 17 December 1992 concerning measures for protection against specified zoonoses and specific zoonotic agents in animals and products of animal origin in order to prevent outbreaks of food-borne infections and intoxications, last amended by Directive 97/22/EC of 22 April 1997.

Mousing, J., P.T. Jensen, C. Holgaard, F. Bager, N. Feld, J.P. Nielsen and S. Beck-Nielsen, 1997. Nationwide *Salmonella* enterica surveillance and control in Danish slaughter swine herds. Prev. Vet. Med. 29, 247-261.

Köfer, J., P. Pless and K. Fuchs, 1999. Results of the Styrian Salmonella Surveillance in pigs. Proc. 3rd International Symposium on the Epidemiology and Control of *Salmonella* in Pork, Washington DC, p. 362-364.

Köfer, J., P. Pless, K. Fuchs and W. Thiel, 2000. Aufbau eines *Salmonella*-Überwachungsprogrammes für die steirische Schweinefleischerzeugung. Wien. Tierärztl. Mschr. 87, 14-20.

Occurrence of *Campylobacter* spp. in styrian broiler flocks

P. Pless, B. Ursinitsch and J. Köfer
Department Of Veterinary Administration i Styria, Zimmerplatzgasse 15, A-8010 Graz, Email: Post@Favw.Stmk.Gv.At

The increasing prevalence of *Campylobacter* infections in human medicine requires in-depth investigations into the occurrence and spread of these pathogens. 60.2% of the 201 Styrian broiler flocks examined produced positive results. *C. jejuni* was isolated from 64.5% of the positive flocks, *C. coli* was detected in 20.7%, both strains in 14% and *C. lari* in 0.8%. The determination of antibiotic resistance showed a further increase in resistance to quinolones, e.g. to ciprofloxacin (*C. jejuni* 51.8%, *C. coli* 58.6%).

Introduction

In human medicine, infections with pathogenic *Campylobacter* spp. have gradually caught up with *Salmonella* infections both in terms of significance and prevalence. Diseases caused by *Campylobacter* spp. have experienced a steep increase in the past few years and now rank second behind salmonellosis (Anonymous, 1999). Antibiotic-resistant *Campylobacter* strains result in additional problems. The Institute of Hygiene at the University of Graz, for instance, detected resistances to ciproflaxin in nearly 40% of human *Campylobacter* isolates (Feierl et al., 2000).
First surveys in Styrian broiler farms carried out in 1998 revealed 45.5% of the flocks to be *Campylobacter*-positive. The determination of resistance patterns showed that 48.8% of the isolated strains had developed a resistance to ciprofloxacin and nalidixic acid, 23.3% and 16.3% to tetracycline and ampicillin respectively. A comparison of the poultry strains with human strains from the same sampling period using pulsed-field gel electrophoresis yielded a genetic matching rate of 27% (Lehner et. al., 2000).

Material and Methods

From February to August 2000, a total of 201 broiler flocks from 73 farms (2-3 fattening periods) situated in the supply area of a large poultry slaughterhouse were examined for the presence of thermophilic *Campylobacter* spp. The samples were taken in the slaughter plant in the course of ante-mortem and meat inspections. Five intestines were taken from each flock and sent to the QA lab of the Department of Veterinary Administration in Graz for examination. In the diagnostic test faeces was streaked directly onto a CCDA medium (Oxoid SR 155). After an incubation period of 48 h at 42°C under microaerophilic conditions, suspected colonies were sub-cultured onto Columbia agar with 5% sheep blood (Becton-Dickinson). The results were confirmed by phase contrast microscopy as well as

catalase and oxidase tests. Differentiation between *C. jejuni*, *C. coli* and *C. lari* was achieved by hippurate hydrolysis and indoxyl acetate hydrolysis tests.
The antibiotic resistance to ampicillin, tetracycline, lincomycin, nalidixic acid, erythromycin and sulfamethoxazole-trimethoprim was tested using the disk diffusion tests. The inhibition zone diameters were evaluated on the basis of the NCCLS protocol (M2-A6, Vol19, No.1/as of Jan.1999).

Results and Discussion

Thermophilic *Campylobacter* spp. was detected in 121 (60.2%) of the 201 broiler flocks from 73 farms examined in the course of 2 to 3 fattening periods. *C. jejuni* was isolated from 64.5% of the positive flocks, *C. coli* was detected in 20.7%, both *C. jejuni* and *C. coli* in 14% and *C. lari* in 0.8%. A total of 278 strains of *C. jejuni* and 111 strains of *C. coli* were used for determining antibiotic resistance. The results are shown in Tables 1 and 2.

Table 1. Test results for resistance to C. jejuni.

	AM	TE	CIP	L	NA	E
Resistant	64 (23.0%)	45 (16.2%)	144 (51.8%)	48 (17.1%)	153 (55.0%)	10 (3.6%)
Intermed.	49 (17.6%)	6 (2.2%)	1 (0.4%)	189 (68,0%)	2 (0.7%)	14 (5,0%)
Suscept.	165 (59.4%)	227 (81.6%)	133 (47.8%)	41 (14.9%)	123 (44.3%)	254 (91.4%)

Table 2. Test results for resistance to C. coli.

	AM	TE	CIP	L	NA	E
Resistant	29 (26.1%)	42 (37.8%)	65 (58.6%)	45 (40.6%)	66 (59.4%)	40 (36.0%)
Intermed.	18 (16.2%)	7 (6.3%)	2 (1.8%)	49 (44.3%)	0 (0.0%)	7 (6.3%)
Suscept.	64 (57.7%)	62 (55.9%)	44 (39.6%)	17 (15.1%)	45 (40.5%)	64 (57.7%)

AM - ampicillin, TE - tetracycline, CIP - ciprofloxacin, L - lincomycin, NA - nalidixic acid, E - erythromycin

The recent investigations produced a percentage of 60.2% of *Campylobacter* spp.-positive flocks, which constitutes a further 15% increase from the results obtained in 1998. At 34.7% the percentage of *C. coli* isolates is very high. The fact that the broiler farms are situated in an area with intensive pig farming may provide an explanation for this.

References

Anonymous (1999): Hygiene-Monitor, Issue 1999/3/00.

Feierl, G., Berghold, C., Fürpass, T. and Marth, E., 1999. Further increase in ciprofloxacin-resistant *Campylobacter jejuni/coli* in Styria, Austria. Clin. Micribiol. Infect. 5 59-60.

Lehner, A., Pless, P., Feierl, G., Schneck, C., Wagner, M., und Brandl, E., 2000. Erhebungen zum Einsatz der Pulsed-Field Gel Elektrophorese als Beitrag zur Abklärung der Epidemiologie von *Campylobacter* spp. Proc. Tagung der Fachgruppe „Epidemiologie und Dokumentation, 6. - 8. Sept., Wien (in print).

An example of the application of HACCP principles and methodology to the safety of raw milk production - monitoring raw milk contamination from udder origin

Joao Niza Ribeiro
Animal Health Department-AGROS, Infesta, Portugal

Introduction

A hazard analysis and critical control point (HACCP) plan for raw milk production in dairy farms needs monitoring tests for the critical control point (CCP) management (monitoring and corrective action taking). The bulk milk tank was considered the CCP where the control of bacterial contamination of udder origin could be achieved, using somatic cell count (SCC) in milk as the monitoring parameter. The verification can be also done with SCC using adequate Statistical Process Control (SPC) tools (Niza-Ribeiro, J. et al., 2000). The California Mastitis Test (CMT) is currently used as a cow side test in udder health monitoring. CMT is either quick and inexpensive and can be performed by the ordinary herdsman. A previous validation of CMT was necessary before its adoption as a monitoring test in HACCP plans. Part of the validation (experiment 1) consisted in the evaluation of the precision characteristics (repeatability and reproducibility) of the test to appraise the possibility of its use by common operators in any farm, as well as the effect of training on the individual performance. The second part of the validation (experiment 2) consisted in investigate the role of the mammary gland in raw milk contamination and the possibility of using CMT scores to identify the quarters shedding milk with heavy bacterial contamination. The results from experiment 1 & 2 were satisfactory but as the ordinary CMT was rather inaccurate to be used directly in the monitorization of BMTSCC, a modification of the method was attempted. The validation of this modified CMT was carried out (experiment 3) to evaluate the test's accuracy to monitor the bulk milk tank (BMT).

Material and Methods

Experiment 1

Series of dilutions of milk with 22 bottles each were prepared with a protocol according to Schneider . A simplified protocol, based in the previous one, was used to prepare series of 9 bottles. The *repeatability* was calculated as the proportion of the duplicates with the same score within all the duplicates in the series. *Reproducibility* was calculated as a proportion of duplicate coincident test scores in the run. Two different panels of operators were challenged: one had 6 farm advisers which performed each 10 trials (test runs) of 22 duplicated samples randomly distributed; in total 2640 tests were done by this panel; the other panel had 35 farmers and three advisers and each one made 3 trials with 9 serial diluted milk samples (nine bottles) in similar conditions to the above mentioned; a total of 1890 tests were done by the farmers. The results were scored in three classes instead of the usual five, as suggested by Schalm.

Experiment 2

Milk samples from 385 quarters samples with known CMT results were aseptically collected from cows of several farms predominantly infected with environmental pathogens. Bacterial counts were done using IDEXX's SimPlate^{TM} methods to estimate the most probable number of microorganisms. The determinations were: the total bacterial count (TBC), the coliform count and the *E. coli* count. For *S. aureus* counts samples from three herds previously known to be heavily infected with this microorganism were collected. In these herds all quarters in milk (873) were tested with CMT and their milk aseptically collected. The number of *S.aureus* in the samples were determined using Baird Parker Agar with RPF. Milk from clinical mastitis was also collected and microrganisms were counted in the way previously mentioned.

Experiment 3

Samples from 133 BMT were taken. Each sample was analyzed in duplicate for SCC in a automated epifluorescence system (COBRA βiocom®). Two different operators scored independently each of the samples for CMT. A ordinary CMT was first done, to score the milk in negative (CMT0), week (CMT1) or strong (CMT2) according to its reaction. Milk with negative reaction was considered under the range of sensitivity of the test, with less then 200 SCC. Milk with CMT1 reaction was diluted in 3:1, 2:1 and 1:1, the three dilutions were submitted to CMT and the result scored. Milk with CMT2 reaction was diluted in 1:1, 1:2 and 1:3 and analyzed as previously. According to the of scores of the three dilutions a SCC value was assigned to the milk. The correlation and a linear regression model were then calculated with the average of the SCC from both methods.
Note : For every experiment the CMT reagent was prepared according to Schalm

Results

Experiment 1

The average of the repeatability was 78% in both panels. The reproducibility was 77% for the farmers and 85% for the advisers. The CMT score negative gave a overall 83% reproducibility. Using single-factor ANOVA it was noticed that the operator does not have a significant interference in the global precision ($P<0.05$) but as the test runs went on, both the advisers and the farmers improved their repeatability.

Experiment 2

It was seen that CMT negative score is strongly associated (P > 90%) with low **TBC** contaminated milk. On the other hand quarters with CMT2 score are more frequently (P> 30%) associated with the higher levels of contamination. About 80% of the 385 samples presented less than 200 cfu/ml and 90% had less than 2000 cfu/ml regardless of their CMT score. Clinical mastitis are more frequently associated with the shedding of high quantities of cfu in the infected milk. ***Staph. aureus*** must be controlled in the case of milk products made of raw milk. The contamination of the tank should remain under 1500 Cfu/ml. Results from the 874 samples cultured show that less than 10% of quarters shed over 500 cfu/ml and only 6.4% of the total quarters sampled had values over 1500 cfu/ml. In the case of Staph. aureus the shedding of bacteria in milk from clinically infected quarters does no play a role so important as the subclinically infected ones with CMT 2 score. On the other contrary clinical mastitis can be important source of **Coliform bacteria** (*E coli* included). Less then 10% of quarters surveyed had more than 20 cfu/ml of Coliform bacteria. Healthy quarters and subclinical infected ones are poor sources of coliform bacteria in raw milk.

Experiment 3

Of the 133 samples only 102 were inside the modified CMT range (200<SCC<900)and therefore used to evaluate the accuracy of CMT as a monitoring test. A significant correlation (R) of 0.76 was found and the linear model was: *SCC = 0.7359 cmtSCC + 21.27* ($P_{\beta=\beta 0}$< E-20). Using the SCC predicted from the CMT scores it is possible to monitor the BMT after each milking and to correct for an eventual excess of milk from sub-clinical mastitis that would keep the BMTSCC in the next milking above the critical limit specified (4 E5 SCC / ml in the European Union). Modified CMT range of sensitivity of the test was considerate satisfactory for BMT control, because farms under 200 BMSCC have no need of corrective action and the critical limit of SCC is within the range. Before the last milking is completed and the milk delivered into the tanker it is always possible to identify the quarters with heaviest contamination and to exclude its milk from the BMT.

Analysis

It was concluded that CMT has good precision characteristics to be used as monitoring test in HACCP plans in dairy farms. Especially important is the good precision characteristics of the CMT negative reaction (> 80%). Its was also noticed the importance of training in the quality of the operator performance.

Quarters shedding high levels of microorganisms have shown sub-clinical mastitis with high CMT scores or clinical involvement. The results from this study suggest that the main sources of microrganisms in raw milk when delivered at the dairy plant are most certainly outside the mammary gland. Frequent mastitis pathogens like *Staph aureus, Streptococci* and *E. coli*, belong to groups of biological hazards with medium or low severity, to which Food Safety Objectives (FSO) can be assigned. FSO are important targets for process control in HACCP plans, and can be assigned for these bacteria. In this context the CMT can be

useful to the control of pathogens from udder origin in the raw milk fulfilling assigned FSO in association with good hygienic practices of milking and storing the milk.
The implementation of HACCP plans in dairy farms, within the context of safe raw milk production in a "farm to table" approach needs the management of the SCC, which is a recognized and widely used udder health parameter. A modification in the method of using the CMT can be useful in the context o BMT monitoring as the results from the experiment suggest. Monitoring the BMT after each milking is a cost effective procedure for the critical the control point management and corrective action at the farm can take place on time, before the milk is pumped into the tanker and delivered to the dairy or directly sold at the farm.

Conclusion

The control of raw milk safety regarding udder pathogens using HACCP principles is possible using CMT and SPC tools. The CMT can be a useful and efficient test in HACCP plans in dairy farms either for monitoring purposes and for corrective action taking. It can be used in the reduction of heavy contamination of raw milk with pathogenic bacteria from infected mammary glands. SPC tools can be used to do the verification of the control measures in place.

References

Niza-Ribeiro, J., Esteves, A., Menezes, J.C. and louzã, a., 2000. the use of some statistical process control tools for monitoring haccp systems in dairy farms. 9th international symposium on epidemiology and economics, denver, august 2000.

schalm, o.w., carrol, e.j. and jain, n.c.. physical & chemical tests for detection of mastitis. in: bovine mastitis. 1st ed., philadelphia, lea & febiger.

schneider, r. and jasper, d.e.. standardization of the california mastitis test. am. j. vet. res. 1964, 25, 109, 1635-1641.

Live dairy cull cow characteristics and associated microbial contamination

J.N. Sofos, K.E. Belk, C. Rossiter, J.O. Reagan and G.C. Smith
Department of Animal Sciences, Colorado State University, Fort Collins, CO 80523-1171, USA

Eighty cull cows from twelve lots of dairy cattle of known breeds were weighed and scored for body, hide, ambulatory and feces condition before slaughter during a 3-day period. Samples of fresh feces, external hide-surface samples (sponge-swabs), and excised carcass tissue samples were analyzed for *Salmonella* and *Escherichia coli* 0157:H7. The carcasses were sampled (excision of 100 cm^2 samples from brisket and round before evisceration) for microbiological analyses of aerobic plate counts (APC), total coliform counts (TCC) and *Escherichia coli* counts (ECC). In this study, most live animal and carcass characteristics had no statistical effect on resulting microbiological populations on carcasses, indicating that activities associated with slaughtering/dressing may have been more important in determining differences in microbiological contamination of carcasses than animal characteristics.

Objectives

The study was performed at a commercial slaughtering/dressing operation to determine if animal characteristics and contamination affected the microbial contamination of resulting carcasses from dairy cows. By examining the potential relationships of condition, cleanliness and other characteristics of animals to microbiological counts on their carcasses, animal and carcass handling/processing protocols might be identified that could reduce carcass contamination.

Materials and Methods

Overall study design

Eighty animals from twelve lots of dairy cattle were individually weighed (live weight) and scored for body condition, hide condition, ambulatory condition and feces condition. The animals were slaughtered over a three-day period, and chilled carcasses were weighed (carcass weight) and graded (carcass grade). Samples were collected from each live animal and each carcass for microbiological analyses. After collection, the samples (in Whirl-Pak™ bags, Nasco, Modesto, CA) were refrigerated and placed in coolers with ice packs for shipment by overnight air express to the laboratory for analysis.

Sampling and evaluation

Each cow was weighed and assigned scores for body condition (1=very lean, 2=some external fat, 3=good condition, 4=some excess fat, 5=fat), hide condition (1=clean, 2=dirty, 3=filthy), ambulatory condition (1=normal, 2=obviously lame, 3=severely lame) and feces condition (1=dry, 2=normal, 3=diarrheal, 4=fluid). A sample of approximately 30g of feces was removed from each animal via palpation from the rectum with the person obtaining the sample wearing a clean, plastic palpation glove. Each fecal sample was placed in a sterile sampling container for shipment to the laboratory. Immediately post-exsanguination and prior to hide removal, each animal's hide was sampled by rehydrating a sterile sponge (Whirl-Pak™, Nasco, Modesto, CA) with 10 ml of sterile double-strength skim milk (Difco Laboratories, Detroit, MI) and aseptically swabbing 100 cm^2 of the brisket area of the hide using a sterile rubber template. The person taking the samples used new sterile gloves for each animal tested. The sponge was placed in a sterile Whirl-Pak™ bag (Nasco) for shipment to the laboratory. Carcass sampling was performed after hide removal (pre-evisceration) on two anatomical carcass sites. The sampling sites on the carcass were the cushion of the round and the brisket, in the area anterior to the navel on the ventral mid-line. Three 100cm^2 portions of the adipose/muscle-tissue surface taken from each of 80 carcasses were aseptically removed from each sampling site of the carcass by use of a sterile rubber template, forceps, and scalpel.

Microbiological analyses

Fresh feces, hide-surface sampling sponges, and one of the three 100cm^2 carcass tissue samples were analyzed for *Salmonella* spp. Fresh feces, hide-surface sampling sponges, and the second of the three 100cm^2 carcass tissue samples were analyzed for *E. coli* 0157:H7. The third 100cm^2 carcass tissue sample from each animal was analyzed for aerobic plate counts (APC), total coliform counts (TCC), and *E. coli* biotype I counts (ECC).

Statistical analysis.

All data were converted to log CFU/cm^2 and analysis of variance, least squares means, standard deviations, and least significant differences for comparison of logarithmic means were used to evaluate the significance of relationships of live animal factors to microbiological contamination of the carcasses.

Results and Conclusions

- None of the fresh feces, hide or carcass samples was found positive for *E. coli* 0157:H7, while *Salmonella* was detected in 13.8% of the hide and 1.9% of the carcass samples.
- Ambulatory condition, hide cleanliness, live animal weight, carcass weight and grade score had no major or consistent effect on carcass microbial counts.
- The only factors that had significant ($P \leq 0.01$) effects on APC and TCC of carcass samples were sampling date and lot number, but it was not possible to determine whether these effects were due to animal characteristics or plant operation differences

during those days. However, since no other animal-associated characteristic had a significant effect on bacterial counts, it may be postulated that the differences were more likely associated with plant operations.

- Nevertheless, highly soiled animals are an important potential source of carcass and plant contamination with pathogenic bacteria. Therefore, poor sanitation, hygiene and manufacturing practices during slaughtering, fabrication and processing can lead to excessively contaminated meat, even when less heavily soiled animals are processed.

Table 1. Effects (Pr > F) of animal characteristics on aerobic plate counts (APC), total coliform counts (TCC) and E. coli *counts (ECC) for both carcass sites combined at pre-evisceration.*

Characteristic	APC	TCC	ECC
Sampling date	0.0001*	0.0030*	0.4636
Lot number	0.0001*	0.0084*	0.2764
Body condition	0.7200	0.7441	0.8238
Feces condition	0.7225	0.7925	0.7306
Live weight	0.8256	0.9676	0.9723
Ambulatory condition	0.3522	0.5776	0.9325
Carcass weight	0.9788	0.9775	0.9867
Animal breed	0.9287	0.4496	0.2311
Carcass grade	0.3014	0.6698	0.8096
Hide condition	0.2736	0.1473	0.3915

Active surveillance for BSE, targeting risk groups

Aline de Koeijer, Bram Schreuder and Mart C.M. de Jong
Quantitative Veterinary Epidemiology, ID-Lelystad, P.O.Box 65, 8200 AB Lelystad, the Netherlands

Introduction

BSE is a disease that is very difficult to diagnose clinically. Post-mortem confirmation is necessary to prove that a diseased animal actually suffers from a BSE infection. Furthermore, early stages of the infection cannot be detected by clinical investigation or by a post-mortem test. These features lead to difficulties in assessing the prevalence of infection.
BSE imposes a risk for human health (Bruce, 1997). Therefore, it is important to reliably assess the prevalence of infection, and to determine risk groups in the cattle population. Subsequently, an active surveillance system can be focussed at groups with higher risk. Thus, the real prevalence in the population can be estimated with minimal surveillance effort.
In 2001, the countries of the European Union will start an active surveillance program for BSE, focussing on two exit routes, with demonstrated higher BSE prevalence (Doherr, 1999). All animals over 24 months of age will be included. This age targeting schedule was chosen because in all BSE cases detected so far only 2 animals below 24 months were found positive.
In the following, we show that further age targeting can improve the efficacy of a BSE surveillance program. Age structured modelling, risk assessment, and analysis of case notification data on incubation period are essential ingredients of this study.

Methods

Given the difficulties in BSE diagnosis, we suggest the following operational definition for BSE prevalence: Prevalence of BSE is defined as the fraction of the population that will give a positive test result in one of the recently developed fast tests. Prevalence by this definition is clearly lower than the prevalence of infected animals, because these tests are not sufficiently sensitive to detect early infection stages. Estimates on the sensitivity of the tests vary a lot. They are assumed to become positive roughly between 1 and 12 months before clinical onset of the disease.
BSE develops very slowly over time, which leads to a typical distribution of the cases over the ages. Furthermore, in a steep increasing phase of an epidemic, the animals with short incubation periods will dominate, whereas in a decreasing phase of an epidemic, the longer incubation periods will become more prevalent among the detected cases.
Therefore, to estimate the prevalence per age group, we first determine whether BSE is in an increasing of decreasing phase. A risk assessment by quantifying local risk factors can lead to an estimate of the so-called basic reproduction ratio of BSE, specific for a time period and population (de Koeijer et al, in prep). Next, using case notification data from

the UK and Switzerland, we estimate the distribution of the incubation period. Knowing the age distribution of cattle in the population observed, we can calculate the age distribution of the BSE prevalence. So far, this is still independent of the actual prevalence of the infection, only the relative frequency per age group can be given. If BSE is prevalent in that population, cases are expected to be found according to that age distribution. With a similar method, we can estimate the distribution of infected animals over the age classes.

Results

When there is sufficient information available on BSE control measures and risk factors in a certain population, the model can predict the age groups in which BSE prevalence is highest. The model calculates the distribution of prevalence over the age groups. The absolute height of the prevalence however, needs to be estimated from the results of the surveillance program, but surveillance at a small age group suffices.
We found that populations which are still in the increasing phase of the epidemic, 4 and 5 year old animals have the highest BSE prevalence, so targeting at these animals would be optimally efficient in that case. However, for countries in a decreasing phase of the BSE epidemic, the highest prevalence is found in 6 and 7 year olds. Age targeting is about 3 times more efficient than random sampling and 2 times more efficient than sampling from animals over 24 months.

Discussion

Combining age targeting and exit route targeting could lead to a far more efficient surveillance system. It may be up to 10 times more efficient than random sampling. However, we should first make sure that the Swiss results on exit route targeting can be extrapolated to other populations. Therefore, testing over all exit routes would be essential to verify the distribution of cases over exit routes and thus it is essential, to reliably estimate the actual prevalence in a population from the results of the surveillance.

References

M.E. Bruce, R.G. Will, J.W. Ironside, I. McConnell, D. Drummond, A. Suttie, L. McCardle, A. Chree, J. Hope, C. Birkett, S. Cousens, H. Fraser and C.J. Bostock, 1997. Transmission to mice indicate that new variant CJD is caused by the BSE agent. Nature 389: 498-501

M. Doherr, B. Oesch, B. Moser, M. vandevelde and D. Heim , 1999. Targeted surveillance for Bovine Spongiforme Encephalopathy (BSE). Veterinary Record, 145: 672

A.A. de Koeijer, B.E.C. Schreuder, J.A.P. Heesterbeek, R.A. Oberthur, J. Wilesmith and M.C.M. de Jong. BSE Risk assessment by calculating the basic reproduction ratio for the infection among cattle. (in prep.)

The cause of variant CJD

Andrea Lowman
National Creutzfeldt-Jakob Disease Surveillance Unit, Research Fellow ,Western General Hospital, Crewe Road, Edinburgh, EH4 2XU

Hypothesis

Variant Creutzfeldt-Jakob disease is due to bovine spongiform encephalopathy transmitted via food the relationship between BSE & vCJD

Agents are identical

Variant CJD is a new disease with a different clinicopathological phenotype to other forms of CJD. A new phenotype is usually due to new agent or a new route of infection.
The new agent is thought to be BSE.
It is hypothesised that Variant CJD and BSE are caused by the same agent strain.

Evidence:
a. Prion protein glycoform analysis (1) (2)
b. Agent strain typing by bioassay characteristics (3)
c. BSE innoculation into macaques (4)

Transmission

a) sheep => cow
The first BSE cows occurred in 1986. At this time it was established practice to feed cattle with meat and bone meal derived from sheep. This may have been contaminated with sheep scrapie, especially as a change in rendering techniques in late 1970's increased the likelihood of infectious material in feed. The use of ruminant protein in ruminant feed was not banned until July 1988.

The epidemic of BSE peaked in 1992 with >3000 cases per month.
This was consistent with the causal theory, as the incubation period for BSE is 5-6 years.

b) cow => man
It is hypothesised that BSE is essentially scrapie in cows. There is currently no evidence that scrapie is a significant risk to humans. However, modification of prions occurs during passage through animal hosts. Therefore scrapie passed through cows may represent a risk to humans.

Evidence that BSE causes variant CJD:

i) Reasonable explanation for occurrence of BSE which cannot be applied directly to variant CJD.
ii) BSE => cattle 1985 Variant CJD => man 1994
Equivalent to minimum incubation period of BSE in humans, parallels kuru and human growth hormone related CJD.
iii) No obvious candidate common source for BSE/vCJD.

How? ? Food

The most likely route of transmission from cattle to humans is ORAL i.e. consumption of infection in food derived from BSE cattle.
Infectivity in prion disease is known to vary between tissues. Natural BSE has high infectivity in brain, spinal cord, dorsal root ganglia and retina. These are not likely to have been directly consumed (with the possible exception of ganglia) but spinal cord may have been used in mechanically recovered meat (MRM)(5).

The proof

Population exposure

A large number of dead/infected cattle entered the food chain up until the change in legislation in 1988-89.
Mechanically recovered meat probably contained bovine vertebral column - therefore, it may have contained significant amounts of spinal cord (5). MRM was used in processed meat products i.e. beefburgers, pies and sausages. Before 1988, infected bovine material was likely to have been included in human food and the human population could have had significant exposure to BSE infection via diet up to late 1989. The number of potential cases is difficult to quantify. (6,7)

Case control study

In the CJD unit we are currently carrying out a case control study. This is the classical epidemiological tool but has obvious limitations (to be discussed further).

Cases are patients classified as probable or definite v CJD (8) when seen in life by CJD unit staff.
Controls are either 'Hospital' (due to cease - possible source of bias) or 'Community' (4 from same General Practice as case, age +/- 4 years, same sex).

In the present case control study there is no confirmatory evidence of particular foodstuffs being the cause of vCJD. However, this is not necessarily surprising because:
1. Small numbers in the study.

2. No specific targets - infectivity could have been present in different items at different times, infectivity could have been present in one food intermittently.
3. Recall problems - asking for information regarding diet in the last 20 years - information gathered might not be reliable; recall bias.
4. Gathering information from a surrogate (as case usually cognitively impaired).

Other factors

There is currently no data to support any other factors in the causation of vCJD. We have looked at:

a. occupations e.g. medical, paramedical, nursing and dentistry
animal laboratories
pharmaceutical laboratories
other research laboratories
animal farming / veterinary medicine
meat industry (butchers/abattoirs/rendering plants)
catering industry
other occupations involving animal products

We have found no evidence to date of any occupational risk for CJD.

b. medical exposure to animal sourced material e.g. medicines (including those of bovine origin withdrawn in 1990), sutures, vaccines.
c. non-occupational exposure to live animals e.g. cats/dogs/ferrets
d. other exposure to animal material
e. transmission via a different route

There is no evidence for any of the above at present time. The recently reported geographical cluster in Leicestershire is currently being investigated. There is no other demographic clustering.

References

1) Collinge J, Siddle KCL, Meads, Ironside J, Hill AF:Molecular analysis of prion strain variation and the aetiology of 'new variant' CJD. Nature 1996;383:685-690
2) Hill AF, Desbruslais M,Joiner S, Siddle KCL, Gowland I, Collinge J: The same prion strain causes vCJD and BSE. Nature 1997;389:448-450
3) Bruce ME, Will RG, Ironside JW, McConnell I, Drummond D, Suttie A, McCardle L, Chree A, Hope J, Birkitt C, Cousens S, Fraser H, Bostock CJ: Transmissions to mice indicate that 'new variant' CJD is caused by the BSE agent. Nature 1997;389:498-501
4) Lasmezas CI, Deslys J-P, Demaimay R, Adjou KT, Lamoury F,Dormont D, Robain J,Ironside JW, Hauw J-J: BSE Transmission to macaques. Nature 1996; 381:743-744
5) Will R: New variant Creutzfeldt-Jakob Disease. Darlington Post Graduate Medical Journal 1998;17:35-42.
6) Cousens SN, Vynnvcky E, Zeidler M, Will RG, Smith PG: Predicting the CJD epidemic in humans. Nature 1997;385:197-198

7) Ghani AC, Ferguson NM, Donnelly CA, Anderson RM: Predicted vCJD mortality in Great Britain. Nature 2000;406:583-584

8) CJD Unit Classification Criteria for variant CJD 2000

9) Eighth Annual Report 1999 of Creutzfeldt-Jakob Disease Surveillance in the UK.

A pilot study for BSE active surveillance in Italy: Preliminary results

P.L. Acutis[1], M. Caramelli[1], C. Casalone[1], E. Bozzetta[1], M. Mazza[1], E. Pavoletti[1], G. Moda[2] and G. Ru[1]
[1]Istituto Zooprofilattico Sperimentale del Piemonte, Liguria e Valle d'Aosta. Centro di Referenza Nazionale per le Encefalopatie Animali, Turin, Italy. [2]Regione Piemonte, Servizio Sanità Pubblica Veterinaria, Turin, Italy.

Introduction

The surveillance of bovine spongiform encephalopathy (BSE) is a matter of general concern, due to the serious implications for human health. Since 1998 an active surveillance system for BSE was started in Switzerland, leading to the identification of cases of infection in animals that would have otherwise been missed. As a consequence, the European Union deemed as priority the need of increasing the efficacy of the BSE surveillance. The new Decision 2000/374/EC, modifying the Decision 98/272/EC, states that from January 2001 all the EU member states must carry out a monitoring plan aimed to test a representative sample of bovines, over 24 months of age, coming from fallen stock or emergency slaughter. Italy will have to test 5000 animals. This new programme is possible because of the availability of 3 so-called "rapid" diagnostic tests evaluated by the European Commission.
In view of the new approach, a pilot study was planned by the Italian Reference Centre for TSE and is now going on in the Piedmont region. The aim of this study is to apply an active surveillance in a limited area, in order to check its feasibility and to gain information useful for a larger scale testing of cattle i.e. procedures for collection and handling of samples, setting up of a rapid test and of the information system. Moreover, the program will increase the knowledge about the epidemiological situation of BSE in the study area.

Materials and Methods

The study started on May the 1st and shall be concluded by the end of the year 2000. The bovines tested are to be selected from two main categories: 1) sub-populations foreseen by the Decision 2000/374/EC (fallen stock, heads from emergency slaughter); 2) sub-populations meeting the requirements of the Italian BSE Surveillance Plan for the current year (foreign cattle or cattle born before the feed ban, or accidentally exposed to MBM). Based on the bovine population of the study area, a sample size of 500 heads to be examined has been determined in order to comply the Decision 2000/374/EC.
The first two months have been spent organising meetings with stakeholders: study staffs, regional health authorities, incinerator's responsible, veterinary officers. Relevance of the project was presented and a general agreement about the logistics and the flow of both samples and information was reached (see Fig.1). It was agreed to destroy all fallen stocks by incineration.

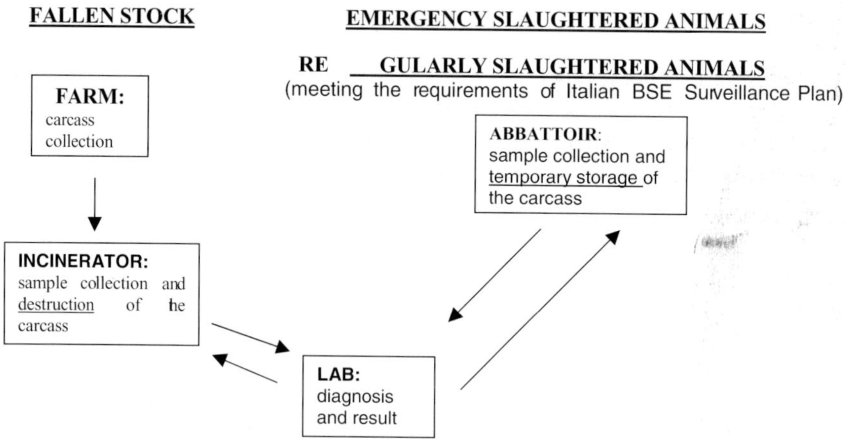

Figure 1. Flow of the samples.

A period of 6 month of sampling was settled from July the 1st with the following timetable: 1) July-September the collection and diagnosis by routine methods of fallen stock samples coming from a smaller area, allowing the setting up of at least one of the rapid test; 2) October-December application of a rapid test on samples of all bovine categories coming from throughout the Region, and summary of on field problems and results.
A preliminary 1-day training about the procedure of collection by means of an *ad hoc* spoon and storage of brain stem specimens was attended by veterinary officers working at various slaughterhouses and at the incinerator for animal waste.
As rapid test, the Prionics-Check test was selected merely because of commercial availability at that moment. As known this test is based on a western blotting procedure for the detection of the disease specific, protease resistant prion protein (PrP_{sc}). Histology, immunohistochemistry (IHC) or, in case of autolytic tissue, Western blot based on SAF extraction are the routine methods used in the first collection period and in order to get diagnosis confirmation of possibly positive sample to the rapid test.

Results

At the moment the first step of the study is ongoing and a total number of 110 brain stem specimens has been collected. As routine method have been used, so far the lab has been able to process and test 73 samples. The median time of diagnosis (from sample received to result issued) is 18 days (3 to 36, minimum to maximum respectively). 70% of the samples reached the lab within 24-36 hours from the death of the animals. All the samples but 4 (94.5%) could be tested by both histology and IHC. The remaining specimens were shipped to the laboratory in an autolytic status and were tested by Western blot. All brain stems resulted negative for PrP_{sc}.
During this study step only fallen stock was considered. The median age of the examined animals is 5 years and 75% of them were over 4 years of age.

Discussion

The first step of the ongoing pilot screening confirm the feasibility of an active approach to TSE surveillance. Main logistic problems have not come out. The organising efforts along with the training and information provided resulted in a full co-operation by all the operators. The expected sample size for the first period was easily reached although poor information was available about monthly rate of fallen stock and its characteristics. The pattern of ages observed point out an highly significant sample of the bovine population in term of risk. The logistics regarding the fallen stock was greatly simplified by restricting the sampling at the incinerator. Minor problems of timely shipment of sample are to be worked out in order to decrease the rate of autolytic specimens while time of diagnosis will be sensibly reduced by the forthcoming adoption of a rapid test.
The experience gained through the pilot study will allow implementing a national active surveillance helpful to disclose the actual epidemiological situation regarding TSEs.

Reference

Moynagh J., Schimmel H.. Tests for BSE evaluated. Nature, vol.400, 8 July 1999.

The use of waste animal fats as feedstock for the production of an environmental friendly fuel for diesel engines (biodiesel)

E. Ahn[1] and M. Mittelbach[2]
[1]BDI Anlagenbau Ges.m.b.H., Eduard-Ast-Straße 1, A-8073 Feldkirchen/Graz, www.BioDiesel-intl.com; [2]Institute of chemistry, Working group renewable ressources, Karl Franzens University Graz, Heinrichstraße 28, A-8010 Graz

With a newly developed process technology it is possible to produce an environmental friendly, high quality diesel fuel out of any kind of waste animal fat. A production plant successfully applying this Austrian technology was built in the USA. Basics of this technology are explained. Possible obstacles to this alternative usage of waste animal fat due to newly proposed European regulations are discussed.

Introduction

The use of waste animal fats as raw material for the animal feed production is increasingly limited by severe regulations. World-wide published scandals, like the BSE-contamination of British beef dramatically showed the dangerous side of this sort of utilisation of waste animal fats. A search for alternative, secure ways of utilisation for these valuable raw materials is necessary more than ever.

Nearly 20 years ago, an industrial scale process was developed at the institute of chemistry of the Karl-Franzens-University Graz, with which vegetable oils could be changed into an environmental friendly, high quality substitute fuel for diesel engines [1]. This special fuel, meanwhile brand-named "BioDiesel", can be used in any kind of diesel engine, without adaptation of the engine. BioDiesel offers a lot of advantages compared to mineral oil diesel when it comes to the impact on the environment [2,3]. Many research studies world-wide proved that the exhaust gas emissions are much lower compared to mineral oil diesel. Being a non-toxic fuel which biodegrades completely within 3 weeks, BioDiesel is most efficiently used in environmental fragile areas (like nature parks, lakes) but also in urban area (City bus fleets).

The institute of chemistry and the company BDI Anlagenbau Ges.m.b.H. have succeeded during a joint research project, sponsored by the Austrian Trade and Industry Research Fund, to improve the production technology, especially in that way, that the variety of feedstock increased to all kinds of vegetable and animal fresh or waste fats/oils. 1991 the results of this research won the "Austrian Innovation Award".

In 1994 the industrial scale production of BioDiesel out of waste cooking oil started. Since many years now, up to 40 buses of the City of Graz run successfully on BioDiesel, produced out of waste cooking oil which is collected in households and restaurants.

The constant improvement of the production technology at BDI lead to the successful construction and start-up of the world-wide first multi-feedstock BioDiesel production plant in Butler / Kentucky (USA) in 1998. In this plant the feedstock ranges from virgin oils, waste cooking oils (with a high content of chicken fat) to waste animal oils from slaughter houses.

Material and methods

Figure 1 shows the basic chemical equation for the production of BioDiesel out of animal oil. During the so-called transesterification triglycerides are transformed with methanol under the influence of a catalyst into fatty acid methyl esters (= BioDiesel) and glycerine. Using a caustic catalyst the reaction can be carried out under ambient conditions.

$$\begin{array}{l} CH_2COOR \\ | \\ CHCOOR \\ | \\ CH_2COOR \end{array} + 3\,CH_3OH \underset{\text{caustic catalyst}}{\overset{40°C,\ 1\ bar}{\Leftrightarrow}} \begin{array}{l} CH_2OH \\ | \\ CHOH \\ | \\ CH_2OH \end{array} + 3\,RCOOCH_3$$

triglyceride + 3 methanol ⇔ glycerine + 3 fatty acid methylester

R = fatty acid chain

Figure 1. Basic chemical equation for the transesterification.

To fulfil the severe limits of the international standards for BioDiesel (f.e. DIN E 51606) the fatty acid methylester has to be purified after the transesterification step in special process steps (figure2).

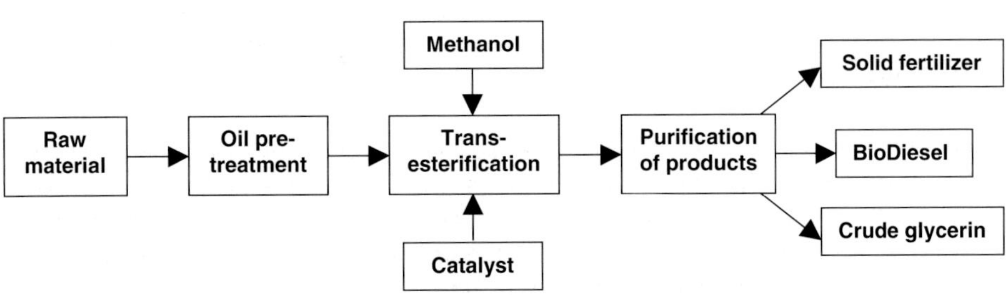

Figure 2. Basic scheme of the BioDiesel production process from BDI Anlagenbau Ges.m.b.H.

Applying the BDI-process technology the side-products during BioDiesel production are crude glycerine of saleable quality and a solid phase, which can be sold to farmers as a valuable fertiliser. The main advantages of the BDI-process are its environmental friendliness (no wastes or waste water during production) and its cost efficiency (due to a world-wide patented process step 100% yield guaranteed).

When using waste animal fats as feedstock additional process steps (oil pre-treatment) are necessary. Within these units the raw material is separated from non-transesterificable material and the sometimes very high content of free fatty acids of these raw material is reduced before the transesterification step.

In the proposal for a regulation of the European parliament and of the council laying down the health rules concerning animal by-products not intended for human consumption several processing methods for animal fats, in particular for specified risk materials (SRM) are specified [4]. Although with these processing methods a sterilisation of the raw material should be guaranteed, the further processing of SRM-derived fats in oleochemical industry is planed to be forbidden. But for BioDiesel production, using industrial standardised process technology, these sterilised fats would be a perfect raw material.
According to the cited proposal, animal fats derived from category 2 material are approved for further processing in an oleochemical plant. But in the case of a subsequent transesterification the proposed regulations demand a minimum reaction temperature of 200°C for 20 minutes. From the point of hygienic this demand is unnecessary because sterilisation already happens during the mandatory, preceding processing methods. Furthermore this extremely high reaction temperature is not state of the art in oleochemical industry any more. Since years, several low-temperature, caustic transesterification processes produce fatty acid methylester of comparatively higher quality. For example the low reaction temperatures have a positive effect on the oxidation stability of the produced fatty acid methylester.
Seeing it from an economic viewpoint, the demand for a minimum reaction temperature of 200°C would mean a drastic restriction in the possibilities of utilisation of waste animal fats for the oleochemical industry. The production of a high-quality BioDiesel, fulfilling the severe international fuel standards, would not be possible under the proposed production conditions, and therefore the chance for the oleochemical industry to produce an alternative, profitable product out of a hazardous material would be given away.

Results and discussion

Due to a newly developed process it is now possible to produce an environmental friendly diesel fuel out of waste animal fats. Even proper sterilised animal fats derived from specified risk materials can be used as feedstock in the BDI-Process technology. Due to the increasingly restrictive regulations concerning the processing and deposition of waste animal fats, the production of BioDiesel out of these materials could be an economical and ecological alternative to the present use of waste animal fats in the feedstock industry.

References

[1] H.Junek; M. Mittelbach, 1986. Austrian patent Nr.386 222B: "Verfahren und Vorrichtung zur Herstellung eines als Kraft- bzw. Brennstoff geeigneten Fettsäuregemisches".

[2] Brussels, Belgium, 10 - 13 October 1999, Proceedings to the 3rd European Motor Biofuels Forum.

[3] Curitiba, Brazil, 19 - 22 July 1998. Proceedings to the International Liquid Biofuels Congress.

[4] Commission of the European Communities, 2000, Proposal for a regulation of the European parliament and of the council laying down the health rules concerning animal by-products not intended for human consumption, SANCO/1226/2000, Brussels.

Animal TSE surveillance system in the province of Styria

P. Wagner, J. Köfer and G. Odörfer
Department of Veterinary Administration in Styria, Zimmerplatzgasse 15, A-8010 Graz, E-mail: post@favw.stmk.gv.at

An animal TSE surveillance system was established in the Province of Styria in accordance with Commission Decision 98/272/EC. The programme places special emphasis on the consistent registration and examination of potentially infected animals and their offspring. For this purpose, a central database was established and a stringent examination and notification scheme was put in place. These measures were augmented by strict control and external validation of the processing procedure in the Styrian rendering plant.

Introduction

With a view to precautionary consumer protection, the Department of Veterinary Administration in Styria makes every effort to protect the population against health risks arising from the potential occurrence of transmissible spongiform encepha-lopathies in animals. For this purpose an extensive TSE monitoring programme was set in place.

Material and methods

Based on Commission Decision 98/272/EC (CEC, 1998) and the corresponding decree of the Federal Chancellery, the Department of Veterinary Administration of the Styrian Government extended the existing epidemiological surveillance system which had been established in 1990. The system is based on the registration of all cattle, sheep and goats (including their offspring) originating from countries with increased TSE risk in a database. In addition to this, farms containing registered animals are entered into a geographical information system. The official veterinarians are responsible for quarterly clinical examinations for symptoms of TSE and the commissioning of histological TSE examinations at the time of slaughter or after the death of the animal. In order to guarantee the safe storage of all carcass pieces in the slaughter plant until the results of the examination are available, the enterprises were provided with special containers. In the Styrian rendering plant a special cold-storage room is available, where those carcasses which must be subjected to follow-up TSE examination are placed in intermediate storage untill final verification. In addition to the official veterinarians' inspections, an external validation by a skilled quality manager from a scientific research institute (Joanneum Research, Graz) of the proper processing in the rendering plant is performed on an annual basis.

Results

As of 30 June 2000, a total of 2.307 cattle and sheep were registered in Styria. Figure 1 shows the exact number of the individual categories of registered animals. In the course of 1999 the official veterinarians in Styria commissioned a total of 352 laboratory diagnostic TSE examinations (Table 1). To date, not a single case of TSE in animals has been detected in Styria.

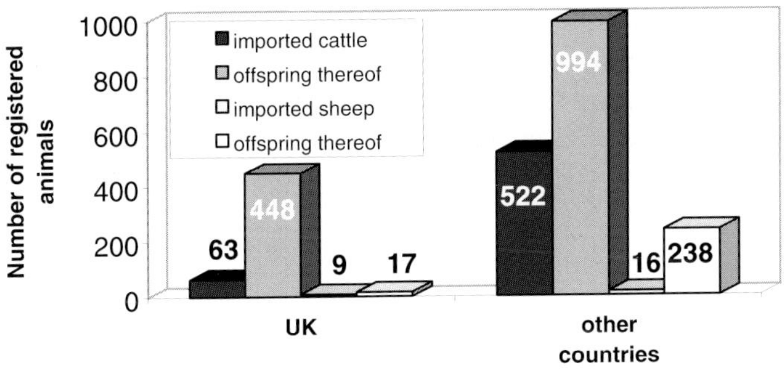

Figure 1. Registered animals in the Province of Styria, as of 30 June 2000.

Table 1. TSE examinations in the Province of Styria, 1999.

Cattle	Sheep	Goats	Total
299	17	9	325

Discussion

For countries outside the United Kingdom, imports of meat and bone meal or live cattle are the major risk factors (SCHREUDER et al., 1997). Therefore any surveillance programme should take special account of these risk factors (O.I.E., 1999). In this respect the consistent registration and monitoring of potentially infected animals and their offspring with regard to BSE is inevitable. TALYOR et al. (1995) proved that some rendering procedures leed to a survival of BSE infectivity. Strict compliance with the required processing parameters in the carcass disposal plant is of primary importance in order to prevent the risk of introducing potentially infected material into the food chain. An additional external validation of the processing procedure provides additional certainty. On the occasion of an inspection visit of the European Commission in October 1998, the inspection team convinced themselves of the efficiency of the measures taken (CEC, 1998).

References

CEC, 1998: Commission Decision 98/272/EC of 23rd April 1998 on the epidemiological monitoring of transmissible spongiform encephalopathies and amendment to Decision 94/474/EC. Official Journal of the European Communities, L 122 of 24.4. 1998, 59-63.

CEC, 1998: Report of a veterinary mission to Austria with regard to the implementation of commission decision 97/735/EC, commission decision 98/272/EC and the follow-up of the implementation of commission decision 96/449/EC. Document XXIV/1480/98/MR-final-EN.

OIE, 1999: International Animal Health Code for Mammals, Birds and Bees, Eight edition, 1999, Chapter 3.2.13. Bovine Spongiform Encephalopathy (BSE).

Schreuder, B.E.C., Wilesmith J.W., Ryan J.B.M., Straub O.C., 1997: Risk of BSE from the import of cattle from the United Kingdom into countries of the European Union. Vet. Rec. 141, 187-190.

Taylor, D.M., Woodgate S.L., Atkinson M.J., 1995: Inactivation of the bovine spongiform encephalopathy agent by rendering procedures. Vet. Rec. 137, 605-610.

Lead, mercury and cadmium burden of the german mud flats (Wattenmeer) does not hamper quality of common shrimp

Hans Marx
Institute for Hygiene and Technology of Food of Animal Origin, Veterinary Faculty, Ludwig-Maximilians-University Munich

Atomic Absorption Spectrometry (AAS) was used to investigate heavy metal contamination of common shrimp *Crangon crangon* from the German mud flats in the North Sea. The shrimp were hauled in two seasons: November and May.
The Lead, Mercury and Cadmium contents of the altogether 32 samples investigated were low: 0.043 ± 0.026 mg Cadmium, 0.019 ± 0.011 mg Lead and 0.033 ± 0.016 mg Mercury (mean ± standard deviation) per kg wet weight. The lowest value detected in Lead was below limit of determination, the highest 0.045 mg per kg. In Cadmium, the contamination spanned 0.016 - 0.091 mg per kg. In Mercury, the highest load amounted to 0.076 mg per kg. All of the samples contained lower levels of the investigated elements than the levels of concern of the Federal Institute for Health Protection of Consumers and Veterinary Medicine, i.e. 0.5 mg Lead per kg, 0.1 mg Cadmium per kg and 0.5 mg Mercury per kg wet weight. Most of the samples showed significantly lower levels of contamination. Therefore, no risk to the consumer arises from the Lead, Mercury and Cadmium contents of the shrimp caught in this area.
The main reason for the over all low levels of contamination might be the short period of feeding of the shrimp in connection with a dramatic growth rate. Primarily, this seems to be the case concerning the low Cadmium burden. WHO surveys indicate that all types of shellfish generally show low levels of Mercury contamination. According to this source, mean values of Mercury contamination in shrimp are below 0.2 mg per kg. In contrast to this, salt water fish are on a high level (shark: 1.8 mg per kg; swordfish: 1.5 mg per kg). With respect to Lead, an active mechanism of secretion must be taken into consideration. This phenomenon of apocrine secretion of Lead into the thoracic extensions of the antennal gland followed by excretion with the urine was described for giant tiger prawn, *Penaeus monodon*.
Besides this, the heavy metal load of the shrimp also depended on the fishing seasons November and May. The difference between the two groups was highly significant (P = 0.01) in the case of Lead and Cadmium and significant (P = 0.05) with respect to Mercury (Student's t-test for independent samples).

References

Buchwald K., 1987. Forstwiss Centralbl 106: 189-206.
Bundesgesundheitsblatt (1996) 5: 193-194.
Dethlefsen V., 1977. Inf Fischwirtsch 24: 161-162.

Falandysz J., 1989. Food Addit Contam 6: 483-488.
Friloux J, Hunt L, Grodner R, Novak AF., 1972. Quick Frozen Foods 34: 79-80.
Guns M, Van Hoeyweghen P, Vyncke W, Clerk R. de, 1992. Rev Agric (Bruss) 45: 731-739.
Llergandi GB, Garcia C, Torres O, Garcia-Roche MO (1987) Alimentaria 24: 43-44.
Madany IM, Wahab AAA, Alawi Z., 1996. Water Air Soil Poll 91: 233-248.
May TW, McKinney GL., 1981. Pestic Monit J 15: 14-38.
Moustafa EK, Moharram YG, El-Sokkary I., 1987. Nahrung 31: 773-776.
Mueller M, Anke M, Hartmann E, Illing-Guenther H., 1996. Food Addit Contam 13: 359-378.
Oehlenschlaeger J., 1986. FIMA-Schriftenreihe 11: 81-92.
Parvaneh V., 1977. Bull Environ Contam Toxicol 18: 778-782.
Rat von Sachverständigen für Umweltfragen. Umweltprobleme der Nordsee, 1980. Stuttgart.
Schulz-Schroeder G, Schering B., 1995. Arch Lebensmittelhyg 46: 40-43.
Sharif AKM, Mustafa AI, Hossain MA, Amin MN, Safiullah S., 1993. Sci Total Environ 133: 193-199.
Tahvonen R, Kumpulainen J., 1996. Food Addit Contam 13: 647-654.
Vijayram K, Geraldine P., 1996. Ecotoxicol Environ Safety 34: 1180-183.
Vogt G, Quinitio ET (1994) Aquat Toxicol 28: 223-241.
WHO, 1986. Mercury in Fish, FAO/WHO Food Contamination and Monitoring Programme for the Codex Committee on Food Additives. WHO/EHE/FOS/86.3.
Yusof AM, Rhaman N, Wood AKH., 1994. Biol Trace Elem Res 43-45: 239-249.
Zahari bin Abd Hamid M, Shafie Emmong M., 1987. J Micronutr Anal 3: 129-135.
Ghoshdastidar N, Chakabarti I., 1991. Indian J Med Res 94: 384-386.

Prevalence of *Clostridium botulinum* type E in finnish wild and farmed fish

Sebastian Hielm, Eija Hyytiä-Trees and Hannu Korkeala
Department of Food and Environmental Hygiene, Veterinary Faculty, Helsinki University

In a recent survey of Finnish sediments, spores of *Clostridium botulinum* type E were found in 81% and 61% of sea and freshwater samples, respectively (Hielm et al., 1998b). Upon growth and subsequent neurotoxin release in foods, this psychrotrophic bacterium is responsible for the potentially fatal foodborne disease botulism. Combined with data from earlier surveys of more southern parts of the Baltic Sea, it is safe to say that the Baltic catchment area in general is among the world's most contaminated fishing grounds in regard to *C. botulinum*. As the environmental pressure in regard to *C. botulinum* type E on fish in this area is quite high, our research group undertook a series of surveys to determine the effect of the high spore levels on the quality of fish harvested in Finland. To estimate the degree of *C. botulinum* contamination in samples, a semi-quantitative PCR analysis employing MPN estimates were used instead of the classical mouse bioassay (Hielm et al., 1996).

Hyytiä et al. (1998) surveyed 438 raw fish samples from intestines, surface and whole fish from five different species and 208 fish roe samples from four different species. The fish species were: rainbow trout (*Onchorhyncus mykiss*), burbot (*Lota lota*), whitefish (*Coregonus lavaretus*), vendace (*Coregonus albula*), and Baltic herring (*Clupea harengus membras*; only whole fish sampled). The results of this survey can be seen in Figure 1 below.

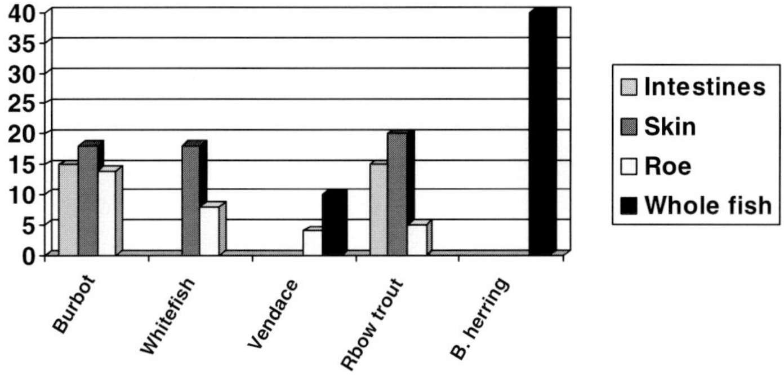

Figure 1. Prevalence (%) of Clostridium botulinum *type E spores in Finnish fish samples.*

Hielm et al. (1998a) surveyed 21 Finnish trout farms for *C. botulinum*. This material included 125 farm sediment samples, and 165 intestinal and 43 skin samples from farmed fish. *C. botulinum* type E spores were found in 68% of the sediment samples, representing all but one (95%) of the farms in this study. Spores were also found in 15% and 5% of the intestinal and skin samples, respectively. The fish farms were also divided into groups depending on the type of farming system they represented: 11 farms were marine net cages and 10 were freshwater ponds situated on the Finnish mainland. Three of the ponds were of a more modern type with concrete walls and equipped with a self-cleaning bottom suction device for the removal of biological discharge. Interestingly, the spore levels of sediments and fish in both marine net cages and traditional earth ponds were on an equally high level (approximately 2000 and 160 spores kg^{-1}, respectively), while the contamination levels of the suction device-equipped farms were significantly lower than these. These findings seem to indicate that the use of such devices, presently developed for the collection of polluting nutrients, would also be indicated in *C. botulinum* -rich waters to improve the safety of harvested fish in regard to botulism.

References

Hielm, S., Björkroth, J., Hyytiä, E. and Korkeala, H., 1998a. Prevalence of *Clostridium botulinum* in Finnish trout farms: PFGE typing reveals extensive genetic diversity between type E isolates. Appl. Env. Microbiol. 64: 4161-4167.

Hielm, S, Hyytiä, E., Andersin, A.-B. and Korkeala, H., 1998b. A high prevalence of *Clostridium botulinum* type E in Finnish freshwater and Baltic Sea sediment samples. J. Appl. Microbiol. 84: 133-137.

Hielm, S, Hyytiä, E., Ridell, J. and Korkeala, H., 1996. Detection of *Clostridium botulinum* in fish and environmental samples using polymerase chain reaction. Int. J. Food Microbiol, 31: 357-365.

Hyytiä, E., Hielm, S., and Korkeala, H., 1998. Prevalence of *Clostridium botulinum* type E in Finnish fish and fishery products. Epidemiol. Infect. 120, 245-250.

Molecular epidemiology of Norwalk-like viruses in Ireland - emerging agents of foodborne disease

Barbara Foley[1], Colin Hill[2], and John G. Morgan[2]
[1]Food Safety Authority of Ireland, Abbey Court, Lr. Abbey Street, Dublin 1, Ireland, [2]Department of Microbiology, National University of Ireland, Cork, Ireland.

It is only in recent years that the role of viruses as aetiologic agents of foodborne illness has emerged. These simple life forms have long been associated with disease in humans but with the advent of molecular biology and molecular epidemiology, scientists are now able to study these agents as foodborne pathogens. The most important foodborne viruses are hepatitis A and the Norwalk-like viruses (NLV) previously referred to as small round structured viruses (SRSV). This paper focuses on the Norwalk-like viruses in particular.

Although Norwalk-like viruses have been recognised worldwide as the single most important aetiological agents of outbreaks of gastroenteritis, at the outset of this study there was no epidemiological information available in Ireland regarding the NLV strains circulating in the country, or their association with foods. The objectives of this study therefore were to determine the molecular epidemiology of NLVs in Ireland for the first time, to develop model assays as indicators of viral contamination of shellfish, particularly bivalve molluscs, and to develop quantitative assays for enumeration of viruses in food and clinical samples.

In order to determine the molecular epidemiology of these viruses, faecal specimens were examined from sporadic and outbreak cases occurring in Ireland over a three-year period. For the sporadic cases, faecal specimens from cases of non-bacterial gastroenteritis were collected from a single Irish hospital. 360 of these samples were tested for the presence of NLV by RT-PCR (reverse-transcription polymerase chain reaction) using a broadly reactive set of primers to detect all viruses (both genogroup G1 and G2). The results of this screening study were that 29 (8%) of these specimens were found to be positive for NLV by RT-PCR. Genotyping of these isolates then determined that the majority of these isolates (25/29) were G2, the remainder being G1. Sub-typing of the G2 isolates then revealed that the most common subtype was P2B. It was then decided to obtain sequence data from eight of these sporadic cases and eight isolates originating from separate outbreaks that had occurred in Ireland over the 3-year period. The latter specimens had been deemed positive for NLV by electron microscopy. A common region within the highly conserved RNA polymerase gene was sequenced and compared between all 16 isolates. The two G1 specimens were shown to be most similar to Southampton virus (98% at the nucleotide level). The remainder of the specimens (all G2) were seen to cluster with Bristol/ Lordsdale virus. We conclude that the Bristol/ Lordsdale virus strain is the most prevalent NLV strain circulating in Ireland and this result correlates with worldwide trends.

The second part of the study was to develop a model system for indication of viral contamination of shellfish, particularly bivalve molluscs. The F+ bacteriophages Q(proved to be very good indicator organisms for the following reasons: firstly they occur in high densities in raw wastewater and sewage treatment effluents, secondly, they have several morphological and biochemical similarities to NLVs (25 nm diameter, icosahedral shape, and single-stranded RNA genome), and thirdly, viable bacteriophage particles can be very easily enumerated using a simple plaque assay. In this study, the sensitivity of detection of Q(from water, faecal specimens and shellfish tissue was assessed using plaque assays and RT-PCR.

The final part of this study involved development of a quantitative-competitive RT-PCR assay for the quantitation of NLVs in clinical, food or environmental samples. This assay was applied to the direct enumeration of NLVs in faecal specimens and the results obtained (of the order of 10^7 virus particles) suggested that previous determinations of NLV titre (10^5 to 10^6) may have been underestimated.

Overall this paper highlights the importance of NLVs as foodborne pathogens in Ireland, and raises many issues with regard to prevention and control measures that impact on public health, food safety, and industry, particularly the shellfish industry.

A case-study of the microbiological condition of a carp pond and the associated fish quality, with particular reference to human pathogens

Dagmar Suppin, Bettina Rippel-Rachlé, Peter Paulsen, Erich Schopf and Frans J.M. Smulders
Institute of Meat Hygiene, Meat Technology and Food Science, University of Veterinary Medicine, Vienna

Introduction

The *per capita* fish consumption in Austria has been slowly but constantly rising since the 1950`ies, especially the consumption of fresh fish. Of the volume consumed the percentage of local sweet water fish is very low, however. In 1997 the self-supply covering Austrian demand was merely 7 % (Austrian Nutrition Report, 1998). EU grants are available to support the establishment of fish-farms. Therefore, Austrian aquaculture is an agricultural branch with potential for expansion. The managers of fish farms currently lack information on the microbiological condition of the food items they produce from aquaculture. Consequently, it is essential to identify the various microorganisms (pathogenic or non-pathogenic) which are associated with fish and aquaculture (HUSS,1994, GONZÀLEZ et al.,1999) and to clearly outline the potential risks for people who consume aquaproducts contaminated with pathogenic microflora such as salmonellae, listeriae and campylobacters (HUSS, 1994, FELDHUSEN, 1999).

Material and Methods

The samples originated from a carp-breeding farm with semi-intensive production. Water samples (700ml) were taken out of 50 cm depth; the temperature during sampling was 11,5°C.

Harvesting procedures included: fishing out, trawling, placing fish in plastic baskets (manual), manual sizing, redistribution of the fish over a wooden slide into a tub with fresh water supply and sacrificing the fish by a neck cut. Subsequently, the fish remained unstripped but well-refrigerated during transport (max. 12 hours) to the lab where they were rinsed thoroughly before stripping. A sample of 7 carps with similar size and weighing approximately 700g were subjected to further microbiological analysis.

The bacterial load on the fish was determined by rinsing the entire fish surface with buffered peptone solution (OXOID, 400 ml per fish). Furthermore, the gills (10g), the intestinal tract (10g) and the water sample (10ml) underwent bacteriological examination (1:10 buffered peptone, OXOID, serial dilutions). The aerobic colony count (Plate -Count

Agar, MERCK), *Pseudomonas* and *Aeromonas* (GSP- Kielwein agar, MERCK), *Enterobacteriaceae* (VRBD- agar after Mossel, MERCK) were determined.
In addition, the prevalence of the following pathogenic organisms was assessed: *Listeria sp.* (Fraser Enrichment Broth, MERCK, Palcalm- agar, MERCK), *Salmonella* (buffered peptone- solution, OXOID, MSRV- agar, OXOID) and *Campylobacter* (Karmali- agar, OXOID). For *Listeria sp.* a further determination on species-level was achieved using the CAMP- test and API- Listeria (bio Meriéux).

Results

Table 1. The microbiological condition of a carp pond, carp-surface, -gills and -guts expressed as log 10 cfu.

number of microorganisms	water (cfu/ml)	surface (cfu/ml)	gills (cfu/g)	guts (cfu/g)
aerobic colony count	$2{,}3 \times 10^{3}$	$1{,}04 \times 10^{5}$	$1{,}9 \times 10^{4}$	$6{,}9 \times 10^{3}$
Enterobacteriaceae	$7{,}1 \times 10^{2}$	6×10^{1}	$1{,}7 \times 10^{2}$	$2{,}3 \times 10^{3}$
Pseudomonas	1×10^{2}	$1{,}3 \times 10^{3}$	$3{,}8 \times 10^{3}$	$6{,}4 \times 10^{1}$
Aeromonas	1×10^{2}	1×10^{2}	$1{,}4 \times 10^{2}$	$4{,}5 \times 10^{3}$

Table 2. Analysis for presence-or-absence of human pathogens in a carp pond, on the fish-surface, and in fish-gills and -gut contant.

Pathogens	water	surface	gills	guts
Salmonella	not detectable in 700 ml	not detectable in 400ml	not detectable in 25g	not detectable in 25g
Listeria	*Listeria seeligeri*	*Listeria seeligeri*	*Listeria seeligeri*	not detectable in 25g
Campylobacter	<1/100ml	<10cfu/ml	<10cfu/g	<10cfu/g

Conclusions

Our results show that upon harvesting the bacteriological condition of fish with regard to pathogenic and non-pathogenic flora is determined by the environment and the water´s microbiological quality. Microorganisms apathogenic to man were consistently found on the skin, the gills and in the intestines of fish; as a rule the fish muscles were essentillly sterile at the time the fish were caught. The clearly increased aerobic-colony-count ($1{,}04 \times 10^{5}$) on the fish surface represents the cumulative result of many manual production steps, which certainly adversely affects the product´s keepability and general quality and will lead to faster spoilage.

The carps intestinal flora included *Aeromonas* as described by Gonzalez et al., 1999.

Salmonella and *Campylobacter* were isolated neither from the water nor from the fish. However, the isolation of *Listeria seeligeri* (apathogenic to man) from water, fish surface and gills indicates that the pond`s environment will offer similarly favourable growth conditions for *Listeria monocytogenes* (pathogenic).
In conclusion, the bacterial microflora of foodstuffs from aquacultured products is closely related to environmental contamination and processing and, obviously, ultimately dependent on consumer behaviour.

References

Österreichischer Ernährungsbericht 1998; Im Auftrag von Bundesministerium f. Arbeit, Gesundheit und Soziales, Bundesministerium für Arbeit, Gesundheit und Soziales; Institut f. Ernährungswissenschaften der Univ. Wien, o.Univ.-Prof.Dr. I.Elmadfa

Huss, H.H., 1994. Assurance of Seafood Quality, FAO Fisheries Technical Paper 334, Rom: Eigenverlag FAO

González, C.J., López-Díaz, T.-M., García-López, M.-L., Prieto, M., and Otero, A., 1999. Bacterial Microflora of Wild Brown Trout, Wild Pike, and Aquacultured Rainbow Trout, Journal of Food Protection, 62, No.11, 1270-1277

Feldhusen, F., 1999. Durch Meerestiere übertragene Erkrankungen, Dtsch.tierärztl.WSchr.106, Heft 8, 309-372

Listeria monocytogenes in finnish raw fish

Annukka Markkula, Tiina Autio and Hannu Korkeala
Department of Food and Environmental Hygiene, Faculty of Veterinary Medicine, University of Helsinki

Listeria monocytogenes is a foodborne pathogen, which can cause a serious illness, listeriosis, to susceptible people, like newborns, elderly, pregnant and immunocompromised people. A meaning of *L. monocytogenes* as a cause of food poisonings have grown last years with ready-to-eat food industry. One reason is the ability of *Listeria* spp. to live and multiply in microaerophilic athmosphere and refrigeration temperatures.

L. monocytogenes have been isolated from many foodtypes. Several listeriosis outbreaks have been linked to ready-to-eat fishery products (Ericsson et al., 1997; Miettinen, 1999). According to Autio et al. (1999) most common biotypes that contaminate fish products come from production environment. However, the role of raw fish in contamination of production lines have not fully been determined. The aim of our study was to examine the prevalence and biodiversity of *L. monocytogenes* in Finnish raw fish to find out the meaning of fish in contamination of fish prosessing plants.

102 rainbow trout samples from fish farms from costal areas of Finland, each containing 2-5 slaughtered rainbow trouts, were examined. Samples contained totally 25 g gut, skin, slime and fin. Samples were homogenized 30 s in Stomacher. Isolation were carried out according to ISO 11290-1 method using Oxoid´s (Basingstoke, UK) media. Identification were carried out with gram-staining, katalase-reaction and API Listeria kit (bioMérieux sa, Lyon, France). Four to six isolates from every *L. monocytogenes* -positive sample were typed by pulsed field gel electrophoresis (PFGE). Brain heart infusion (Difco, Detroit, Michigan, USA) was inoculated with a single colony grown 24 h on sheep blood agar. DNA isolation and electrophoresis were carried out according to Autio et al. (1999) with the following modifications. Restriction enzymes *Apa*I and *Asc*I (New England Biolabs) were used. The pulse time ramped from 1 to 35 s for 18 h. The gels were stained with ethidium bromide and photographed under UV transillumination. Low range PFG Marker (New England Biolabs) were used for fragment size determination.

Three (3 %) samples were positive for *L. monocytogenes.* Other *Listeria* spp. were isolated from seven (7 %) samples, of which five (5 %) were *L. innocua*. Only one *Listeria* sp. were isolated from a single sample. PFGEs of the isolated *L. monocytogenes* strains with *Apa*I and *Asc*I both yelded three restriction patterns, dividing the isolates into three different pulsotypes. All the *L. monocytogenes* -positive samples were contaminated with only one pulsotype and pulsotypes of every sample were different.

The prevalence of *Listeria* spp. and *L. monocytogenes* in Finnish raw fish grown in brackish water were 10 % and 3 %, respectively. Our results correlate well to previous studies about prevalences in fishes caught from cold sea waters (Autio et al., 1999; Eklund et al., 1994; Masuda et al., 1992; Rørvik et al., 1995; Vaz-Velho et al., 1998). In tropical waters prevalences have been lower (Adesiyun, 1993; Karungasar et al., 1992; Manoj et al., 1991), probably behause of the high background flora that inhibits the growth of *Listeria* spp. (Ben Embarek, 1994) The differences in salt content between brackish and seawater does not seem to influence *Listeria* prevalenses.

Only one pulsotype could be found in a single sample and pulsotypes in every *L. monocytogenes* -positive sample were different. Isolates of every sample were of different origin. The probable sources of contamination of raw fish are water, slaughterhouse environment (Eklund et al. 1994; Rørvik et al., 1995) and ice bags (Autio, 1999). In raw fish the prevalense of *L. monocytogenes* is low and biodiversity is high when compared to fish products (Autio, 1999). Behause of great biodiversity of *L. monocytogenes* strains, raw fish is a potential source of *L. monocytogenes* contamination in fish prosessing plants.

References

Adesiyun, A.A., 1993. Prevalence of *Listeria* spp., *Campylobacter* spp., *Salmonella* spp., *Yersinia* spp. and toxigenic *Eschterichia coli* on meat and seafoods in Trinidad. Food Microbiol. 10:395-403.

Autio, T., Hielm, S., Miettinen, M., Sjöberg, A-M., Aarnisalo K., Björkroth, J., Mattila-Sandholm, T. and Korkeala, H., 1999. Sources of *Listeria monocytogenes* contamination in a cold-smoked rainbow trout prosessing plant detected by pulsed-field gel electrophoresis typing. Appl. Environ. Microbiol. 65:150-155.

Ben Embarek, P.K., 1994. Presence, detection and growth of *Listeria monocytogenes* in seafoods: a review. Int. J. Food Microbiol. 23:17-34.

Eklund, M.W., Poysky, F.T., Paranjpye, R.N., Lashbrook, L.C., Peterson, M.E. & Pelroy, G.A., 1994. Incidense and sources of *Listeria monocytogenes* in cold-smoked fishery products and prosessing plants. J. Food Prot. 58(5):502-508.

Ericsson, H., Eklöw, A., Danielsson-Tham, M.-L., Longarevic, S., Mentzing, L.-O., Persson, I., Unnerstad, H. and Tham, W., 1997 An outbreak of listeriosis suspected to have been caused by rainbow trout. In: Methods Based on Molecular Biology for Charasteerication and Detection of *Listeria monocytogenes.* Ph.D. thesis. Swedish University of Agricultural Sciences, Uppsala, Sweden.

Karungasar, I., Segar, K., Karungasar, I., and Goebel, W., 1992. Incidence of *Listeria* spp., in tropicalseafoods. In: *Listeria* 1992. Abstract 155. 11th Int. Symp. Problems of Listeriosis. 11-14 May 1992. Copenhagen.

Manoj, Y.B., Rosalind, G.M., Karungasar, I. and Karungasar, I., 1991. *Listeria* spp. in fish and fish-handling areas, Mangalore, India. Asian Fish. Sci. 4:119-122.

Masuda, T., Iwaya, M., Miura, H., Kokubo, Y. and Maruyama, T., 1992. Occurrence of *Listeria* species in fresh seafood. J. Food Hyg. Soc. Japan 33:599-602.

Miettinen, M.K., Siitonen, A., Heiskanen, P., Haajanen, H., Bjorkrooth, K.J. & Korkeala, H., 1999. Molecular epidemiology of an outbreak of febrile gastroenteritis caused by *Listeria monocytogenes* in cold-smoked rainbow trout. J. Clin. Microbiol. 37(7):2358-2360.

Rørvik, L.M., Caugant, D.A. and Yndestad, M., 1995. Contamination pattern of *Listeria monocytogenes* and other *Listeria* spp. in a salmon slaughterhouse and smoked salmon prosessing plant. Int. J. Food Microbiol. 25:19-27.

Vaz-Velho, M., Duarte, G. and Gibbs, P., 1998. Note. Occurance of *Listeria* spp. in salmon-trout (*Onchorhyncus mykiss*) and salmon (*Salmo salar*). Food Sci. Tecnol. Int. 4:121-125.

Environment-related factors influencing end-aquaculture product quality, focusing on antibiotics residues at the pre-harvest phase

E.P. Papapanagiotou
Laboratory of Food Hygiene, Faculty of Veterinary Medicine, Aristotle University of Thessaloniki, 54006 Thessaloniki, Greece. Tel. & Fax +30 31 999833, e-mail: batzias@vet.auth.gr

Various environmental hazards can influence the quality of the final product from aquaculture. These hazards must be identified at the pre-harvest level so as to secure that the end product of the aquaculture process will be safe for the consumer. Biological hazards that can influence the pre-harvest quality of the aquacultured fish products can be classified into:
a) Parasites. Proper cooking or freezing prevents transmission of ingested human pathogenic parasites, b) Bacteria. Any potential microbial safety risk can be significantly reduced by cooking, proper sanitation in processing, handling and storage and personal hygiene, c) Viruses. d) Other biological hazards include algal toxins which may pose a human health risk. Potential risks can be reduced by understanding the natural phenomena of biotoxins, planning carefully for site selection and monitoring for known toxin-producing algae. Public awareness, coastal engineering and classification of waters can also protect public health.(FAO/NACA/WHO, 1999).

Chemical hazards can be present in products from aquaculture through exposure to compounds used in the aquaculture industry and can cause acute and chronic pollution of the water. The chemical hazards can be classified into the following categories:
a) Agrochemicals. b) Chemotherapeutants (antimicrobial agents, parasiticides, hormones), c) Metals. ii) antifoulants and molluscicides, iii) Pollutants d) Feed ingredients, additives and contaminants e) Organic pollutants, PCBs. (FAO/NACA/WHO, 1999).

Concerning the chemotherapeutants, the pre-harvest monitoring of the presence of veterinary drug residues in the final product is definitely of great importance in safeguarding the public health and strengthening the consumer confidence towards safe and high-quality products. Foodborne diseases associated with aquaculture products can be largely prevented and controlled through appropriate food-safety measures. Responsibility for food safety associated with products from aquaculture is shared between governments, the fish-farming and processing industries and consumers. The principles of the HACCP system should form an integral part of the farm to table food safety continuum, thereby increasing consumer confidence in products from aquaculture and expanding market opportunities.

The use of drugs in the aquaculture industry is without risks once the withdrawal times of the drugs are followed. Withdrawal times have not been set for all drugs and all fish species that receive drug treatments, and the usual-incorrect-procedure is to extrapolate results from residue and kinetic studies from one fish species to another, rather than implementing studies on each species separately for each drug. Where antibiotics are incorrectly used or withdrawal times are not observed, residues in fish flesh may pose a potential threat to consumers, especially those who are hypersensitive, leading to possible allergic reactions. Safe levels of residues for some of the more common antibiotics used in aquaculture have been recommended by the Joint FAO/WHO Expert Committee on Food Additives (JECFA, 1996).

Potentiated sulfonamides in the form of commercially available broad spectrum antibacterial preparations have long been used in the aquaculture industry all over the world. The two existing drug formulations are Tribrissen® (combination of Sulfadiazine and Trimethoprim at a 5:1 ratio) and Romet-30® (combination of Sulfadimethoxine and Ormetoprim at a ratio of 5:1). Tribrissen® is widely used in the aquaculture industry in the EU whereas Romet-30® is widely used in the USA and Canada.

One excellent example of the effect of an environmental factor namely temperature, is the great difference in withdrawal times of the same drug preparations in different fish species that we have shown in residue studies in order to determine the withdrawal time of Tribrissen® and Romet-30® in gilthead sea bream (*Sparus aurata*, L.) and compare these results with the corresponding results from salmonids. The difference between sea bream and the other finfish species in terms of withdrawal times is best illustrated by the following results from other researchers and our own.

Salte and Leistol (1983) suggested a 60-day withholding period for rainbow trout in salt water at 10°C and treated with Tribrissen® (TMP 5 mg/kg and SDZ 25 mg/kg) for 10 consecutive days, whereas Jacobsen (1989) suggested a withholding period of 60-days for freshwater rainbow trout treated with TMP at a water temperature of 8°C and 40 days for temperatures above this. In Norway, the official withholding period for Atlantic salmon treated with Tribrissen® is 40 days for water temperatures above 12°C, 40-90 days for temperatures of 8-12°C and 90 days for temperatures below 8°C.

The significantly higher temperature range in which the Mediterranean fish species are cultured along with the species differences provide the excellent strategic advantage that enables much shorter withdrawal times to be set for the two aforementioned drug formulations. As far as Tribrissen® is concerned both SDZ and TMP were not detectable from muscle plus skin, liver, gills and fat by the third day after the end of medication whereas in kidney TMP persisted for over 6 days (Papapanagiotou, *et al.*, 1999). As for Romet-30® both SDM and OMP were not detectable in muscle plus skin and liver by the first day following the cessation of treatment, whereas OMP persisted for over 5 days in kidney (Papapanagiotou, *et al.*, in press). The same pattern for OMP and TMP persistence in the kidney was consistent for both residue studies performed in gilthead sea bream after in-feed administration. and in agreement with the results in Atlantic salmon (Samuelsen, 1997). Tribrissen® and Romet-30® were administered to gilthead sea bream incorporated in the feed for 7 and 5 days respectively at a water temperature of 25°C and at a dose rate of 75 and 50 mg/kg biomass/day, respectively. As a result of the faster elimination rates in gilthead sea bream, in comparison with salmonids, recorded for Tribrissen® and Romet-30® shorter withdrawal times can be set in favour of the Mediterranean mariculture industry, in relation to the commercially available potentiated sulfonamides.

References

Food Safety Issues Associated with Products from Aquaculture (1999). Joint FAO/NACA/WHO Study Group, WHO, Geneva.

Jacobsen M.D., 1989. Withdrawal times of freshwater rainbow trout, *Salmo gairdneri* Richardson, after treatment with oxolinic acid, oxytetracycline and trimethoprim. *J. Fish Dis.*, 12: 29-36.

JECFA., 1996. *Toxicological evaluation of Certain Veterinary Drug Residues in Food.* Prepared by the 45th Meeting of the Joint FAO/WHO Expert Committee on Food Additives. WHO Food Additive Series 36, WHO, Geneva.

Papapanagiotou E.P., Iossifidou E.G., Psomas I.E. and Photis G., 1999. Tissue distribution and residue depletion of Sulfadiazine and Trimethoprim after oral administration to cultured gilthead sea bream (*Sparus aurata*, L.) In: Proceedings of the 26th World Veterinary Congress, 23-26 September 1999, Lyon, France.

Salte R. and Leistol K., 1983. Drug withdrawal from farmed fish. Depletion of oxytetracycline, sulfadiazine and trimethoprim from muscular tissue of rainbow trout (*Salmo gairdneri*) *Acta Vet. Scand.*, 24: 418-430.

Samuelsen O.B., Pursell L., Smith P. and Ervik A., 1997. Multiple dose pharmacokinetic study of Romet30 in Atlantic salmon (*Salmo salar*) and in vitro antibacterial activity against *Aeromonas salmonicida, Aquaculture,* **152**, pp 13-24.

A comparative assessment of the use of ammonium-ferric(III)-hexacyanoferrate(II) (AFCF) and bentonite from polish geological deposits in reducing the radiocaesium transfer to hen eggs

Jaroslaw Rachubik
Laboratory of Radiological Protection and Isotope Studies, National Veterinary Research Institute, Al. Partyzantow 57, 24-100 Pulawy, Poland, e-mail: rachubik@piwet.pulawy.pl

Introduction

Radiocaesium is one of the most important radionuclides which is transferred through the soil-plant-animal food chain and may be accumulated in meat, milk and eggs consumed by men. Consumption of contaminated eggs may be enhanced by the fact that in many Polish rural regions hens are kept free outdoors and thus can ingest polluted plants, grains and soil elements. Therefore, there is a need to have appropriate countermeasures for reducing radioactive contamination of farm animals and farm animal products. They range from alterations in agricultural management (sheltering, deep ploughing) to administration of chemical substances reducing the absorption of radionuclides from the digestive tract or increasing their elimination from the animal organisms. Ammonium-ferric(III)-hexacyanoferrate(II) (AFCF or Giese salt), as other hexacyanoferrate derivatives, has the advantage that it can be fed in a very low quantity and is practically not absorbed from the alimentary tract. On the other hand, bentonite is easily available and inexpensive. This substance is a mixture of clay minerals with the main component montmorylonite determining its biological activity. Since the decontamination effectiveness of bentonite depends on its chemical composition, which may differ among countries or even among regions, it was reasonable to use bentonite from Polish geological deposits. The objective of the present work was to compare the efficiency of AFCF and bentonite in decreasing the radiocaesium transfer to hen eggs.

Materials and Methods

Hundred and five laying hens were used. Each group of animals (15 per group) was kept outdoors in a separated pen. Caesium chloride ($^{137}CsCl$) was supplied by the Nuclear Energy Institute in Swierk, Poland. Ammonium-ferric(III)-hexacyanoferrate(II) was purchased from Riedel-deHaën AG, Germany. Radioactivity of yolk, albumen and whole eggs (mixed yolk and albumen) was measured in a scintillation (NaI/Tl) counter (Polon, Poland) using vessels of the standard Marinelli geometry prepared in our laboratory.

Hens were fed a standard chow and watered *ad libitum*. After a week of acclimatisation birds were randomly divided into eight groups (two control groups and six experimental groups). All groups were treated daily with a dose of radiocaesium (~3.8 kBq per bird) in drinking water for 14 days. The control hens were fed in the same manner as during the acclimatisation period. Animals in the experimental groups were given a chow supplemented with a decontamination agent (0.2% of AFCF or 10% of bentonite) for 14 consecutive days. The schedule of giving the caesium binder was as follows: groups I and II received hexacyanoferrate or bentonite along with the radiocaesium administration, groups III and IV were fed with the enriched chows from day 8 of the radioisotope treatment, groups V and VI from the day after the last radionuclide dosage, respectively.

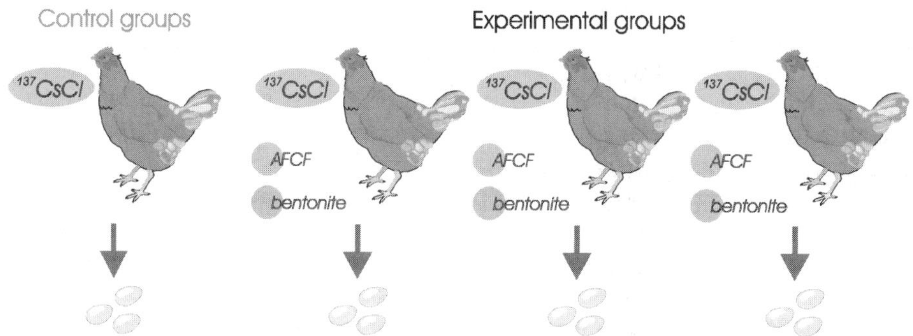

Figure 1. Schematic representation of the experimental design on hens.

Results

Radiocaesium in albumen.

The results are presented in Figs 2 and 5. The concentrations of albumen radiocaesium in the experimental groups peaked on day 2 in group I, on day 6 in groups III and V, on day 8 in group II, on day 6 in group IV, and on day 12 in group VI, respectively. The results obtained indicate a very high degree of AFCF protection in group I (up to 96.1% in comparison with control on day 12 of the experiment) for this egg constituent. Giving bentonite simultaneously with radionuclide treatment lowered the radioactivity concentration by more than 50% in comparison with that in the controls during the entire experiment. In groups III-VI reduction was lower, although still remarkable, especially in groups III and IV from the day of the first binder administration.

Radiocaesium in yolk.

The results are presented in Figs 3 and 6. Radiocaesium concentrations in yolk were generally lower than those in albumen. In all groups treated with the hexacyanoferrate enriched diet radioactivities peaked on day 8 of the experiment and in hens treated with the bentonite supplemented chow radionuclide concentrations reached the maximal values on day 8 in group II, on day 10 in group IV, and on day 14 in group VI, respectively. The highest decontamination efficacy of AFCF (up to 94.7% in comparison with control on day 16 of the study) was observed in group I. In groups III and V the reduction rate was slightly slower than that in egg white. In bentonite treated groups, similarly as in albumen, the most pronounced reduction of the ^{137}Cs concentration was noted in group II. In groups IV and VI the decontamination efficiency of this caesium binder was lower than that in group II, especially in group VI receiving the clay mineral after the last radiocaesium administration.

Radiocaesium in whole eggs.

The results are presented in Figs 4 and 7. Radiocaesium concentrations in whole eggs were of intermediate values between those in albumen and yolk. In all groups treated with the hexacyanoferrate enriched diet radioactivities peaked on day 6 and in animals treated with bentonite supplemented chow those peaked on day 6 in group II and IV, and on day 10 in group VI, respectively. Similarly as in albumen and yolk, in whole eggs the highest effectiveness in lowering the radionuclide transfer was noticed in groups I and II which were receiving AFCF or bentonite along with ^{137}Cs.

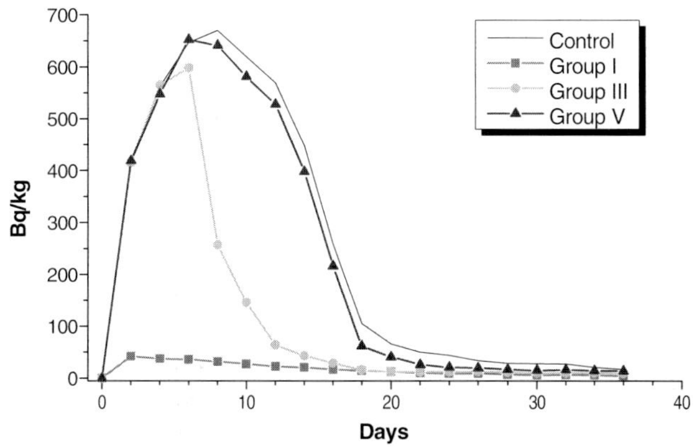

Figure 2. The radioactivity concentration in albumen (AFCF).

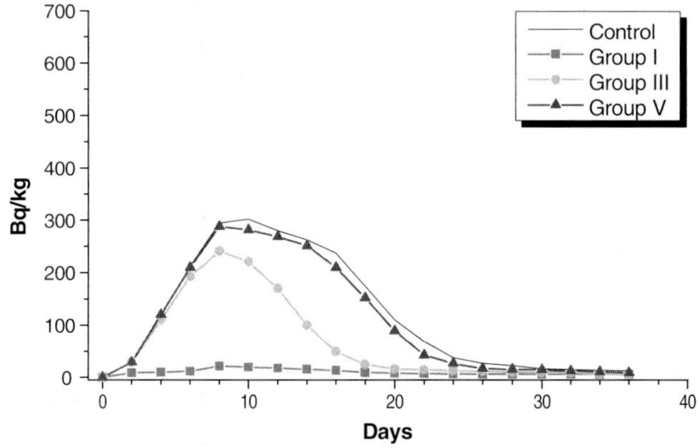

Figure 3. The radioactivity concentration in yolk (AFCF).

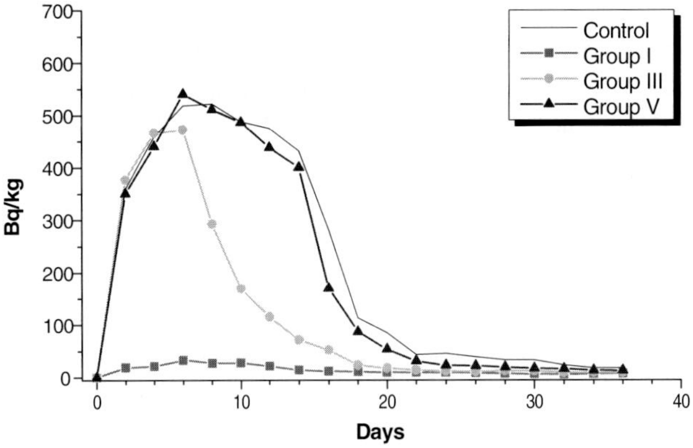

Figure 4. The radioactivity concentration in whole eggs (AFCF).

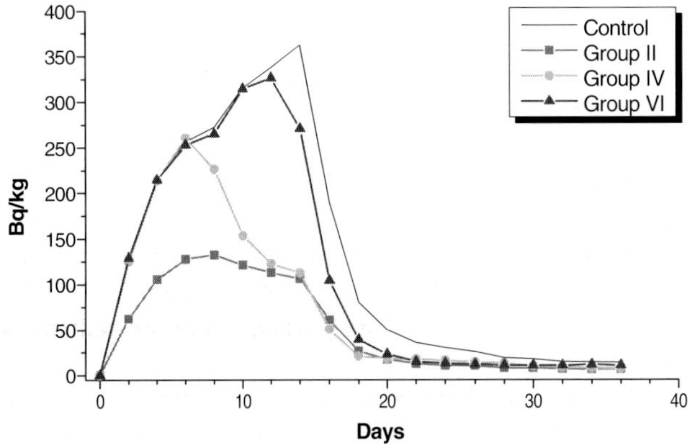

Figure 5. The radioactivity concentration in albumen (bentonite).

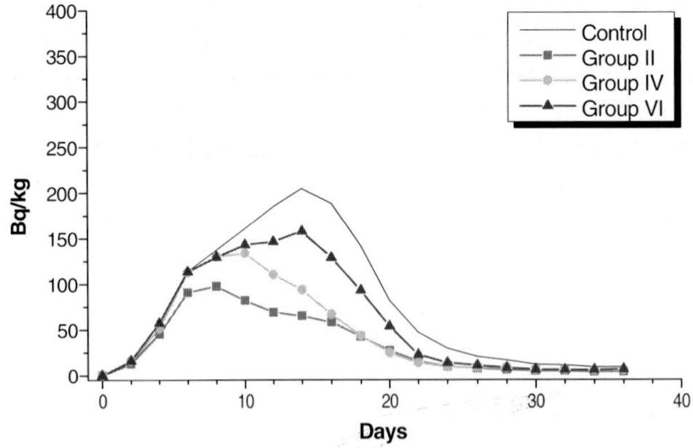

Figure 6. The radioactivity concentration in yolk (bentonite).

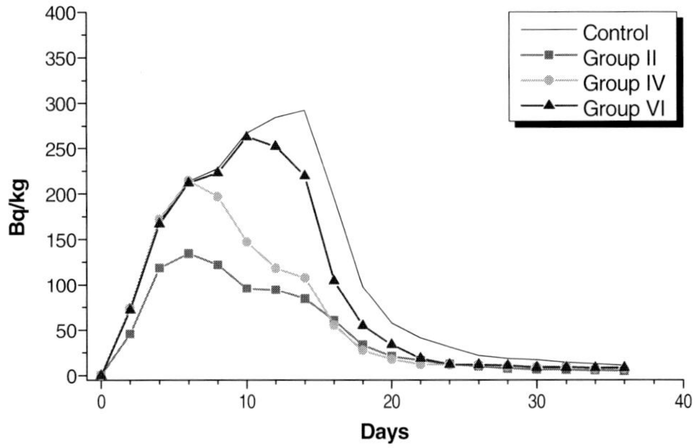

Figure 7. The radioactivity concentration in whole eggs (bentonite).

Discussion

Eggs, beside milk and meat, are an important source of animal origin proteins and are frequently consumed by men. Thus, they can substantially contribute to the total radioactivity burden of human organisms. So, it is necessary to set up appropriate measures for reducing the radionuclide transfer to edible egg components. This radionuclide exhibited the higher affinity for egg white. We found the inhibited ^{137}Cs transfer to hen eggs by even more than 95% for AFCF. Bentonite from Polish geological deposits was less effective than model compound - AFCF. The main factor influencing the efficiency of a decontaminating substance seems to be the time interval between radioactive pollution and the beginning of the treatment with a radionuclide binder. The delayed treatment with radionuclide binders was substantially less effective in reducing the ^{137}Cs transfer to egg components. It is a great advantage to use hexacyanoferrate in hens because of the very small amount needed to achieve a decontamination effect. However, there are some pros to use bentonite. This decontamination agent is widely distributed in farms (a commonly feed addition for ruminants) and easy to obtain. It was necessary to study bentonite from Polish deposits because its decontamination efficacy depends on the clay origin region. This variation in reduction the radiocaesium transfer to animal organisms might be due to the use of varying quality of the material according to the age (collapsed mineral structures) or unspecified different forms such as Ca, Na or other derivatives of bentonite. In conclusion, our work demonstrated that using AFCF as a caesium binder is more effective in reduction of the radiocaesium transfer to hen eggs. However, in Poland bentonite seems to be an fairly efficient alternative to AFCF because of wide distribution and easy availability.

The two component system PHOP/PHOQ is required for full chloramphenicol resistance in the multiresistant *Salmonella typhimurium* DT104

Friederike Hilbert[1] and Eduardo A. Groisman[2]
[1]Institute for Meat Hygiene, Meat Technology and Food Science, University of Veterinary Medicine Vienna. F.H. was supported by an Erwin-Schroedinger fellowship from the Austrian Science Fund, [2]Department of Molecular Microbiology and Howard Hughes Medical Institute, Washington, University School of Medicine, St. Louis, Missouri, 63110-1093, USA

Abstract

Salmonella enterica serovar *typhimurium* phage type DT104 with multiple antibiotic resistance to streptomycin, chloramphenicol, tetracycline, ampicillin and sulfonamides is becoming an increasing cause of disease in humans, cattle and pigs. The *floR* (*cmlA*) gene, conferring chloramphenicol and florfenicol resistance shows structural similarity to multidrug efflux transporters driven by the proton motive force. We show now that the resistance to chloramphenicol mediated by the *floR* (*cmlA*) gene needs a functional PhoP-PhoQ regulatory system. We hypothesise that PhoP-PhoQ induced changes in the bacterial membrane are required for an effective antiport of this multidrug efflux systems and to establish a sufficient proton motive force of the transmembrane electrochemical proton gradient.

Introduction

Salmonella enterica serovar *typhimurium* phage type DT104 has become a dominant phage type to cause disease in humans as well as in cattle and pigs. An increasing number of these strains exhibit multiple antibiotic resistance to streptomycin, chloramphenicol (including florfenicol), tetracycline, ampicillin and sulfonamides. Antibiotic therapy of salmonellosis is necessary for treating severe infections mainly occurring in immunocompromised, elderly, children and to prevent chronic disease. Hence, this antibiotic resistance pattern is an important concern to public health authorities. The *floR* (*cmlA*) gene has been analysed in more detail, representing a new gene with high identity to *cmlA* in *Pseudomonas aeruginosa*, a gene coding for a chloramphenicol exporter protein. In addition the *floR* (*cmlA*) gene product provides resistance to florfenicol, an antibiotic of the same family as chloramphenicol. The FloR (CmlA) protein reveals structural similarity to multidrug efflux systems, driven by the proton motive force of the transmembrane electrochemical proton gradient. Their function is most likely based on a multidrug - proton antiport mechanism.

The PhoP-PhoQ two-component regulatory system in *Salmonella enterica* is known to regulate a variety of genes transcriptionally. Most important for controlling genes substantial for virulence in *Salmonella*, the PhoP-PhoQ system acts as a global regulatory system in a variety of enteric species.

In this study we show that chloramphenicol resistance mediated by the *floR* (*cmlA*) gene in *Salmonella enterica* serovar *typhimurium* phage type DT104 needs a functional PhoP-PhoQ regulatory system. PhoP is also essential in *E. coli* for chloramphenicol resistance carrying the *floR* (*cmlA*) chloramphenicol resistance gene on a plasmid. We hypothesise that PhoP-PhoQ induced changes in the bacterial membrane are required for an effective antiport of this multidrug efflux systems and to establish a sufficient proton motive force of the transmembrane electrochemical proton gradient.

Results

Chloramphenicol resistance in *Salmonella enterica* serovar *typhimurium* phage type DT104.

We examined wild type isolates EG12401, a human isolate, and H3380 for chloramphenicol resistance by agar diffusion and broth dilution testing (NCCL standards). H3380 carries the *floR* (*cmlA*) gene, a putative chloramphenicol exporter protein. In EG12401, a streptomycin, chloramphenicol (including florfenicol), tetracycline, ampicillin and sulfonamide resistant isolate of *Salmonella enterica* serovar *typhimurium* phage type DT104, we confirmed the presence of the *floR* (*cmlA*) gene by PCR using primer derived from the sequence of H3380 and sequencing of the gene.

PhoP/PhoQ is required for full chloramphenicol resistance.

PhoP - PhoQ regulates resistance to a number of compounds including bile salts, antimicrobial peptides, like defensins, magainins, cecropins and peptide antibiotics such as polymyxin B. Mutants of EG12401 and H3380 were analysed for resistance to chloramphenicol by disk susceptibility testing and microdilution susceptibility testing. *phoP* mutants exhibited only half the resistance of the wild type (Table 1.) in both strains.

Table 1. Chloramphenicol resistance of wild types and PhoP mutants

strain	MIC_{90} in MH-broth µg/ml Cm
EG12401	180
EG12409	80
H3380	200
H3380 $PhoP^-$	125
MC1061+pFH10	120
FS1000+pFH10	80
14028s+pFH10	250
MS7953+pFH10	170

A plasmid harboring a 1.6 kb fragment including the floR (cmlA) gene confers resistance to chloramphenicol.

We constructed pFH10 with a 1.6 kb DNA fragment including the putative *floR* (*cmlA*) gene sequence. This plasmid conferred chloramphenicol resistance upon *E. coli* strain JM109 as well as to *Salmonella enterica* serovar Typhimurium 14028s at levels higher than the parent strain (Table 1.). Expression of chloramphenicol resistance driven by the plasmid was also dependent on a functional PhoP/Q system. This was not only seen in *Salmonella* but also in *E. coli*. FS1000 a *phoP* mutant of *E. coli* strain MC1061 developed almost no resistance to chloramphenicol when transformed with the plasmid pFH10 whereas MC1061 was highly resistant.

PhoP/Q dependent loss of the chloramphenicol resistance does not occur on transcriptional level.

To examine if the chloramphenicol resistance is regulated at the transcriptional level, we investigated expression of the *floR* (*cmlA*) gene by primer extension analysis in the wild type EG12401 and the *phoP* mutant of EG12401 (=EG12409).Quantitation, revealed the very same intensity for both strains. Therefore the PhoP/Q system does not appear to regulate the *floR* (*cmlA*) gene at the transcriptional level.

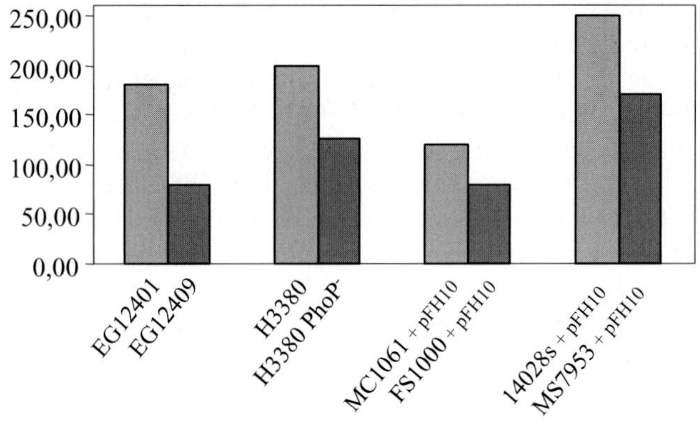

Figure 1. MIC90 in MH-broth µg/ml Cm.

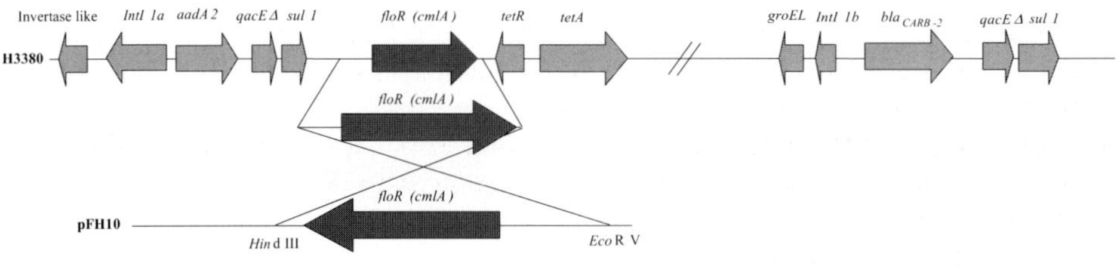

Figure 2.

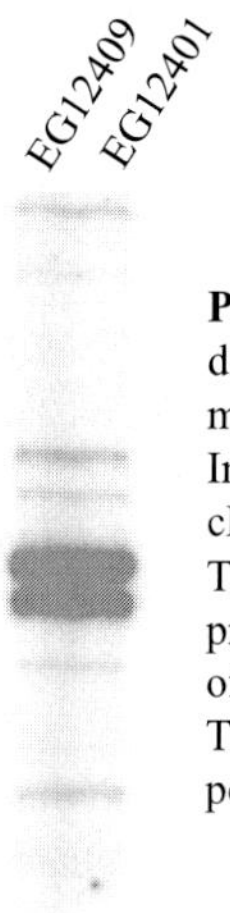

Primer extension: EG12401 and EG12409 were grown in LB overnight, diluted 1:100 in LB, respectively. Total RNA was extracted from cells at midexponential phase cultures (A_{600} 0.6) with Trizol (Life Technologies, Inc.) according to the manufacturerís specification. The Primer, used for cDNA synthesis is 26 bp downstream of the *floR* (*cmlA*) initiation codon. The cDNA synthesis was performed with 40 g of total RNA, 2 pmol of primer labeled with T4 polynucleotide kinase, and [γ-^{32}P]ATP and 50 units of SuperScript^{TM}II RNaseH^{-} reverse transcriptase (Life Technologies, Inc.). The extension products were analyzed by electrophoresis in a 6% polyacrylamide, 7.5 M urea gel.

Figure 3.

Discussion

Salmonella enterica serovar *typhimurium* phage type DT104 is a multiresistant *Salmonella* strain with a broad host specificity, infecting humans, cattle, pigs, and chicken. The molecular analyses of this resistance gene cluster revealed the existence of a new gene conferring chloramphenicol and florfenicol resistance, two antibiotics of the same group, the phenicols. This gene shows sequence similarity to the family of secondary transporters most likely driven by the proton motive force ($\Delta\mu_H$+). In this study we have shown that the *floR* (*cmlA*) gene is dependent on a functional PhoP/Q two component system both in *Salmonella* and *E. coli*. It is known that the PhoP/Q system induces changes in the membrane and the lipid A fraction. As chloramphenicol did not penetrate the membrane of a PhoP mutant to an higher extent than the wild type we hypothesise that a PhoP mutant has a lower ability to induce a sufficient membrane potential for transporters driven by the proton motive force. This could also explain the reduced bile salt resistance of PhoP mutants as proton dependent efflux pumps are known to be responsible for bile salt resistance in *E. coli*

References

Arcangioli, M.A., Leroy-Setrin, S., Martel, J.L., Chaslus-Dancla, E., 2000. Evolution of chloramphenicol resistance, with emergence of cross-resistance to florfenicol, in bovine *Salmonella typhimurium* strains implicates definitive phage type (DT) 104. J. Med. Microbiol. 49:103-110.

Briggs, C.E., Fratamico, P.M., 1999. Molecular characterization of an antibiotic resistance gene cluster of *Salmonella typhimurium* DT104. Antimicrob. Agents Chemother. 43:846-849.

Garcia Vescovi, E., F.C. Soncini, Groisman, E.A, 1996. Mg^{2+} as an extracellular signal: environmental regulation of Salmonella virulence. Cell 84:165-174.

Glynn, M.K., C. Bopp, W. Dewitt, P. Dabney, M. Mokhtar, Angulo, F.J., 1998. Emergence of multidrug-resistant *Salmonella enterica* serotype *typhimurium* DT104 infections in the United States. N. Engl. J. Med. 338:1333-1338.

Groisman, E.A., Heffron, F., 1995. Regulation of *Salmonella* virulence by two component regulatory systems, p.319-332. *In* J.A. Hoch and T.J.Silhavy (ed), Two component signal transduction. ASM Press, Washington, D.C.

Groisman, E.A., Heffron, F., Solomon, F., 1992. Molecular genetic analysis of the *Escherichia coli phoP* locus. J Bacteriol. 174:486-491.

Gunn, J.S., Lim, K.B., Krueger, J., Kim, K., Guo, L., Hackett, M., Miller, S.I., 1998. PmrA-PmrB-regulated genes necessary for 4-aminoarabinose lipid A modification and polymyxin resistance. Mol Microbiol. 27:1171-1182.

Guo, L., Lim, K.B., Poduje, C.M., Daniel, M., Gunn, J.S., Hackett, M., Miller, S.I., 1998. Lipid A acylation and bacterial resistance against vertebrate antimicrobial peptides. Cell. 16:189-198.

Oyston, P.C., Dorrell, N., Williams, K., Li, S.R., Green, M., Titball, R.W., Wren, B.W., 2000. The response regulator PhoP is important for survival under conditions of macrophage-induced stress and virulence *inYersinia pestis*. Infect. Immun. 68:3419-3425.

Paulsen, I.T., Brown, M.H., Skurray, R.A., 1996. Proton-dependent multidrug efflux systems. Microbiol. Rev. 60:575-608.

Poppe, C., Smart, N., Khakhria, R., Johnson, W., Spika, J., Prescott, J., 1998. *Salmonella typhimurium* DT104: a virulent and drug-resistant pathogen. Can. Vet. J. 39:559-565.

Ridley, A., Threlfall, E.J., 1998. Molecular epidemiology of antibiotic resistance genes in multiresistant epidemic *Salmonella typhimurium* DT 104. Microb Drug Resist., Summer 4:113-118.

Soncini, F.C., Garcia Vescovi, E., Solomon, F., Groisman, E.A., 1996. Molecular basis of the magnesium deprivation response in Salmonella typhimurium: identification of PhoP-regulated genes. J. Bacteriol. 178: 5092-5099.

Thanassi, D.G., Cheng, L.W., Nikaido, H., 1997. Active efflux of bile salts by *Escherichia coli*. J. Bacteriol. 179:2512-2518.

Threlfall, E.J., Frost, J.A., Ward, L.R., Rowe, B., 1994. Epidemic in cattle and humans of *Salmonella typhimurium* DT 104 with chromosomally integrated multiple drug resistance. Vet. Rec. 134:577.

Van Velkinburgh, J.C., Gunn, J.S., 1999. PhoP-PhoQ-regulated loci are required for enhanced bile resistance in *Salmonella spp*. Infect Immun. 67:1614-1622.

Biographies

Henk Beumer graduated in Food Technology in 1963 at the Wageningen University, The Netherlands. From 1971 to 1976 he was involved in UNDP funded and bilateral development projects on fish processing in Peru. In 1976 he joined TNO Nutrition and Food Research Institute, the Netherlands, where he was in charge of the Institute's research in Feed Technology, which was later on integrated in the Department of Animal Nutrition. From 1988-1992 he co-ordinated an UNDP/UNIDO funded training and consultancy project aimed at establishing a Centre for Research, Development, Quality Control and Training for the food industry in Egypt. In 1998 he joined the Institute's Food Science and Technology Department and in 1999 the Risk Management and Microbiology Department. At present he is a senior consultant in risk management and food safety issues.

Raymond Bradley qualified as a veterinarian from the Royal Veterinary College (RVC) in London 1959. He spent ten years in mixed veterinary practice in Hampshire, Devon and Surrey before joining the Pathology Department of the Central Veterinary Laboratory of the Ministry of Agriculture Fisheries and Food (MAFF) in 1969. He obtained a Masters Degree in Veterinary Pathology, also from the RVC, in 1972. His main research interests were skeletal and cardiac muscle diseases of farm animals, and the ultrastructural pathology of central nervous and muscle tissue in farm animal diseases. He became Head of the Pathology Department in 1983. In 1986, BSE was discovered and as Head of the lead Department he became the Project Leader for the BSE research programme. In 1991 he became BSE Co-ordinator for MAFF. He retired in 1995 to become a Private BSE Consultant and maintains a national and international presence and interest in regard to BSE, its control and elimination.

Marion Brank studied Biology at the University of Vienna, Austria, where she received in 1997 the M.Sc. degree for her work on the construction of recombinant *Escherichia coli* with membrane-anchored antigen fragments of pseudorabies virus for use as a non-living vaccine. Subsequently, she joined the Institute of Bacteriology, Mycology and Hygiene at the University of Veterinary Medicine Vienna as graduate research assistant and is currently completing her Ph.D. thesis on the molecular characterization of immunogenic surface proteins of *Mycoplasma bovis.*

Christine Citti received her education at the University of Bordeaux II, France. In 1988 she obtained her diploma (DEA) and in 1992 her Ph.D. degree in Viral and Microbial Pathology. Following a three-year period as Postdoctoral Fellow at the Deaprtment of Molecular Microbiology and Immunology at the University of Missouri-Columbia, USA, Dr. Citti joined the Institute of Bacteriology, Mycology and Hygiene at the University of Veterinary Medicine Vienna, Austria, first as Research Associate and later as University Assistant and Research Group Leader. Her main research interests focus on mycoplasma membrane proteins and their consequences as surface components mediating various interactions with host cells and the immune system.

John Daniel Collins is Professor - and Head of Department of Farm Animal Clinical Studies at the Faculty of Veterinary Medicine, National University of Ireland- University College Dublin. He is the Chair of the Management Committee of the Centre for Food Safety at University College Dublin. He is also Director of the Veterinary Epidemiology and Tuberculosis Investigation Unit at the College. Professor Collins is a member of the Board of the Food Safety Authority of Ireland and Secretary of the European College of Veterinary Public Health (ECVPH).

Willem Daelman received his DVM degree at Ghent University, Belgium in 1973. From 1973 through 1976 he was assistant Professor in Hygiene of Foods of Animal Origin at that University. From 1976 through 1979 he was affiliated with the Veterinary Service of the Belgian Ministry of Agriculture, and from 1979 through 1982 he served as an administrator of the EC Veterinary Inspection Services (DG VI) and from 1982 through 1998 at the Unit Veterinary Legislation (DG VI). Since 1998 Dr. Daelman is Deputy head of the unit DG Health and Consumers Protection, Directorate D: Food Safety - production and distribution chain, Unit D2 Biological risks. His current activities include recast of the Community food hygiene legislation and recast of the Community legislation on food controls.

Johanna Fink-Gremmels studied veterinary medicine at the Veterinary School in Hannover, Germany. She qualified as veterinarian in 1974 and in the same year received her doctor's degree based on a thesis in biochemistry. After 2 years in general veterinary practice she returned to Hannover's University to specialize in veterinary pharmacology and toxicology. Upon receiving specialist status in 1980 she was appointed assistant professor in the Department of Veterinary Pharmacology and Toxicology in 1981. In 1985 she accepted an appointment as Scientific Director at the German Federal Institute for Meat Research, as head of the Department of Food Toxicology. In 1991 she was appointed at the Utrecht University, Faculty of Veterinary Medicine as professor in veterinary pharmacology and clinical toxicology. Presently she is head of the Department of Veterinary Pharmacology, Pharmacy and Toxicology at this Faculty. In addition, she serves in various international organizations, among others as President of the ECVPT, Honorary member of EAVPT and Chairperson of the EU's Scientific Committee of Veterinary Measures Relating to Public Health (SCVPH) and member of the Scientific Steering Committee (DG Sanco).

Walter H. Günzburg studied biology at the University of Birmingham, England. He obtained his Ph.D. in 1984 at the Institute of Genetics, Research Centre Karlsruhe and the Imperial Cancer Research Fund, London, England. After some years of postdoctoral experience in Switzerland and the USA, he received his Habilitation in Genetics at the Ludwig-Maximilians University, Munich in 1992. He was appointed Professor and Chairman of the Institute of Virology at the University of Veterinary Sciences, Vienna in 1996. A Faculty member of European Academy for the Study of the Consequences of Scientific and Technical Advances, member of the Somatic Gene Therapy Commission of the German Federal Medical Association, Section Editor of Current Opinion in Molecular Therapy, Editorial Board member for Gene Therapy and for Trends in Molecular Medicine, he is also Board Member and Consultant to Austrian Nordic Biotherapeutics AG. His current fields of interest include the regulation of retroviral gene expression, the gene therapy of solid tumours and the development of retroviral vectors.

Andreas Hensel graduated in Veterinary Medicine at the Tierärztliche Hochschule Hannover (TiHo, Germany) in 1985. He continued his studies ("Aufbaustudium" in microbiology) and received a doctorate degree in 1998. From 1998-1990 he was research scientist at the Institute of Microbiology and Infectious Animal Diseases at this Faculty. In 1990 he joined the Institute of Bacteriology and Animal Hygiene at the University of Veterinary Medicine in Vienna and received a Ph.D. Degree at Utrecht University in 1994. From 1994 until 1997 he became a senior scientist am Institute of Microbiology and Genetics, Biocenter, University of Vienna. In 1997 he finished his "Habilitation" degree at the TiHo in Hannover. Since 1998 he is Full Professor of Animal Hygiene and Control of Infectious Animal Diseases and head of the Institute of Animal Hygiene and Public Veterinary Health at the University of Leipzig.

Sebastian Hielm graduated as veterinarian from Helsinki, Finland in 1994. Since then, he has been working at the Department of Food and Environmental Hygiene, Faculty of Veterinary Medicine, University of Helsinki. He received a Ph.D. in Food Hygiene in 1999. He is assistant professor and head of the botulism laboratory at the Department of Food and Environmental Hygiene since 2000. Currently he is involved in a three-year project involved in applying risk analysis theory to the development of HACCP programs in the Finnish food industry.

Dieter Klein studied veterinary medicine at the Ludwig-Maximillians-University in Munich, Germany. After two years research work at the Institute of Experimental Surgery and the Clinics for Internal Medicine at the University in Munich, he received the degree of DVM for his work on hyperacute rejection in xenotransplantation in 1994. Subsequently, he joined the Institute of Molecular Virology at the Research Center for Environment and Health in Oberschleißheim, Germany, to work on retroviral vectors for gene therapy. In 1996 he continued his work in this field at the Biotech company Bavarian Nordic in Neuherberg, Germany, before moving in 1997 to Vienna to become Lecturer at the Institute of Virology at the University of Veterinary Sciences. In 2000 he received his Habilitation in Virology. His main research interests include molecular virology of retroviruses, retroviral vectors for gene therapy and quantification of microorganisms using molecular methods.

Hannu Korkeala graduated as veterinarian from Helsinki, Finland in 1975. Since then, he has been working at the Department of Food and Environmental Hygiene, Faculty of Veterinary Medicine, University of Helsinki. He received a Ph.D. in Veterinary Medicine in 1980 and a M.Sc. degree in Social Sciences in 1985. He is professor in Food Hygiene since 1990 and Head of the Department of Food and Environmental Hygiene. His current research interests include diagnostics, pathogenesis and epidemiology of foodborne pathogens.

Miia Lindström qualified as DVM in 1998 at the Faculty of Veterinary Medicine, University of Helsinki, Finland. She is currently working on her PhD on diagnostics and inactivation of *Clostridium botulinum* in food industry at the Department of Food and Environmental Hygiene in the veterinary faculty.

Heinrich Neubauer studied veterinary medicine from 1987 to 1992 in Munich. In 1995 he finished his education in Munich with the degree Dr. med vet. From then on he worked in the field of molecular epidemiology and diagnostics of bacterial infections at the Institute for Microbiology, SanAkBw, Munich.

Jos Noordhuizen is Professor in Ruminant Health Sciences at the Faculty of Veterinary Medicine in Utrecht, The Netherlands and head of the Ruminant Health Group. He is also acknowledged veterinary specialist in the area of bovine herd health care. He is member of the Scientific Committee on Animal Health and Welfare of the European Commission. He was founder of the Dutch Association for Veterinary Epidemiology & Economics (VEEC) in 1988 and coordinator of a European Network on Veterinary Epidemiology and Economics (EPIDECON). His research interests are in the fields of dairy herd health and production management at dairy farm level. He has published several books and papers on these subjects.

Servé Notermans received his academic training at Wageningen University, The Netherlands . In 1972 he graduated from this University both in food technoloy and food microbiology. In 1975 he received his PhD in food microbiology, whereafter he was employed at the National Institute of Public Health, The Netherlands, being made responsible for the food microbiological rsearch. Since the late 1990's Dr. Notermans is employed as a Principle Scientists at the Nutrition and Food Research Institute at Zeist The Netherlands. His research interests cover food safety, more in particular HACCP, risk analysis and preservation technology.

Larry G. Paisley received a DVM degree from Washington State University in 1970 and a MS degree in Veterinary Obstetrics and Gynecology at the University of Minnesota in 1972. After two years of general large animal clinical practice in Montana, he returned to Washington State University where he taught until 1985. After teaching at Ross University in the West Indies 1985 through 1989 he entered a Ph.D. program at the Royal Veterinary Agricultural University, Copenhagen, Denmark. After obtaining the PhD in 1991 he worked for USDA:APHIS:VS in Puerto Rico and New York, the National Veterinary Institute, Oslo, Norway and is presently employed at the Danish Veterinary Institute, Copenhagen, Denmark. His main research interests are risk analysis and statistical and simulation modeling of animal health surveillance and other animal health problems particularly TSEs.

Renate Rosengarten received her education at the University of Hannover and the School of Veterinary Medicine Hannover, Germany. She obtained her Ph.D. degree in Microbiology in 1985, and her DVM degree and Ph.D. degrees in Infectious Diseases of Animals in 1993. Following a five-year period as Postdoctoral Research Associate at the School of Veterinary Medicine Hannover and the University of Missouri-Columbia (USA), Dr. Rosengarten accepted a position as Research Group Leader at the Institute of Microbiology and Infectious Diseases of Animals at the School of Veterinary Medicine Hannover in 1990. From 1993 to 1995 she joined the Department of Membrane and Ultrastructure Research at the Hebrew University-Hadassah Medical School in Jerusalem, Israel, as a Visiting Scientist. In 1996 she was appointed as Professor of Bacteriology and Hygiene and Chair of the Institute of Bacteriology, Mycology and Hygiene at the University of Veterinary Medicine Vienna, Austria.

Eystein Skjerve graduated at the Norwegian College of Veterinary Medicine in 1983, and received his PhD in 1987. Since then he has been Associate Professor of Food Hygiene at the same Institution, with the exception of a period of sabbatical leave spent at Utrecht University, The Netherlands, 1989-1990, and the University of California, Davis (USA), 1997-1998. His main area of work has been the epidemiology of different foodborne infections, including Toxoplasma gondii, Taenia saginata, Trichinella spriralis, Yersinia enterocolitica, Salmonella, Campylobacter, Listeria monocytogenes, E.coli 0157:H7 and enteroviruses. In recent years Dr. Skjerve particularly focuses on mathemathical modelling of the microbial ecological systems, linked to animal husbandry and foods of animal origin.

John N. Sofos received his B.S. degree in Agriculture from the Aristotle University of Thessaloniki, Greece in 1971, his M.S. degree in Animal Science forms the University of Minnesota in 1975, and in 1979 he got his Ph.D. degree in Food Science (Meat Microbiology) at the same University. Following a two-year period as a Research Associate at the University of Minnesota, Dr. Sofos accepted a position as Assistant Professor at the Department of Animal Science of Colorado State University in 1980. In 1984 he became Associate Professor at this department and was jointly appointed to the Departments of Food Science and Human Nutrition. Since 1987, Dr. Sofos is Full Professor at both these departments. He also serves as scientific adviser to the United States Food and Drug Administration in Denver, Colorado. His main research interests include the areas of sources, ecology and extent of bacterial pathogen contamination of meat and other foods; procedures to reduce contamination and to inactivate or inhibit bacterial pathogens; resistance of microorganisms to preservation procedures; stress adaptation of pathogenic bacteria; and, methods of sampling and detection of bacteria in foods.

Preben W. Willeberg studied Veterinary Medicine at the Royal Veteinary and Agricultural University (RVAU) at Copenhagen from 1960 - 1967. In September 1972 he received the PhD, - and in 1977 the Dr. Vet.Sc. degree from that same University. May 1995 Dr. Willeberg was awarded a Dr.med.vet.hc. by the Faculty of Veterinary Medicine, University of Helsinki. From 1985 through 1999 Dr. Willeberg was Professor of State Veterinary Medicine and Epidemiology at RVAU and since October 1999 he serves as Denmark's Chief Veterinary Officer for the Danish Veterinary and Food Administration of the Ministry of Food, Agriculture and Fisheries.

Index

D

O

P